Multimodal Intelligent Sensing in Modern Applications

IEEE Press
445 Hoes Lane
Piscataway, NJ 08854

Multimodal Intelligent Sensing in Modern Applications

Edited by

Masood Ur Rehman
James Watt School of Engineering, University of Glasgow, Glasgow, UK

Ahmed Zoha
James Watt School of Engineering, University of Glasgow, Glasgow, UK

Muhammad Ali Jamshed
James Watt School of Engineering, University of Glasgow, Glasgow, UK

Naeem Ramzan
School of Computing, Engineering and Physical Sciences, University of the West of Scotland, Paisley, UK

IEEE Press

WILEY

Published by John Wiley & Sons, Inc., Hoboken, New Jersey.
Published simultaneously in Canada.

Library of Congress Cataloging-in-Publication Data Applied for:

Hardback ISBN: 9781394257713

Cover Design: Wiley
Cover Image: © MF3d/Getty Images

Set in 9.5/12.5pt STIXTwoText by Straive, Pondicherry, India

Dedication

To Allah, the most merciful and compassionate, who guides us through every challenge in life. His love and grace sustain us always and we dedicate this book to him, seeking his continued guidance and blessings.

To my parents, Khalil and Ilfaz, who have nurtured me with unconditional love, sacrifice, and wisdom. To my brothers, Habib and Waheed for their wholehearted assistance. To my wife, Faiza, for her being a great support system through thick and thin. To my son, Musaab, for filling my life with immense love and joy.

Masood Ur Rehman

To my wife, Mariyam, for her unwavering support, love, and patience throughout the journey of editing this book. Her encouragement has been invaluable. To my daughters, Zainab and Yusra, thank you for your joy and the inspiration you bring into my life every day. This accomplishment would not have been possible without the strength and motivation you all gave me.

Ahmed Zoha

To my biggest inspiration, my parents, Jamshed Iqbal and Nuzhut Jamshed
my support system, wife: Aqsa Tariq, and my son: Zohaan Ali.

Muhammad Ali Jamshed

To my parents for all their love and for raising me in a way I am proud of. To my wife Nasira for her resolute support. To my daughters Saba, Bisma, and Hadiya for cherishing and elating my life with all the bliss and love.

Naeem Ramzan

Contents

About the Editors

Masood Ur Rehman received a BSc degree in electronics and telecommunication engineering from UET, Lahore, Pakistan in 2004 and an MSc and PhD in electronic engineering from Queen Mary University of London, UK, in 2006 and 2010, respectively. He worked at QMUL as a postdoctoral research assistant till 2012 before joining the Centre for Wireless Research at the University of Bedfordshire as a Lecturer. He served briefly at the University of Essex and then moved to the James Watt School of Engineering at the University of Glasgow as an Assistant Professor in 2019. He currently works as an Associate Professor at the University of Glasgow. His research interests include compact antenna design for 6G, Industry 5.0, and Global Navigation Satellite Systems; flexible, wearable and implantable sensors and systems; bio-electromagnetics and exposure of biological tissues to RF; mmWave and nano-communications for body-centric networks and wireless sensor networks in industrial automation, healthcare and environmental monitoring, and device-to-device and human-to-human communications. He has worked on several projects supported by industrial partners and research councils. He has contributed to a patent and authored/coauthored 7 books, 13 book chapters, and over 200 technical articles in leading journals and peer-reviewed conferences. He is a Fellow of the Higher Education Academy (UK), a Senior Member of the IEEE, a Member of the IET and BioEM and part of the technical program committees and organizing committees of several international conferences, workshops, and special sessions. He is a committee member of IEEE APS/SC WG P145, IEEE APS Best Paper Award committee and Pearson's focus group on formative assessment. He is acting as an editor of PeerJ Computer Science, associate editor of IEEE Sensors Journal, IEEE Journal of Electromagnetics, RF and Microwaves in Medicine and Biology, IEEE Access, IET Electronics Letters and Microwave & Optical Technology Letters, topic editor for MDPI Sensors, editorial advisor to Cambridge Scholars Publishing, and lead guest editor of numerous special issues of renowned journals. He is chair of the IEEE UKRI Section Young Professionals Affinity Group, vice-chair of the IEEE UKRI Section APS/

MTTS Joint Chapter and acting/acted as the Communications Chair for EuCAP2024, Workshop Chair for Workshop on Sustainable and Intelligent Green Internet of Things for 6G and Beyond in IEEE ICC 2024, IEEE GLOBECOM 2024, IEEE VTC-S 2023, IEEE GLOBECOM 2023, and TPC chair for UCET 2020 and BodyNets 2021 conferences.

Ahmed Zoha is an Associate Professor at the James Watt School of Engineering, University of Glasgow, and a globally recognized expert in artificial intelligence, machine learning, and smart energy systems. With a PhD from the 6G/5GIC Centre at the University of Surrey and over 15 years of experience, he has contributed to cutting-edge research in AI-driven 5G networks, healthcare technologies, and smart energy monitoring, earning prestigious accolades including IEEE Best Paper Awards and endorsement as a UK exceptional talent by the Royal Academy of Engineering. He serves as an Associate Editor for the *Journal of Big Data and Frontiers in Communication and Networks*, as well as a Guest Editor for several Q1 journals. His work, widely cited and recognized in both academia and industry, supports scalable technological solutions that benefit vulnerable populations and advance the Sustainable Development Goals (SDGs).

Muhammad Ali Jamshed has been with the University of Glasgow since 2021. He is endorsed by the Royal Academy of Engineering under the exceptional talent category and was nominated for the Departmental Prize for Excellence in Research in 2019 and 2020 at the University of Surrey. He is a Fellow of the Royal Society of Arts, a Fellow of Higher Education Academy UK, a Senior Member of the IEEE, an Editor of IEEE Wireless Communication Letters, and an Associate Editor of IEEE Sensor Journal, IEEE Communication Standard Magazine and IEEE IoT Magazine. He is Co-inventor of one patent and has more than 70 publications in top-tier international journals, including IEEE Transactions and Magazines, flagship IEEE COMSOC conferences, such as ICC, VTC, INFOCOM, WCNC, etc. He is an Editor of four books. He has been the Lead Guest Editor for Special Issues in IEEE Wireless Communications Magazine (2024–2025), IEEE Communications Standards Magazine (2024–2025), and IEEE Internet of Things Magazine (2023–2024). He has been serving as General Chair for over 10 workshops at IEEE international conferences such as IEEE GLOBECOM 2024, IEEE VTC Fall 2024, IEEE MECOM 2024, IEEE CAMAD 2023, IEEE WCNC 2023, IEEE PIMRC 2022–2023, IEEE GLOBECOM 2023, IEEE VTC Spring 2022–2023, IEEE CAMAD 2023, and IEEE CAMAD 2019. He has been a TPC Member at IEEE ICC 2022–2024 IEEE VTC Spring 2024, IEEE GLOBECOM 2023, and Sessions Chair at IEEE VTC Spring 2022 and IEEE WCNC 2019 and 2023. He is a founding member of the IEEE Workshop on Sustainable and Intelligent Green Internet of Things.

Naeem Ramzan received an MSc degree in telecommunication from the University of Brest, France, in 2004, and a PhD in electronics engineering from Queen Mary University of London, London, UK, in 2008. Currently, he is a Full Professor of Artificial Intelligence and Computer Engineering and the Director of the Artificial Intelligence, Virtual Communication, Network (AVCN) Institute at the University of West of Scotland. Before that, he was a senior research fellow and lecturer at Queen Mary University of London from 2008 to 2012. He is a Fellow of the Royal Society of Edinburgh, a Senior Member of the IEEE, Senior Fellow of the Higher Education Academy (HEA), Cochair of MPEG HEVC verification (AHG5) group and a voting member of the British Standard Institution (BSI). In addition, he holds key roles in the Video Quality Expert Group (VQEG) such as Co-chair of the Ultra High Definition (UltraHD) group; Co-chair of the Visually Lossless Quality Analysis (VLQA) group; and Co-chair of the Psycho-Physiological Quality Assessment (PsyPhyQA). He has been a lead researcher in various nationally or EU-sponsored multimillion-funded international research projects. His research interests are cross-disciplinary and industry-focused and include video processing, analysis and communication, video quality evaluation, Brain-inspired multimodal cognitive technology, Big Data analytics, Affective computing, IoT/smart environments, natural multimodal human–computer interaction, eHealth/connected Health. He has a global collaborative research network spanning both academia and key industrial players. He has been the Lead supervisor/supervisor for about 30 postdoctoral research fellows and PhD research students. He has published over 250 articles in peer-reviewed journals, conferences, and book chapters including standardized contributions. His paper was awarded Best Paper Award 2016 of the IEEE Transaction on Circuit and System for Video Technology and three conference papers were selected for Best Student Paper awards in 2015/2016. He was awarded the Scottish Knowledge Exchange Champion award in 2023 and 2020 and only academics in Scotland got this award twice. He received STARS (Staff Appreciation and Recognition Scheme) award in 2014 and 2016 for "Outstanding Research and Knowledge Exchange" (University of the West of Scotland) and was awarded the Contribution Reward in 2009 and 2011 for outstanding research and teaching activities (Queen Mary University of London). He has chaired/cochaired/organized more than 25 workshops, special sessions, and tracks at International conferences.

List of Contributors

Qammer H. Abbasi
James Watt School of Engineering
University of Glasgow
Glasgow, UK

Abdul Jabbar
James Watt School of Engineering
University of Glasgow
Glasgow, UK

Iftikhar Ahmad
James Watt School of Engineering
University of Glasgow
Glasgow, UK

Saima Gulzar Ahmad
Department of Computer Science
COMSATS University Islamabad,
Wah Campus
Wah Cantt, Pakistan

Nadeem Ajum
Department of Software Engineering
Capital University of Science and
Technology
Islamabad, Pakistan

Romina Soledad Albornoz-De Luise
Departament d'Informàtica
Universitat de València
València, Spain

Kamran Ali
Department of Computer Science
Middlesex University
London, UK

Pablo Arnau-González
Departament d'Informàtica
Universitat de València
València, Spain

Fahad Ayaz
James Watt School of Engineering
University of Glasgow
Glasgow, UK

Kashif Ayyub
Department of Computer Science
COMSATS University Islamabad,
Wah Campus
Wah Cantt, Pakistan

Rami Ghannam
James Watt School of Engineering
University of Glasgow
Glasgow, UK

Muhammad Hanif
Department of Computer Science
COMSATS University Islamabad,
Wah Campus
Wah Cantt, Pakistan

Bushra Haq
Department of Computer Science
Balochistan University of Information
Technology, Engineering and
Management Sciences
Quetta, Pakistan

Sajjad Hussain
James Watt School of Engineering
University of Glasgow
Glasgow, UK

Shagufta Iftikhar
Department of Software Engineering
Capital University of Science and
Technology
Islamabad, Pakistan

Muhammad Ali Imran
James Watt School of Engineering
University of Glasgow
Glasgow, UK

Tassawar Iqbal
Department of Computer Science
COMSATS University Islamabad
Wah Campus
Wah Cantt, Pakistan

Muhammad Ali Jamshed
James Watt School of Engineering
University of Glasgow
Glasgow, UK

Sana Ullah Jan
School of Computing Engineering
and the Built Environment
Edinburgh Napier University, UK

Tahera Kalsoom
Manchester Fashion Institute
Manchester Metropolitan University
Manchester, UK

Mumraiz Khan Kasi
Department of Computer Science
Balochistan University of Information
Technology
Engineering and Management Sciences
Quetta, Pakistan

Ahsan Raza Khan
James Watt School of Engineering
University of Glasgow
Glasgow, UK

Dost Muhammad Khan
Department of Information Technology
The Islamia University of Bahawalpur
Bahawalpur, Pakistan

Sara Khosravi
James Watt School of Engineering
University of Glasgow
Glasgow, UK

Nasira Kirn
School of Computing, Engineering
and Physical Sciences
University of the West of Scotland
Paisley, UK

Haobo Li
James Watt School of Engineering
University of Glasgow
Glasgow, UK

Habib Ullah Manzoor
James Watt School of Engineering
University of Glasgow
Glasgow, UK

Ehsan Ullah Munir
Department of Computer Science
COMSATS University Islamabad
Wah Campus
Wah Cantt, Pakistan

Hassan Rabah
Institut Jean Lamour
Université de Lorraine
France

Rao Naveed Bin Rais
Artificial Intelligence Research Center (AIRC), Ajman University
Ajman, UAE

Naeem Ramzan
School of Computing, Engineering and Physical Sciences
University of the West of Scotland
Paisley, UK

Muhammad Maaz Rehan
Department of Computer Science
COMSATS University Islamabad
Wah Campus
Wah Cantt, Pakistan

Masood Ur Rehman
James Watt School of Engineering
University of Glasgow
Glasgow, UK

Omer Riaz
Department of Information Technology
The Islamia University of Bahawalpur
Bahawalpur, Pakistan

Najia Saher
Department of Information Technology
The Islamia University of Bahawalpur
Bahawalpur, Pakistan

Ana Serrano-Mamolar
Departamento de Lenguajes y Sistemas, Universidad de Burgos
Burgos, Spain

Syed Ahmed Shah
Department of Computer Science, Balochistan University of Information Technology, Engineering and Management Sciences
Quetta, Pakistan

Rizwan Shahid
NHS Trust, Lincoln County Hospital
Lincoln, UK

Sergi Solera-Monforte
Departament d'Informàtica
Universitat de València
València, Spain

Muhammad Suleman
Department of Information Technology
The Islamia University of Bahawalpur
Bahawalpur, Pakistan

Yuyan Wu
Departament d'Informàtica, Universitat de València, València
Spain

Ahmed Zoha
James Watt School of Engineering
University of Glasgow
Glasgow, UK

Preface

In recent years, there has been a significant surge in the utilization of the Internet of Things (IoT) and wireless sensors to meet escalating demands for high data rates, low latency, and ultrareliable communication in 5G/6G systems. Various strategies are under development utilizing intelligent sensing platforms to address these growing needs. Using multiple sensors has proven to be an effective approach to enhance reliability, efficiency, and user experience in diverse application scenarios across healthcare, transportation, environmental monitoring, industrial automation, and entertainment industries.

Incorporating data from multiple modalities such as visual images, radiation levels, texture details, and behavioral patterns captured by sensors like light, temperature, humidity, vision, and motion enhances the information assimilation process, leading to precise decision-making. While offering immense benefits, multi-modal sensing presents significant challenges such as energy efficiency, mobility, reliability, interference mitigation, reliability, security, and real-time processing requirements.

In an interconnected world, seamless integration of IoT with intelligent sensing platforms is essential to deliver transformative solutions. The customization and diversification of intelligent sensors require efficient techniques for designing and integrating sensors, as well as extracting valuable insights from vast amounts of multimodal data. This involves leveraging advanced sensor design, robust big data analytics, and stringent security measures to promote sustainability, spur innovation, and explore new possibilities.

To date, there is a lack of comprehensive literature that addresses design, implementation, and analytical techniques for multimodal intelligent sensing. A dedicated book focusing on these crucial aspects will not only bridge this gap but also educate readers on the key aspects of efficient sensor networks, laying the groundwork for future advancements in a smart and interconnected world. This book is a structured effort in this direction that explores the cutting-edge advancements and challenges in the realm of multimodal sensing, discussing both software and hardware solutions. It covers a broad spectrum of topics in multimodal intelligent sensing for a range of applications, bringing together experts from various disciplines including wireless communications, signal processing, and sensor design.

Key topics discussed include sensor design, deployment efficiency, energy management, data fusion, and information extraction through machine learning, deep learning, and federated learning to showcase the latest developments in this dynamic field. By considering challenges and future prospects, the book caters to a diverse readership within the scientific community.

Chapter 1 delves into the realm of multimodal intelligent sensing, uncovering the various sensor types and the integration of multiple sensors for enhanced capabilities. The chapter also explores the applications of multimodal sensing in different sectors and the challenges and opportunities that come with this dynamic field.

Chapter 2 focuses on antennas for wireless sensors, emphasizing their crucial role as the sensory gateway for wireless networks. Readers will gain insights into fundamental antenna parameters, fabrication methods, and the different types of antennas utilized in sensing networks.

Chapter 3 discusses the sensor design for environmental monitoring, shedding light on combating deforestation through wireless sensor networks and the design considerations involved in creating effective systems for environmental conservation.

Chapter 4 explores the applicability of wireless sensors for multimodal health monitoring, detailing the use of wearable and implantable sensors, multimodal healthcare sensing devices, and the use of AI methods in healthcare systems for enhanced monitoring and diagnosis.

Chapter 5 shifts gears to sensor design for industrial automation, highlighting the role of multimodal sensing in revolutionizing industrial processes. From RF sensors to vision sensors, this chapter explores the design considerations and challenges faced in implementing multimodal sensing in industrial settings.

Chapters 6 and 7 probe hybrid neuromorphic federated learning for activity recognition and multimodal beam prediction in drone communication networks, respectively, showcasing the integration of cutting-edge technologies to enhance sensing capabilities in diverse applications.

Chapter 8 studies the domain of mind-wandering through multiple wearable sensors with the aim of detecting mind-wandering episodes through deep learning and paving ways to improve the detection of students' learning and concentration levels.

Chapters 9 and 10 investigate the advancements in remote infant monitoring systems and telehealth patient monitoring, emphasizing the design of multimodal wearable systems and secure telehealth systems for improved patient care and monitoring.

Chapter 11 explores the ethical considerations involved in the development of multimodal intelligent tutoring systems, respectively, highlighting the balance between innovation and ethical responsibility in the deployment of such systems.

Lastly, Chapter 12 paves the way for the future of multimodal intelligent sensing in the deep learning era, outlining the challenges, perspectives, and ethical considerations that will shape the evolution of sensor technology in the years to come.

This book serves as a comprehensive guide for researchers, engineers, and professionals interested in the advancements and applications of multimodal intelligent sensing systems. Each chapter offers insights, practical considerations, and future directions to inspire further innovation and exploration in this vibrant field.

Masood Ur Rehman
James Watt School of Engineering,
University of Glasgow, Glasgow, UK

Ahmed Zoha
James Watt School of Engineering,
University of Glasgow, Glasgow, UK

Muhammad Ali Jamshed
James Watt School of Engineering,
University of Glasgow, Glasgow, UK

Naeem Ramzan
School of Computing, Engineering and Physical Sciences,
University of the West of Scotland, Paisley, UK

1

Advances in Multi-modal Intelligent Sensing

Masood Ur Rehman[1], Muhammad Ali Jamshed[1], and Tahera Kalsoom[2]

[1] *James Watt School of Engineering, University of Glasgow, Glasgow, UK*
[2] *Manchester Fashion Institute, Manchester Metropolitan University, Manchester, UK*

Intelligent sensing systems play a crucial role in various fields, enabling the acquisition of valuable data for analysis, monitoring, and decision-making processes. Multi-modal intelligent sensing, with its capability to gather information from multiple sensor types and sensing parameters, has emerged as a powerful tool in diverse applications. This chapter aims to provide a comprehensive overview of multi-modal intelligent sensing, offering insights into the diverse sensor types, sensing parameters, application scenarios, and data analysis tools associated with this rapidly evolving field.

1.1 Multi-modal Intelligent Sensing

Multi-modal intelligent sensing refers to the use of multiple sensor types, such as optical, acoustic, thermal, and chemical sensors, to capture and analyze different aspects of the environment or a system. By combining data from various sensors, multi-modal sensing systems can provide a more comprehensive and holistic view of the monitored phenomenon, leading to better insights and decision-making [1].

A simple example is a system designed to monitor a remote environment, such as a home or office, with a view to enhancing security. Such a system might use a variety of sensing devices including visible light cameras, LiDAR, infra-red motion detectors, and contact microphones, each capturing unique types of data as shown in Figure 1.1. The behavior of this environment could be quite complex,

Multimodal Intelligent Sensing in Modern Applications, First Edition.
Edited by Masood Ur Rehman, Ahmed Zoha, Muhammad Ali Jamshed, and Naeem Ramzan.

Published 2025 by John Wiley & Sons, Inc.

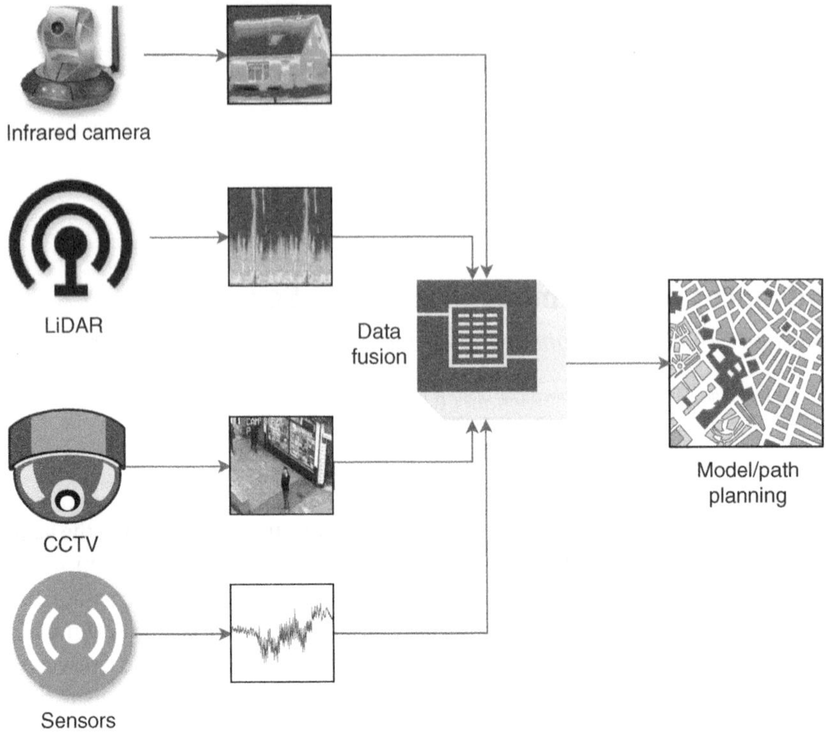

Figure 1.1 Multi-modal intelligent sensing.

for example, the sound of a door opening followed by an increase in infra-red activity and ending with the turning off of a light, could be automatically interpreted as a sequence involving an entry into the environment by an unwanted visitor (door opening), followed by movement to a specific location (increase in infra-red activity), and culminating in an attempt to allay suspicion by simulating the turning off of a light. The analysis of this complex behavior would be greatly enhanced if the system could automatically determine spatio-temporal relationships between events and classify these events into a taxonomy based on their threat to security. This could be achieved by using the information provided by the various sensing devices, organizing this information into one unified data structure through data fusion, with each type of data being a mode, and building a symbolic or semantic model of the observed events through automatic learning and reasoning using methods from artificial intelligence. The intelligent agent, influenced by this understanding, then executes actions or generates outputs accordingly [2].

The components and functionalities of this simple example can be tailored to various applications by selecting appropriate sensors, data fusion methods, and intelligent processing techniques. In recent times, sensing devices like digital cameras, microphones, and range finders have become ubiquitous. Single-mode intelligent sensing refers to the automatic extraction of information from the data produced by these devices using processing and analysis techniques. There is great demand to build systems that can interact more richly with humans in their environment, for applications such as smart environments, augmented reality, and human–computer interfaces [3]. Systems that can emulate or augment human perception by automatically integrating and processing information from different modes (multi-modal sensing) are a natural next step in this quest.

1.2 Sensors for Multi-modal Intelligent Sensing

1.2.1 Sensor Types

The success of multi-modal intelligent sensing systems relies on the selection and integration of appropriate sensor types that can capture different aspects of the phenomenon under observation. Sensors have been known to guide humans since the Han Dynasty [4]. The seismometer is a sensor developed in the second century by Zhang Heng (a Chinese astronomer and mathematician), used to sense earthquakes [5], and the auxanometer and the crescograph are the sensors used to measure the growth in plants [6, 7]. Whereas the galvanometer developed by Hans Christian Ørsted in 1820 (a Danish chemist and physicist) is used to measure the flow of electric current [8]. Moreover, the actinometer developed in 1825 by Sir John Frederick William Herschel (an English astronomer and mathematician) is used to measure thermal power and radiations [9]. Based on the applications, scenarios, and the environment, the sensor(s) can be of different types. For instance, to measure sound/noise levels in a given environment, acoustic sensors, for example, hydrophone and microphone, are used [10]. Similarly, the thermometer is a popular sensor, used to measure the temperature [11]. Different sensor types are presented in Figure 1.2.

Nowadays, the sensors are playing a vital role in our daily routine tasks, for example, tactile sensors used in elevators, lamps that are brightened or dim by a touch of hand, fire alarms, motion detectors, etc., are a few examples. The advancements in micro-machinery have enhanced the expansion of sensors beyond their traditional types as indicated in Figure 1.2. The magnetic, angular rate, and gravity, known as MARG, is one of the classic examples of such type of expansion, which is used to estimate the altitude of an aircraft [13]. The microelectromechanical systems (MEMS) technology has enabled the manufacturing of these sensors on a microscopic scale [14]. The microsensors developed using

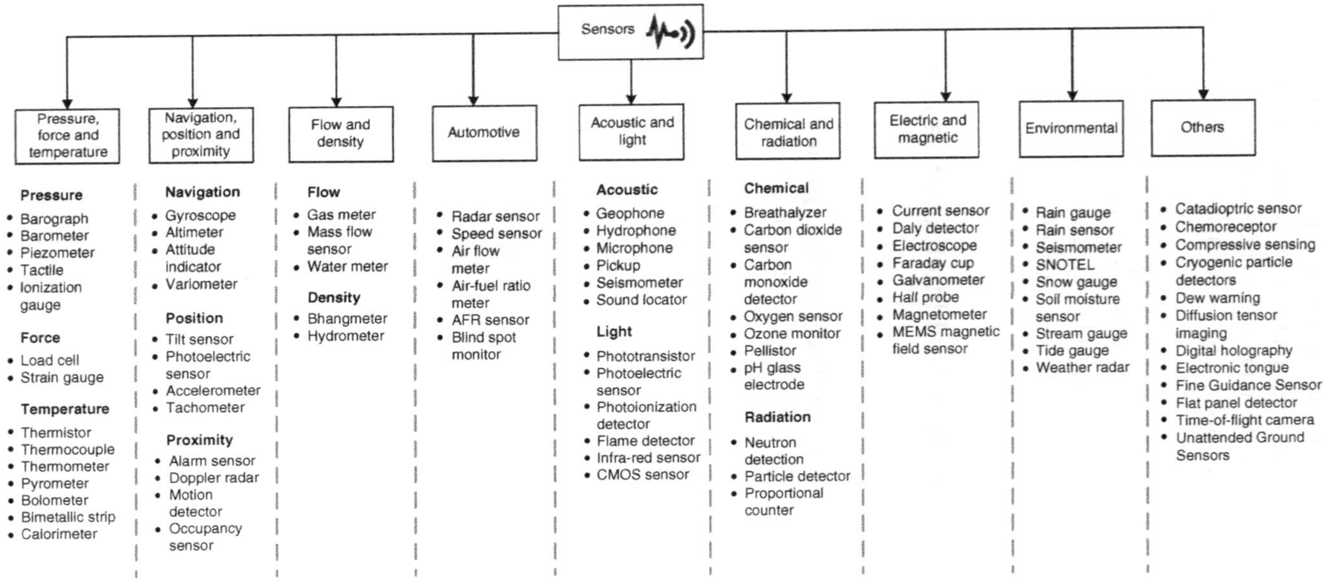

Figure 1.2 Key types of sensors. *Source:* Ref. [12]/IEEE.

different microscopic approaches are comparatively more accurate and faster than the older sensors. Due to the increased demand for rapid, reliable, and affordable data access, the lost cost and easy to use disposable sensors have gained more importance [15].

With the evolution of telecommunication technology, especially, with the advent of wireless communication, more efforts have been put to integrate the sensors with the wireless antennas, that are capable of transmitting sensing information over large distances. The integration of wireless communication technology with the sensors has enabled humans to gain useful information from hard-to-reach areas. In the current era of fifth generation (5G) of mobile communication and beyond, the addition of massive machine type communication (mMTC) and ultra-reliable low latency communications (URLLC) use cases have increased the connectivity and reliability of Internet of Things (IoT) devices [16, 17]. It is envisioned that the number of IoT devices will reach 75 billion by 2025, which will significantly increase the popularity of sensors in wireless communication domain [18]. This increase in the popularity of IoT devices will create a plethora of challenges to meet the end-user requirements. In order to tackle such challenges, first, there is a need to understand the physics behind these sensors, which will result in a more fruitful integration with the wireless domain and in return provide a meaningful platform to address these challenges effectively.

Each sensor type has its own strengths and limitations, and the selection of sensors for a multi-modal intelligent sensing system should be based on the specific requirements and characteristics of the application being monitored. By combining different sensor types, the system can leverage their respective advantages and mitigate their limitations, leading to a more robust and comprehensive sensing solution.

1.2.2 Integration of Multiple Sensor Types for Enhanced Sensing Capabilities

Integrating multiple sensor types in a sensing system can provide enhanced capabilities by leveraging the strengths of each sensor and compensating for their individual limitations. It can enhance capabilities, improve performance, and provide a more comprehensive and accurate representation of the monitored environment. By carefully selecting and integrating sensors with complementary strengths, the system can better adapt to dynamic conditions and deliver more valuable insights for various applications.

1.2.2.1 Advantages of Multiple Sensor Integration

Some key advantages of integrating multiple sensor types in a multi-modal sensing system include the following.

Redundancy and Reliability

By having multiple sensors of different types that measure the same parameters, the system can cross-check and validate the data, increasing reliability and reducing the risk of errors or false readings [19].

Complementary Information

Different sensor types capture different aspects of the environment or object being monitored. Combining optical, thermal, and acoustic sensors, for example, can provide a more comprehensive picture of the surroundings, enabling better analysis and decision-making [20].

Enhanced Accuracy and Precision

Integrating sensors with different resolutions and sensing capabilities can lead to more accurate and precise measurements. For instance, combining a high-resolution optical sensor with a thermal sensor can provide detailed visual information along with temperature data for better object identification and tracking [21].

Robustness to Environmental Conditions

Different sensor types may perform better in specific environmental conditions. By integrating sensors with complementary strengths, the system can maintain functionality across a wider range of operating conditions, such as varying light levels, temperatures, or noise levels [22].

Expanded Detection Capabilities

Combining sensors with different modalities can expand the range of detectable signals or anomalies. For example, integrating optical, thermal, and acoustic sensors can enable the system to detect objects based on visual appearance, heat signatures, and sound signals simultaneously [23].

Improved Contextual Understanding

By combining multiple sensor types, the system can gather a multifaceted view of the environment, leading to a better understanding of the context in which events are occurring. This can enable more informed decision-making and response strategies [24].

1.2.2.2 Key Considerations for Multiple Sensor Integration

Integrating multiple sensor types for multi-modal sensing is a complex task and requires careful consideration of various factors. Some of the key aspects that must be thoughtfully contemplated are illustrated in Figure 1.3 and discussed below.

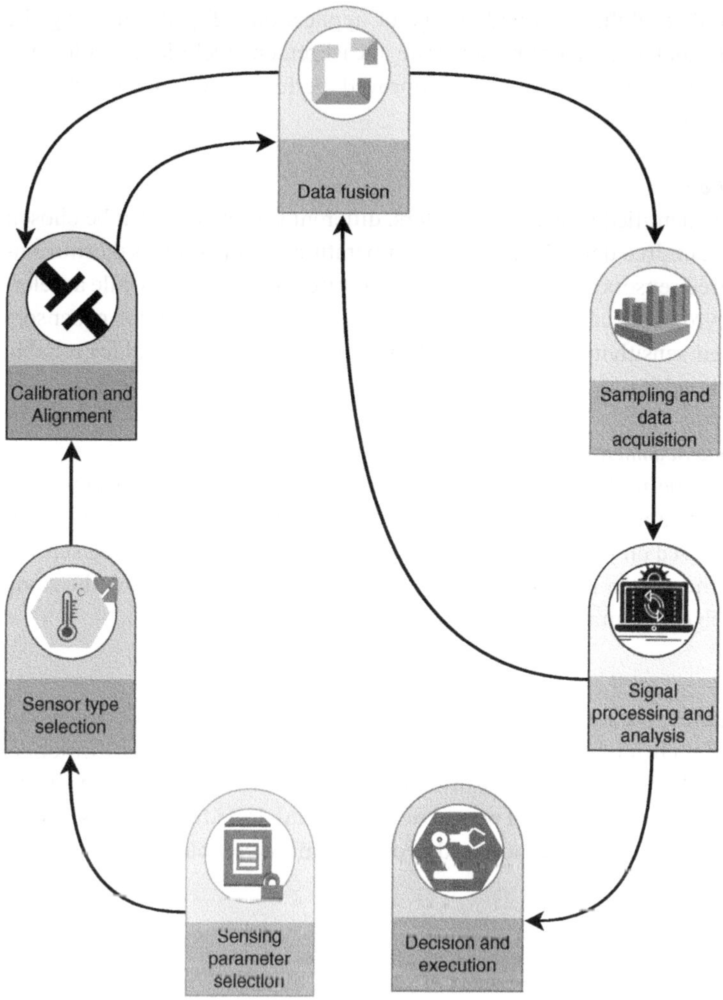

Figure 1.3 Key considerations in multiple sensor integration.

Sensing Parameters

Before integrating sensors, it is essential to identify the specific parameters or variables that need to be measured or monitored. A wide variety of physical quantities such as temperature, pressure, humidity, motion, light intensity, sound, gases and chemicals, proximity and presence, force and strain, and environmental parameters such as moisture level, pH, air quality, radiation level, and magnetic field can be sensed through various types of sensors. Understanding and monitoring these sensing parameters are essential for ensuring system performance, safety, efficiency, and reliability in a wide range of applications [25]. By choosing the right sensors and implementing accurate measurement techniques, informed decisions, optimized processes, and improved performance can be achieved through multi-modal sensing.

Sensor Selection

Based on the identified sensing parameters, different sensor types can be chosen to capture the desired data. For example, temperature sensors such as thermocouples or infrared sensors can be used for temperature measurement, while accelerometers or gyroscopes can be used for motion sensing. Careful selection of sensors based on their sensitivity, accuracy, resolution, and range is essential for reliable data acquisition [26].

Calibration and Alignment

When integrating multiple sensor types, it is crucial to calibrate each sensor to ensure accurate and consistent readings. Calibration involves adjusting the sensor's output based on known reference values [27]. Additionally, aligning sensors properly in the system is important to ensure that they are measuring the same location or object accurately.

Data Fusion

Integrating multiple sensors often involves data fusion, which is the process of combining data from different sensors to improve overall system performance. This can be achieved through sensor fusion algorithms such as Kalman filters, Bayesian inference, or artificial neural networks [26]. Data fusion helps reduce uncertainties, enhance accuracy, and provide a more complete picture of the environment.

Sampling and Data Acquisition

Data acquisition involves the process of sampling, digitizing, and storing sensor data for further processing and analysis. The sampling rate, resolution, and timing synchronization are important considerations in data acquisition. Depending on the application, continuous sampling or event-triggered sampling may be required to capture relevant information [26].

Signal Processing and Analysis

Once the sensor data is acquired, signal processing techniques can be applied to extract meaningful information, detect patterns, or identify anomalies [28]. Signal processing methods such as noise reduction, filtering, feature extraction, and pattern recognition can be used to process the sensor data and derive useful insights.

1.2.2.3 Concurrent Data Acquisition Methods

Concurrent data acquisition from different sensors involves collecting, integrating, and analyzing data from multiple sensors simultaneously to obtain a comprehensive view of the system or environment being monitored. The selection of the appropriate method depends on the specific requirements of the monitoring system, the nature of the sensors involved, and the desired level of synchronization and accuracy in data acquisition. Apart from data fusion, some methods commonly used for concurrent data acquisition from different sensors are the following.

Multiplexing

Multiplexing involves switching between different sensors to sample data sequentially. Analog multiplexers or digital multiplexing systems can be used to connect multiple sensors to a single data acquisition system, allowing data to be collected from each sensor in rapid succession [29]. Time-division multiplexing (TDM) (each sensor is allocated a specific time slot during which it can transmit its data to the data acquisition system enabling better synchronization, as shown in Figure 1.4) or frequency-division multiplexing (FDM) (assigning specific frequency bands to different sensors for data transmission (Figure 1.5) allowing true concurrent data acquisition without interference but at the cost of increased complexity requiring dedicated hardware) are two widely used methods [30].

Parallel Processing

Parallel processing involves acquiring data from multiple sensors simultaneously by using multiple data acquisition channels or systems. Each sensor is connected to a dedicated channel, allowing for independent data collection from different sensors in parallel. This offers the highest level of concurrency but requires more hardware resources [31].

Distributed Data Acquisition

Distributed data acquisition systems consist of multiple nodes or modules, each connected to one or more sensors. These nodes communicate with a central control unit or data aggregator to coordinate data acquisition from different sensors in real time. This approach is well suited for large-scale, geographically distributed sensor deployments [31].

Synchronous Sampling

Synchronous sampling involves triggering all sensors simultaneously to capture data at the same time instance. This method ensures that data from different sensors is synchronized to prevent timing discrepancies and ensure accurate data comparison crucial for applications requiring precise temporal correlation between different modalities (e.g., audio and video synchronization) [32].

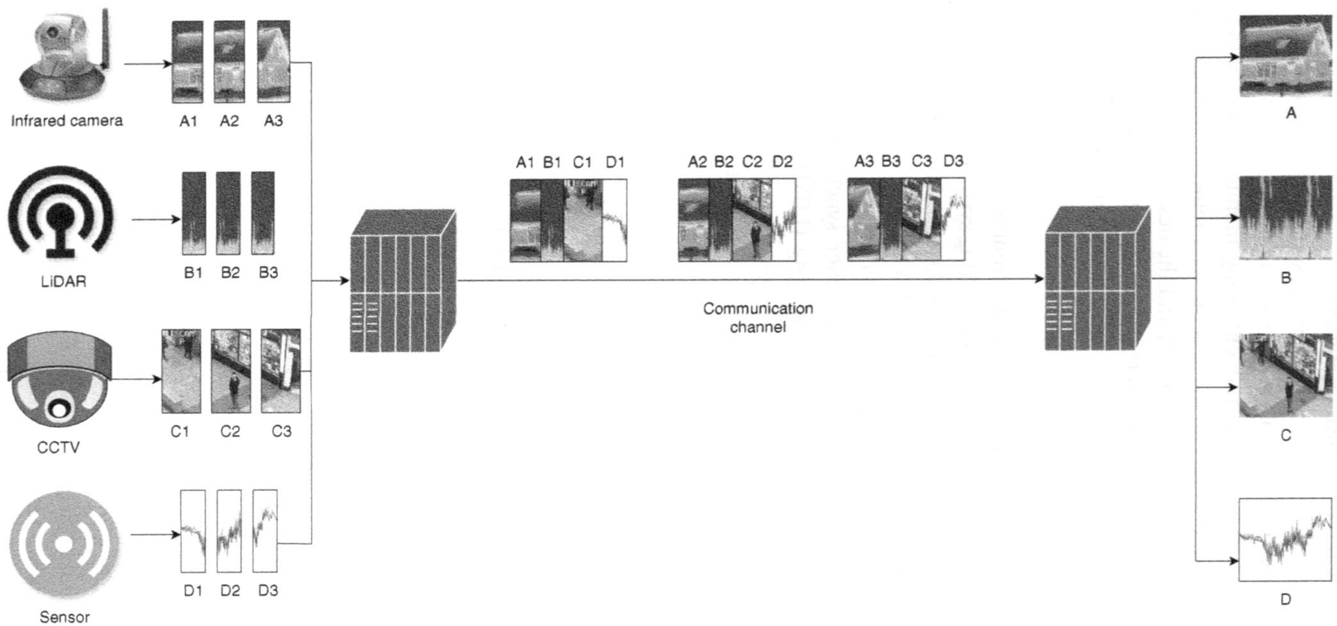

Figure 1.4 Time-division multiplexing.

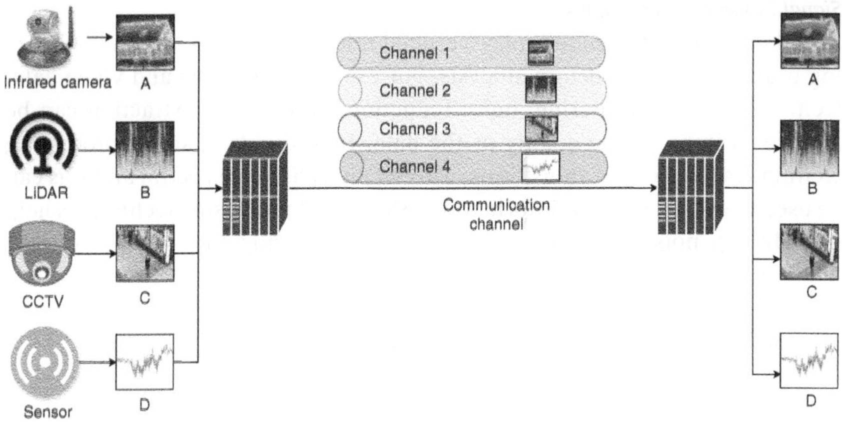

Figure 1.5 Frequency-division multiplexing.

Sensor Networks

Sensor networks consist of interconnected sensors that communicate with each other and a centralized data collection point. Wireless sensor networks (WSNs) and IoT devices enable concurrent data acquisition from multiple sensors distributed across a wide area [33].

Data Bus Systems

Using a common data bus or communication protocol such as controller area network (CAN) or Ethernet can facilitate concurrent data acquisition from different sensors. This method allows for standardized communication between sensors and data acquisition systems [34].

Real-Time Data Processing

Employing real-time data processing techniques such as data streaming, event-driven programming, or edge computing enables immediate analysis of data collected from multiple sensors. Real-time processing helps in making timely decisions based on the acquired sensor data [35].

1.2.2.4 Data Analysis Tools for Multi-modal Sensing

Multi-modal sensing involves the simultaneous collection of data from different sources or types of sensors, such as images, audio, motion, and temperature. Analyzing multi-modal data requires sophisticated tools and techniques to effectively extract meaningful insights. Data analysis tools for multi-modal sensing encompass a range of techniques from signal processing to machine learning, data fusion, and visualization.

Signal Processing Techniques

Signal processing techniques play a crucial role in pre-processing and enhancing multi-modal data before analysis. For example, in image and video data, techniques such as filtering, noise reduction, and feature extraction can be applied to improve the quality of the data. In audio data, signal processing techniques like Fourier transforms, wavelet analysis, and spectral analysis can be used to extract relevant information [36]. Signal processing techniques help in reducing noise, extracting features, and preparing the data for further analysis.

Machine Learning and Artificial Intelligence Algorithms

Machine learning and artificial intelligence algorithms are essential for analyzing multi-modal data and extracting patterns and insights. These algorithms can be used for tasks such as classification, clustering, regression, anomaly detection, and prediction [37]. Techniques like deep learning, neural networks, support vector machines, and random forests can be applied to multi-modal data to uncover hidden relationships and patterns [38]. Machine learning algorithms can learn from the data and make predictions or decisions based on the patterns they find, enabling the automation of complex data analysis tasks.

Data Fusion and Integration Methods

Data fusion and integration involve combining information from multiple sensors or data sources to provide a holistic view of the data. Fusion techniques can be used to merge data at different levels, such as feature-level fusion, decision-level fusion, or sensor-level fusion [39]. By integrating data from diverse sources, data fusion techniques help in reducing redundancy, improving the accuracy of analysis, and providing a more comprehensive understanding of the data. Methods like Bayesian inference, Kalman filtering, and Dempster-Shafer theory can be applied for data fusion in multi-modal sensing scenarios [40].

Visualization and Interpretation of Multi-modal Data

Visualization and interpretation of multi-modal data play a crucial role in understanding the relationships and patterns within the data. Visualizing data from different modalities together can provide a more comprehensive view and facilitate the identification of correlations and insights. Techniques like dimensional reduction, scatter plots, heat maps, and clustering visualizations can be used to represent multi-modal data in a human-readable format [41]. Interpretation of multi-modal data involves extracting meaningful insights, identifying trends, anomalies, and patterns, and communicating the findings effectively to stakeholders.

1.2.2.5 Considerations for Data Fusion and Synchronization

Data fusion and synchronization are crucial aspects of concurrent data acquisition from different sensors to ensure accurate and reliable data processing. Following are some of the key factors that should be deliberated for data fusion and synchronization to make multi-modal sensing effective and efficient.

Sensor Calibration and Synchronization

One of the major challenges in multi-modal sensing is ensuring that sensors are accurately calibrated and synchronized. Sensor calibration involves adjusting sensor readings to account for factors such as systematic error, drift, noise, or bias, which can impact the accuracy and reliability of the data [42]. Synchronization is essential to ensure that data from different sensors are aligned in time enabling accurate fusion. Strategies for addressing calibration and synchronization issues include using calibration algorithms, calibration targets, time-stamping, and synchronization protocols to improve the quality and consistency of the data. Utilizing a common time reference, such as a global positioning system (GPS) signal or a high-precision clock, can ensure that data from various sensors is timestamped consistently [43].

Data Alignment

Ensuring that data samples from different sensors correspond to the same time period is essential for successful data fusion. Aligning data streams by timestamp or aligning data sequences through interpolation techniques can help in merging data from disparate sources [44].

Data Integration

Data fusion involves combining information from multiple sensors to create a unified representation of the monitored system. Using fusion algorithms, such as Kalman filters, particle filters, or Bayesian inference, can help in reconciling conflicting or complementary data from different sensors [45].

Error Handling

Dealing with sensor errors, outliers, or missing data is crucial for maintaining data integrity during fusion. Implementing outlier detection algorithms, interpolation methods for missing values, and error correction techniques can improve the robustness of the data fusion process [46].

Redundancy and Reliability

Incorporating redundancy in sensor data can enhance the reliability of the fusion process by cross-validating measurements from different sensors. Redundant sensors or sensor networks can provide backup data in case of sensor failures or discrepancies [47].

Communication Protocols

Standardized communication protocols and data formats facilitate seamless data exchange between sensors, data acquisition systems, and data processing units. Utilizing protocols like MQTT, OPC UA, or Modbus can streamline data transmission and synchronization efforts [48].

Computational Efficiency

Optimizing data fusion algorithms for efficiency and scalability is essential for processing large volumes of sensor data in real time. Implementing parallel processing techniques, distributed computing architectures, or edge computing solutions can enhance computational performance [49].

Data Security and Privacy

Maintaining the security and privacy of sensor data during fusion is critical to prevent unauthorized access, data breaches, or tampering. Implementing encryption, access controls, and data anonymization techniques can safeguard sensitive sensor data [50].

System Integration

Integrating data fusion processes with the overall system architecture, control systems, visualization tools, or decision-making frameworks is essential for leveraging fused sensor data effectively [51]. Ensuring interoperability and compatibility with existing systems is key to maximizing the impact of data fusion efforts.

1.3 Applications of Multi-modal Intelligent Sensing

The use of multi-modal sensing is rapidly evolving, and nowadays sensors are considered an integral part in performing routine tasks. With the addition of mMTC as a use case in 5G/6G, the popularity and feasibility of using wireless sensors in IoT domain have gained a lot of interest, and it is predicted that the number of connected IoT devices will reach 125 billion by 2030 [52].

Multi-modal intelligent sensing has various application scenarios across different industries and domains making human life more secure, increasing industrial production efficiency, helping towards global warming, etc. Some of the key application scenarios of multi-modal intelligent sensing are discussed below.

1.3.1 Healthcare and Medical Monitoring

The medical sector is experiencing a transformation thanks to multi-modal intelligent sensing that fosters a holistic approach to healthcare. By leveraging data from diverse sources (implanted/wearable/environment-embedded sensors), this technology empowers patients to manage their health, improves treatment outcomes, and facilitates better decision-making by healthcare professionals [53]. It personalizes healthcare, fostering patient engagement and well-being. This technology gathers data from various sensors (vital signs, imaging, and activity trackers) to provide a comprehensive picture of a patient's health. Healthcare providers can remotely track patients' health in real time using wearable devices, smart sensors, and mobile apps. This allows for early detection of potential issues and timely interventions. Combining data from different sources (imaging, genetics, and biomarkers) improves the accuracy of disease detection and diagnosis [54]. This leads to earlier interventions and personalized treatment plans.

Multi-modal sensing allows the analysis of vast datasets including genetics, lifestyle, and environment. This personalized approach tailors interventions and preventive strategies to each patient's unique needs. Surgeons benefit from real-time feedback using data from surgical instruments, imaging devices, and physiological sensors. This results in enhanced surgical precision and improved patient outcomes [55]. Tracking patient progress, movement patterns, and optimizing treatment plans becomes possible with motion sensors, wearables, and biofeedback systems. This leads to improved recovery and rehabilitation. Smart pill bottles, wearables, and digital health platforms track medication adherence and outcomes. This helps healthcare providers identify and address adherence challenges proactively [56].

Orthopedics can monitor the physical conditions of bones in real time using implantable sensors [57]. Body-worn sensors can be used to treat cardiovascular patients effectively [58]. It is estimated that three million elderly people are brought to Accidents & Emergency (A&E) for fall-related injuries every year in the United States [59]. The environment-embedded WSs also play a key role in monitoring the health of patients. The environment embedded sensors can be employed to monitor the health of the elderly, in mobile as well as static conditions [60].

1.3.2 Automotive and Transportation Systems

Multi-modal intelligent sensing is transforming the automotive and transportation landscape. It contributes to the development of smarter, safer, and more efficient transportation systems, leading to a transformed transportation experience for both passengers and operators [61]. By fusing data from various sensors (cameras, LiDAR, and radar), vehicles gain a 360° view of their surroundings, leading to enhanced safety [62]. Advanced driver assistance systems (ADAS) utilize these sensors to provide real-time information, enabling features like lane departure warning and automatic emergency braking, ultimately reducing accidents [63].

Multi-modal sensing is the backbone of self-driving cars. By combining LiDAR, cameras, radars, and GPS data, autonomous vehicles navigate complex environments and make crucial decisions in real time [64]. Sensor data from vehicles and infrastructure helps optimize traffic flow. Intelligent traffic lights and vehicle-to-infrastructure communication systems reduce congestion, improve travel times, and enhance road safety. Sensor integration allows tracking of fuel consumption, engine health, and tire pressure, ensuring optimal vehicle operation and timely maintenance.

Sensors personalize the travel experience by adjusting climate control, lighting, and entertainment based on passenger preferences [65]. Air quality, noise, and emissions can be monitored using multi-modal sensing. This data empowers policymakers to make informed decisions for cleaner and more sustainable transportation solutions.

1.3.3 Environmental Monitoring and Conservation

Multi-modal intelligent sensing allows for comprehensive environmental monitoring, data analysis, and implementation of effective conservation measures. It empowers researchers, policymakers, and conservationists to address environmental challenges, safeguard ecosystems, and promote sustainable development for future generations [66, 67]. Sensor networks, weather stations, and satellite imagery track air pollution, identify emission sources, and assess their impact such as the Breathe London project [68]. Similarly, sensors, acoustic devices, and satellite observations monitor water bodies, detect contaminants, and assess aquatic health. Data on temperature, pH, and oxygen levels helps identify pollution sources and protect water resources.

Camera traps, acoustic sensors, and satellite tracking devices monitor wildlife populations, endangered species, and migration patterns. This data helps prioritize conservation efforts, establish protected areas, and prevent extinction [69]. Moreover, soil sensors, hyperspectral imaging, and drone surveys assess soil properties, moisture levels, and degradation. Data on composition, nutrients, and erosion helps develop sustainable agricultural practices and improve soil fertility [70].

The impact of deforestation has also raised the need to measure the growth of trees, plants, etc., in real time as they directly impact the level of oxygen in the environment. A fast bacteria detection method in olive tree can be adaptive to other kinds of trees as well [71]. A company, Nature 4.0, is purely dedicated to develop innovative IoT products to save the environment from various adverse effects [72]. The TreeTalker (TT+) is one of their products that measures the water consumption, biomass growth, etc., in a tree. Unmanned aerial vehicles (UAVs), satellite imaging, and ground sensors monitor forests, detect deforestation, and assess fire risks. Real-time data on forest cover, tree health, and fire hotspots enables forest conservation, wildfire prevention, and sustainable forest management [73]. Climate models, satellite observations, and ground sensors analyze trends in temperature, precipitation, sea level, and greenhouse gas emissions. This data informs climate policy decisions, helps mitigate climate risks, and promotes climate-resilient practices.

1.3.4 Smart Cities and Infrastructure Management

The vision of cities that are efficient, sustainable, and responsive to residents' needs is becoming a reality with multi-modal intelligent sensing technologies (Figure 1.6). Sensing technologies create efficient, sustainable, and resilient urban environments with optimized resource management practices and a higher quality of life for residents. Traffic sensors, GPS trackers, and smart cameras optimize traffic flow, reduce congestion, and improve public transit. Real-time data on vehicles, pedestrians, and road conditions empowers intelligent transportation systems for efficient urban mobility [74, 75].

Smart meters, energy monitoring devices, and renewable energy sources optimize energy use and promote sustainability. Data on consumption, peak demand,

Figure 1.6 Realization of smart city concept through multi-modal sensing.

and renewable generation allows for implementing smart grids, reducing energy costs, and minimizing a city's carbon footprint [76]. A continuous Wireless Sensor (WS) monitoring setup can save energy consumption by up to 18%, in comparison to traditional manual periodic check-ups [75]. Water quality sensors, leak detection systems, and irrigation controllers ensure efficient water resource management. Data on water quality, consumption, and distribution networks helps identify leaks, prevent wastage, and sustain water resources for city residents [77, 78]. Waste sensors, smart bins, and sensor-equipped collection vehicles optimize waste collection and promote recycling. Data on waste volumes, collection frequencies, and recycling rates allows for optimizing collection schedules, reducing costs, and improving waste disposal practices [79].

Building automation systems, occupancy sensors, and energy management platforms optimize energy use and improve occupant comfort. Data on temperature, lighting, and occupancy patterns enables smart building solutions that reduce energy consumption and enhance indoor environment quality [75, 80]. Sensor data, satellite imagery, and weather monitoring devices improve disaster preparedness and response. This real-time information helps track disasters, assess damage, and coordinate rescue efforts, mitigating the impact of natural disasters on urban communities [81].

1.3.5 Industrial Automation

In industrial settings, multi-modal intelligent sensing is used for condition monitoring, predictive maintenance, and process optimization. By integrating data from sensors measuring temperature, vibration, pressure, and other parameters, manufacturers can identify equipment failures, optimize production processes, and minimize downtime [82]. By integrating data from sensors capturing customer behavior, foot traffic, and product interaction, retailers can enhance the shopping experience, optimize store layouts, and target promotions more effectively. For instance, the inventory can be managed effectively by using the WSs and can eliminate the fear of over-stocking, replenishing the required products on time, removing burden on productions, etc. [83]. Ocado, one of the leading online supermarkets in the United Kingdom, is effectively exploiting IoT-based wireless sensing in increasing the efficiency of their warehouses [84]. Similarly, the environmental parameters in any industrial area can be measured, such as raising an alarm for evacuation on detection of any gas leakage, liquid, and other harmful stuff, hence saving human life [85].

1.4 Challenges and Opportunities in Multi-modal Sensing

Multi-modal sensing offers incredible opportunities for gathering diverse and rich data from multiple sources. However, along with its advantages, there are several challenges that need to be addressed to effectively harness the full potential of multi-modal sensing. Key challenges faced by the multi-modal sensing systems and strategies to address them are discussed below.

1.4.1 Data Security and Privacy

Data security and privacy are significant concerns in multi-modal sensing, as the collection and analysis of sensitive data from multiple sources can pose risks to individuals' privacy and confidentiality. It is essential to prioritize the protection of personal data and ensure that appropriate security measures are in place to safeguard sensitive information in multi-modal sensing applications [86].

Numerous efforts are being made to solve the security and privacy issues in multi-modal sensing systems. Strategies for mitigating data security and privacy concerns include data encryption, secure communication protocols, access control mechanisms, data anonymization techniques, and compliance with regulatory frameworks such as GDPR. Radio frequency identification (RFID) and new 5G standards are aiming to solve privacy concerns at the hardware level [87]. Whereas, at the software level, blockchain and key management systems are proving to be effective solutions against security threats [88]. As the energy constraints also limit the use of more powerful security protocols as sensors, lightweight security algorithms based on intrusion detection can be used to provide energy-efficient security and privacy in multi-modal sensing [89]. The trust-based schemes are also useful in securing the user data at sensor nodes but consume more energy in comparison to lightweight algorithms. New technologies, such as blockchain, have also proven to be much effective in enhancing security [90]. Nevertheless, there still exist challenges in achieving optimal performance in terms of security and privacy and gaining the trust of end users, which imposes a significant impact on the economic growth of IoT-based sensors.

1.4.2 Interoperability and Standardization

The rapid increase in the popularity of multi-modal sensing with the rise of IoT devices has significantly increased the level of heterogeneous traffic. Interoperability and standardization are critical challenges in multi-modal sensing, as data from different sensors and modalities may vary in format, structure, or communication protocols. Currently, there are more than three hundred different types of IoT platforms available in the market, and each of the platforms is working under a closed ecosystem [91]. Ensuring interoperability between sensors and systems is essential for seamless data integration and analysis. Its absence is highlighted as an economic threat and it is stated that, solely, the presence of interoperability between different IoT platforms can increase the potential benefits of IoT devices by 40% [92].

Standardization efforts can help establish common data formats, communication interfaces, and metadata standards to facilitate interoperability across different platforms and devices. Strategies for addressing interoperability and standardization challenges include adopting industry standards, developing interoperable data formats, and promoting open-source solutions to enable seamless data exchange and collaboration in multi-modal sensing environments. The use of multi-layer trust management is considered an effective way to overcome interoperability [93], but it adds an additional trust computational burden on resource-constrained sensor nodes. Block-chain-based trust management can reduce the trust computational load of sensor nodes [94]. The use of the semantic graph representation approach is yet another way to mitigate the issues related to interoperability in multi-modal sensing [95]. The absence of interoperability reduces the scalability of sensor nodes. Despite the abundance of research in the literature on addressing interoperability, it is still an open-ended question that has not been fully resolved.

1.4.3 Energy Efficiency and Power Management

Energy efficiency and power management are key considerations in multi-modal sensing, particularly in applications where sensors are deployed in resource-constrained environments or operate on battery-powered devices. It is expected that the sensors can perform the intended operations for long periods by just relying on a limited power supply. Replacement of batteries for each sensor is cost-prohibitive or even for some hostile environment applications, it is impossible. The simultaneous operation of multiple sensors can consume significant energy, leading to shorter battery life and increased maintenance costs.

The energy consumption requirements of sensors vary from application-to-application and each of the sensors is designed accordingly [1]. With the increase in the demand for IoT applications, achieving the required amount of energy efficiency is becoming a challenging task. Several strategies for optimizing energy efficiency and power management in multi-modal sensing are being proposed. The use of energy-efficient scheduling, routing, and clustering algorithms is one of the simplest, yet effective ways to increase the lifetime of sensor nodes [96]. Energy harvesting is another popular way to enhance the lifetime of sensor nodes, but requires additional electronics and, hence, increases the cost and size of the sensor nodes [97]. The use of compressed sensing has shown to be much effective in prolonging the lifetime of sensor nodes but works effectively for smaller coverage areas [98]. Although the software-based techniques to reduce the overall energy consumption in multi-modal sensing systems such as data aggregation are well researched, there exists a significant gap from the hardware perspective.

1.4.4 Coverage

Covering a sensing region optimally is one of the critical challenges faced by multi-modal sensing systems. Coverage plays a critical role in keeping the energy consumption within the required limits to maintain the Quality of Service (QoS) and to enhance the lifetime of sensing nodes [99]. The physical coverage area of a sensor is defined by the capability of that sensor to sense information in its surrounding radius [12]. Virtual force, Voronoi-based, vector-based, Bee protocol, etc., are some of the examples of algorithms used to optimize the coverage of a sensor node [100]. The topology by which each sensor node is deployed in a multi-modal sensing system is also an effective method to improve the coverage of each node [101]. However, the increase in the use of wireless sensors in delay-tolerant applications, such as healthcare and livestock monitoring, has significantly increased the challenges associated in achieving the optimal coverage.

1.4.5 Summary

Multi-modal sensing plays a crucial role in futuristic smart living scenarios by offering numerous benefits such as redundancy, reliability, enhanced accuracy, robustness to environmental conditions, expanded detection capabilities, and improved contextual understanding. This chapter extensively covers the essential components of multi-modal sensing systems. The effectiveness of a multi-modal sensing system relies on the seamless integration of diverse sensors. This integration necessitates a thoughtful consideration of the required sensing parameters, appropriate sensor types, sensor calibration, alignment to prevent data flooding and detection errors, data fusion from various sensors, data sampling, acquisition, signal processing, and analysis for informed decision-making. The use of multiple data sources also requires efficient concurrent data acquisition methods such as multiplexing, parallel processing, distributed data acquisition, synchronous sampling, sensor networks, data bus systems, and real-time data processing. The chapter emphasizes the importance of data fusion and analysis tools for multi-modal sensing. It delves into the key applications of multi-modal sensing in healthcare, automotive and transportation systems, environmental smart cities, and industrial automation using state-of-the-art technologies. The chapter concludes with a detailed discussion on critical issues like data security, privacy, interoperability, standardization, energy efficiency, power management, and range coverage, along with potential solutions to facilitate the widespread deployment of multi-modal systems.

References

1 Cheng, X., Zhang, H., Zhang, J. et al. (2024). Intelligent multi-modal sensing-communication integration: synesthesia of machines. *IEEE Communications Surveys & Tutorials* 26 (1): 258–301.

2 Karle, P., Fent, F., Huch, S. et al. (2023). Multi-modal sensor fusion and object tracking for autonomous racing. *IEEE Transactions on Intelligent Vehicles* 8 (7): 3871–3883.

3 Wang, X., Sun, Z., Chehri, A. et al. (2024). Deep learning and multi-modal fusion for real-time multi-object tracking: algorithms, challenges, datasets, and comparative study. *Information Fusion* 105: 102247.

4 Needham, J. Science and civilization in China. Volume I, Introductory orientations. *Philosophy* 30 (115): 1955.

5 Sleeswyk, A.W. and Sivin, N. (1983). Dragons and toads. The Chinese seismoscope of AD132. *Chinese Science* 1–19.

6 Olymbios, C.M. (1973). Physiological studies on the growth and development of the carrot (*Daucus carota* L). PhD dissertation. University of London.

7 Bose, J.C. and Das, G. (1919). Researches on growth and movement in plants by means of the high magnification crescograph. *Proceedings of the Royal Society of London. Series B, Containing Papers of a Biological Character* 90 (631): 364–400.

8 Keithley, J.F. (1999). *The Story of Electrical and Magnetic Measurements: From 500 BC to the 1940s*. Wiley.

9 Gouy, M. and Rigollot (1889). On an electrochemical actinometer. *The London, Edinburgh, and Dublin Philosophical Magazine and Journal of Science* 27 (166): 288–288.

10 Sessler, G. (1991). Acoustic sensors. *Sensors and Actuators A: Physical* 26 (1–3): 323–330.

11 Santos, J.L. (2021). Optical sensors for industry 4.0. *IEEE Journal of Selected Topics in Quantum Electronics* 27 (6): 1–11.

12 Jamshed, M.A., Ali, K., Abbasi, Q.H. et al. (2022). Challenges, applications, and future of wireless sensors in Internet of Things: a review. *IEEE Sensors Journal* 22 (6): 5482–5494.

13 Cookson, J.L. (2010). A method for testing the dynamic accuracy of micro-electro-mechanical systems (mems) magnetic, angular rate, and gravity (MARG) sensors for inertial navigation systems (INS) and human motion tracking applications. Naval Postgraduate School Monterey Ca, Technical Report.

14 Gabriel, K., Jarvis, J., and Trimmer, W. (1988). *Small Machines, Large Opportunities: A Report on the Emerging Field of Microdynamics*. Murray Hill, New Jersey, USA: National Science Foundation, AT&T Bell Laboratories.

15 Dincer, C., Bruch, R., Costa-Rama, E. et al. (2019). Disposable sensors in diagnostics, food, and environmental monitoring. *Advanced Materials* 31 (30): 1806739.

16 Jamshed, M.A. and Rehman, M.U. (2021). MIMO and IRS: Enablers for energy efficient wireless sensor networks. *IEEE IoT Newsletter* September.

17 Pokhrel, S.R., Ding, J., Park, J. et al. (2020). Towards enabling critical mMTC: a review of URLLC within mMTC. *IEEE Access* 8: 131796–131813.

18 Statista, I. (2018). Internet Of Things (IoT) connected devices installed base worldwide from 2015 to 2025 (in billions). https://www.statista.com/statistics/471264/iot-number-of-connected-devicesworldwide/ (accessed 17 May 2020).

19 Gravina, R., Alinia, P., Ghasemzadeh, H., and Fortino, G. (2017). Multi-sensor fusion in body sensor networks: State-of-the-art and research challenges. *Information Fusion* 35: 68–80.

20 Chitta, K., Prakash, A., Jaeger, B. et al. (2022). TransFuser: imitation with transformer-based sensor fusion for autonomous driving. *IEEE Transactions on Pattern Analysis and Machine Intelligence* 45: 12878–12895.

21 Darsena, D., Gelli, G., Iudice, I., and Verde, F. (2023). Sensing technologies for crowd management, adaptation, and information dissemination in public transportation systems: a review. *IEEE Sensors Journal* 23 (1): 68–87.

22 Carminati, M. (2018). Trends and paradigms in the development of miniaturized sensors for environmental monitoring. *2018 IEEE International Conference on Environmental Engineering (EE)*, Milan, Italy (12–14 March 2018).

23 Sören Richard Stahlschmidt, Benjamin Ulfenborg, Jane Synnergren, Multimodal deep learning for biomedical data fusion: a review, *Briefings in Bioinformatics*, Volume 23, Issue 2, March 2022, bbab569.

24 Palmerini, L., Reggi, L., Bonci, T. et al. (2023). Mobility recorded by wearable devices and gold standards: the Mobilise-D procedure for data standardization. *Scientific Data* 10: 38.

25 Wilson, J.S. (ed.) (2004). *Sensor Technology Handbook*. Newnes/Elsevier.

26 Liu, Y., Lin, Y.-L., Kyung, C.-M., and Yasuura, H. (ed.) (2020). *Smart Sensors and Systems: Technology Advancement and Application Demonstrations*. Springer Cham.

27 Hering, E. and Schönfelder, G. (ed.) (2022). *Sensors in Science and Technology: Functionality and Application Areas*. Springer Wiesbaden.

28 Zheng, L., Lops, M., Eldar, Y.C., and Wang, X. (2019). Radar and communication coexistence: an overview: a review of recent methods. *IEEE Signal Processing Magzine* 36 (5): 85–99.

29 Babić, D., Pul, M., Umiljanović, L. and Vranješ, M. (2020). Automotive video data gathering and reproduction tool. *2020 International Conference on Smart Systems and Technologies (SST)*, Osijek, Croatia (14–16 October 2020).

30 Ding, Z., Liu, Y., Choi, J. et al. (2017). Application of non-orthogonal multiple access in LTE and 5G networks. *IEEE Communications Magazine* 55 (2): 185–191.

31 Ilyas, K. and Ullah, I. (2016). A state estimation and fusion algorithm for high-speed low-altitude targets. *2016 19th International Multi-Topic Conference (INMIC)*, Islamabad, Pakistan (5–6 December 2016).

32 Wang, C.-X., Renzo, M.D., Stanczak, S. et al. (2020). Artificial intelligence enabled wireless networking for 5G and beyond: recent advances and future challenges. *IEEE Wireless Communications* 27 (1): 16–23.

33 Tubaishat, M. and Madria, S. (2003). Sensor networks: an overview. *IEEE Potentials* 22 (2): 20–23.

34 Shao, C., Asano, S., Muroyama, M. et al. (2018). Connection of 48 sensors on the same serial bus line for human-inspired event-driven tactile sensation covering wide spatial range over 2 meters at millimeter resolution. *2018 IEEE Micro Electro Mechanical Systems (MEMS)*, Belfast, UK (21–25 January 2018), pp. 866–869.

35 Hamid, S., Bawany, N.Z., Sodhro, A.H. et al. (2022). A systematic review and IoMT based big data framework for COVID-19 prevention and detection. *Electronics* 11 (17): 2777.

36 Zhang, J.A., Liu, F., Masouros, C. et al. (2021). An overview of signal processing techniques for joint communication and radar sensing. *IEEE Journal of Selected Topics Signal Process* 15 (6): 1295–1315.

37 Tian, Y., Pan, G., and Alouini, M.-S. (2021). Applying deep-learning-based computer vision to wireless communications: methodologies opportunities and challenges. *IEEE Open Journal of Communications Society* 2: 132–143.

38 Salehi, B. et al. (2022). Deep learning on multimodal sensor data at the wireless edge for vehicular network. *IEEE Transactions of Vehicular Technology* 71 (7): 7639–7655.

39 Yang, Y., Gao, F., Xing, C. et al. (2021). Deep multimodal learning: merging sensory data for massive MIMO channel prediction. *IEEE Journal on Selected Areas in Communications* 39 (7): 1885–1898.

40 Gu, S., Zhang, Y., Tang, J. et al. (2019). Integrating dense LiDAR-camera road detection maps by a multimodal CRF model. *IEEE Transactions on Vehicular Technology* 68 (12): 11635–11645.

41 Huang, Z. and Cheng, X. (2022). A 3-D non-stationary model for beyond 5G and 6G vehicle-to-vehicle mmWave massive MIMO channels. *IEEE Transactions on Intelligent Transportation Systems* 23 (7): 8260–8276.

42 Chen, B., Varshney, P. K., Zulch, P. et al. (2020). Heterogeneous sensor fusion with out of sync data. *2020 IEEE Aerospace Conference*, Big Sky, MT, USA (7–14 March 2020), pp. 1–6.

43 Zhou, L., Feng, Z., Wang, H., and Guo, Q. (2023). MIUIC: a human-computer collaborative multimodal intention-understanding algorithm incorporating comfort analysis. *International Journal of Human–Computer Interaction* 40 (19): 6077–6090.

44 Wan, S. and Ding, J. (2024). A recommendation method based on multi-source heterogeneous hypergraphs and contrastive learning. *IEEE Access* 12: 70001–70016.

45 Sun, D., Junjie, H., Huifeng, W. et al. (2024). A comprehensive survey on collaborative data-access enablers in the IIoT. *ACM Computing Surveys* 56 (2): Article 50.

46 Gao, F., Tan, S., Shi, H., and Zheng, M. (2021). A status-relevant blocks fusion approach for operational status monitoring. *Engineering Applications of Artificial Intelligence* 106.

47 Asif-Ur-Rahman, M., Afsana, F., Mahmud, M. et al. (2019). Toward a heterogeneous mist, fog, and cloud-based framework for the internet of healthcare things. *IEEE Internet of Things Journal* 6 (3): 4049–4062.

48 Swedlow, J.R., Kankaanpää, P., Sarkans, U. et al. (2021). A global view of standards for open image data formats and repositories. *Nature Methods* 18: 1440–1446.

49 Ding, W., Jing, X., Yan, Z., and Yang, L.T. (2019). A survey on data fusion in Internet of Things: towards secure and privacy-preserving fusion. *Information Fusion* 51: 129–144.

50 Chen, J. and Ran, X. (2019). Deep learning with edge computing: a review. *Proceedings of the IEEE* 107 (8): 1655–1674.

51 Dalla Mura, M., Prasad, S., Pacifici, F. et al. (2015). Challenges and opportunities of multimodality and data fusion in remote sensing. *Proceedings of the IEEE* 103 (9): 1585–1601.

52 Markit, I. (2017). The Internet of Things: a movement, not a market. *Critical IoT Insights* 1–9.

53 Soenksen, L.R., Ma, Y., Zeng, C. et al. (2022). Integrated multimodal artificial intelligence framework for healthcare applications. *npj Digital Medicine* 5: 149.

54 Tian, S., Yang, W., Le Grange, J.M. et al. (2019). Smart healthcare: making medical care more intelligent. *Global Health Journal* 3 (3): 62–65.

55 Ginsburg, G.S. and Phillips, K.A. (2018). Precision medicine: from science to value. *Health Affairs (Millwood)* 37 (5): 694–701.

56 Smuck, M., Odonkor, C.A., Wilt, J.K. et al. (2021). The emerging clinical role of wearables: factors for successful implementation in healthcare. *npj Digital Medicine* 4: 45.

57 Karipott, S.S., Nelson, B.D., Guldberg, R.E., and Ong, K.G. (2018). Clinical potential of implantable wireless sensors for orthopaedic treatments. *Expert Review of Medical Devices* 15 (4): 255–264.

58 Gogate, U. and Bakal, J. (2018). Healthcare monitoring system based on wireless sensor network for cardiac patients. *Biomedical & Pharmacology Journal* 11 (3): 1681.

59 Al-khafajiy, M., Baker, T., Chalmers, C. et al. (2019). Remote health monitoring of elderly through wearable sensors. *Multimed Tools Appl* 78: 24681–24706.

60 Vaidehi, V., Vardhini, M., Yogeshwaran, H. et al. (2013). Agent based health monitoring of elderly people in indoor environments using wireless sensor networks. *Procedia Computer Science* 19: 64–71.

61 Huang, Z., Lv, C., Xing, Y., and Wu, J. (2021). Multi-modal sensor fusion-based deep neural network for end-to-end autonomous driving with scene understanding. *IEEE Sensors Journal* 21 (10): 11781–11790.

62 A. Asvadi, L. Garrote, C. Premebida, P. Peixoto and U. J. Nunes, "Multimodal vehicle detection: fusing 3D-LiDAR and color camera data", *Pattern Recognition Letters*, vol. 115, pp. 20–29, Nov. 2018.

63 Yue, L., Abdel-Aty, M.A., Wu, Y., and Farid, A. (2020). The practical effectiveness of advanced driver assistance systems at different roadway facilities: system limitation, adoption, and usage. *IEEE Transactions on Intelligent Transportation Systems* 21 (9): 3859–3870.

64 Nawaz, M., Tang, J.K.-T., Bibi, K. et al. (2024). Robust cognitive capability in autonomous driving using sensor fusion techniques: a survey. *IEEE Transactions on Intelligent Transportation Systems* 25 (5): 3228–3243.

65 Al-Ezaly, E., El-Bakry, M., H., Abo-Elfetoh, A. et al. (2023). An innovative traffic light recognition method using vehicular ad-hoc networks. *Scientific Reports* 13: 4009.

66 Lanzolla, A. and Spadavecchia, M. (2021). Wireless sensor networks for environmental monitoring. *Sensors* 21: 1172.

67 Miller, M., Kisiel, A., Cembrowska-Lech, D. et al. (2023). IoT in water quality monitoring-are we really here? *Sensors (Basel)* 23 (2): 960.

68 Breathe London. https://www.breathelondon.org/ (accessed 16 May 2024).

69 Seymour, A.C., Dale, J., Hammill, M. et al. (2017). Automated detection and enumeration of marine wildlife using unmanned aircraft systems (UAS) and thermal imagery. *Scientific Reports* 7: 45127.

70 Victor, N., Maddikunta, P.K.R., Mary, D.R.K. et al. (2024). Remote sensing for agriculture in the era of industry 5.0 – a survey. *IEEE Journal of Selected Topics in Applied Earth Observations and Remote Sensing* 17: 5920–5945.

71 Di Nisio, A., Adamo, F., Acciani, G., and Attivissimo, F. (2020). Fast detection of olive trees affected by *Xylella fastidiosa* from UAVs using multispectral imaging. *Sensors* 20 (17): 4915.

72 Friess, N., Bendix, J., Brandle, M. et al. (2019). Introducing nature 4.0: a sensor network for environmental monitoring in the marburg open forest. *Biodiversity Information Science and Standards* 2.

73 Rubí, J.N.S., de Carvalho, P.H.P., and Gondim, P.R.L. (2022). Forestry 4.0 and Industry 4.0: Use case on wildfire behavior predictions. *Computers and Electrical Engineering* 102.

74 Elassy, M., Al-Hattab, M., Takruri, M., and Badawi, S. (2024). Intelligent transportation systems for sustainable smart cities. *Transportation Engineering* 16.

75 Sirajuddin, A., Abbas, S.M., and Zia, H. (ed.) (2020). *Smart Cities – Opportunities and Challenges*. Springer.

76 Pandiyan, P., Saravanan, S., Usha, K. et al. (2023). Technological advancements toward smart energy management in smart cities. *Energy Reports* 10: 648–677.

77 Kamyab, H., Khademi, T., Chelliapan, S. et al. (2023). The latest innovative avenues for the utilization of artificial Intelligence and big data analytics in water resource management. *Results Engineering* 20.

78 Kassim, A.M. and Al-Sharif, L. (ed.) (2021). *Smart Cities: Their Framework and Applications*. IntechOpen.

79 Mousavi, S., Hosseinzadeh, A., and Golzary, A. (2023). Challenges, recent development, and opportunities of smart waste collection: a review. *Science of the Total Environment* 886.

80 Al Mughairi, M., Beach, T., and Rezgui, Y. (2023). Post-occupancy evaluation for enhancing building performance and automation deployment. *Journal of Building Engineering* 77.

81 Ghaffarian, S., Taghikhah, F.R., and Maier, H.R. (2023). Explainable artificial intelligence in disaster risk management: achievements and prospective futures. *International Journal of Disaster Risk Reduction* 98.

82 Sandberg, C., McCoy, K., Jim, H., and Koppitsch, H. (2018). The application of a continuous leak detection system to pipelines and associated equipment. *IEEE Transactions on Industry Applications* 25.

83 V. C. Gungor and G. P. Hancke, "Industrial Wireless Sensor Networks: Challenges, Design Principles, and Technical Approaches," *IEEE Transactions on Industrial Electronics*, vol. 56, no. 10, pp. 4258–4265, Oct. 2009.

84 Sarah Shaharuddin, Khairul Nizam Abdul Maulud, Syed Ahmad Fadhli Syed Abdul Rahman, Adi Irfan Che Ani, Biswajeet Pradhan, The role of IoT sensor in smart building context for indoor fire hazard scenario: A systematic review of interdisciplinary articles, *Internet of Things*, Volume 22, 2023.

85 Puccinelli, D. and Haenggi, M. (2005). Wireless sensor networks: applications and challenges of ubiquitous sensing. *IEEE Circuits and Systems Magazine* 5 (3): 19–31.

86 Wang, Y., Su, Z., Zhang, N. et al. (2023). A survey on metaverse: fundamentals, security, and privacy. *IEEE Communications Surveys & Tutorials* 25 (1): 319–352.

87 Liyanage, M., Ahmad, I., Abro, A.B. et al. (2018). *A Comprehensive Guide to 5G Security*. Wiley.

88 Arul, R., Al-Otaibi, Y.D., Alnumay, W.S. et al. (2024). Multi-modal secure healthcare data dissemination framework using blockchain in IoMT. *Personal and Ubiquitous Computing* 28: 3–15.

89 Rani, R., Kumar, S., and Dohare, U. (2019). Trust evaluation for light weight security in sensor enabled Internet of Things: game theory oriented approach. *IEEE Internet of Things Journal* 6 (5): 8421–8432.

90 Ferrag, M.A., Shu, L., Yang, X. et al. (2020). Security and privacy for green IoT-based agriculture: review, blockchain solutions, and challenges. *IEEE Access* 8: 32031–32053.

91 Noura, M., Atiquzzaman, M., and Gaedke, M. (2019). Interoperability in Internet of Things: taxonomies and open challenges. *Mobile Networks and Applications* 24 (3): 796–809.

92 James, M., Chui, M., Bisson, P. et al. (2015). The Internet of Things: mapping the value beyond the hype. *McKinsey Global Institute* 3.

93 Abbasi, M.A., Memon, Z.A., Durrani, N.M. et al. (2021). A multi-layer trust-based middleware framework for handling interoperability issues in heterogeneous IoTs. *Cluster Computing* 1–28.

94 Wu, X. and Liang, J. (2021). A blockchain-based trust management method for Internet of Things. *Pervasive and Mobile Computing* 72: 101330.

95 Di Martino, B. and Esposito, A. (2021). Semantic techniques to support IOT interoperability. In: *Semantic IoT: Theory and Applications: Interoperability, Provenance and Beyond* (ed. R. Pandey, M. Paprzycki, N. Srivastava, et al.), 229. Springer.

96 Jamshed, M.A., Amjad, O., and Zeydan, E. (2017). Multicore energy efficient scheduling with energy harvesting for wireless multimedia sensor networks. *2017 International Multi-topic Conference (INMIC)*, Lahore, Pakistan (24–26 November 2017). IEEE, 1–5.

97 W. Chen, Z. Wang, D. Ding, X. Yi and Q. -L. Han, "Distributed state estimation over wireless sensor networks with energy harvesting sensors," in *IEEE Transactions on Cybernetics*, vol. 53, no. 5, pp. 3311–3324, May 2023.

98 Karakus, C., Gurbuz, A.C., and Tavli, B. (2013). Analysis of energy efficiency of compressive sensing in wireless sensor networks. *IEEE Sensors Journal* 13 (5): 1999–2008.

99 Zhu, C., Zheng, C., Shu, L., and Han, G. (2012). A survey on coverage and connectivity issues in wireless sensor networks. *Journal of Network and Computer Applications* 35 (2): 619–632.

100 He, S., Shi, K., Liu, C. et al. (2022). Collaborative sensing in Internet of Things: a comprehensive survey. *IEEE Communications Surveys & Tutorials* 24 (3): 1435–1474.

101 Farsi, M., Elhosseini, M.A., Badawy, M. et al. (2019). Deployment techniques in wireless sensor networks, coverage and connectivity: a survey. *IEEE Access* 7: 28940–28954.

2

Antennas for Wireless Sensors

Abdul Jabbar, Muhammad Ali Jamshed, and Masood Ur Rehman

James Watt School of Engineering, University of Glasgow, Glasgow, UK

The increasing need for effective and reliable wireless sensing systems is driven by technological advancements in numerous fields such as environmental monitoring, healthcare, agriculture, smart factories, industrial automation, and infrastructure management. Wireless sensor networks (WSNs) are one of the key enablers of multi-modal sensing systems and depend significantly on efficient and reliable antennas. Antennas serve as the means for sending and receiving data in different settings, and they are crucial for facilitating uninterrupted communication within these networks. This chapter provides an overview of the antennas, their design principles, uses, and the recent breakthroughs that are driving innovation in WSNs.

2.1 Wireless Sensors: Definition and Architecture

A modern-day sensor/sensor node is a device capable of gathering changes in its vicinity, sometimes capable of pre-processing those changes and then transmitting them to other nodes or directly to the gateway node [1]. A sensor node can also be termed as a mote, and a mote can be a node but a node is not always a mote [2]. The transmitting medium, over which a sensor node transmits its sensing information, can be a wired or a wireless channel [3]. We will discuss the sensors using wireless medium where antennas play an important role.

Multimodal Intelligent Sensing in Modern Applications, First Edition.
Edited by Masood Ur Rehman, Ahmed Zoha, Muhammad Ali Jamshed, and Naeem Ramzan.

Published 2025 by John Wiley & Sons, Inc.

2.1.1 Wireless Sensor Node Architecture

A classical architecture of a wireless sensor (WS) node is shown in Figure 2.1 where a WS node consists of five fixed modules and one adjustable module. The fixed modules consist of a fixed power supply, which can also be a battery, with or without rechargeable capabilities. A controller controls the functionality of a WS node by processing the data and performing operational and maintenance activities. A memory module is often added to a WS node, based on the application requirements and to perform programming (algorithms implementation) tasks. A transceiver is used for communicating with other WS nodes and to transmit sensory information. The analogue-to-digital converter is used to convert the analogue signal produced by the sensors into a digital form, to make it understandable for the controller. The adjustable module in a WS node is sensor(s), which are entirely application specific. A detailed description of each of these modules is provided in the following.

- **Power Supply:** The power supply is a critical component of any WS node and can be termed as the heart of it. There are some applications, where these WS nodes can be equipped with a continuous power supply, like for instance, in measuring the thermal heat in data centers, etc. [5]. Mostly, the WS nodes are deployed in hard-to-reach locations and are equipped with removable and rechargeable batteries. The lifetime of these batteries is a critical aspect of any WS node. In the development phase of a WS node, it is needed to ensure that there is enough power supply for a wireless mote to perform the intended sensing operation(s), as it is really hard and costly to replace them very often. Hence,

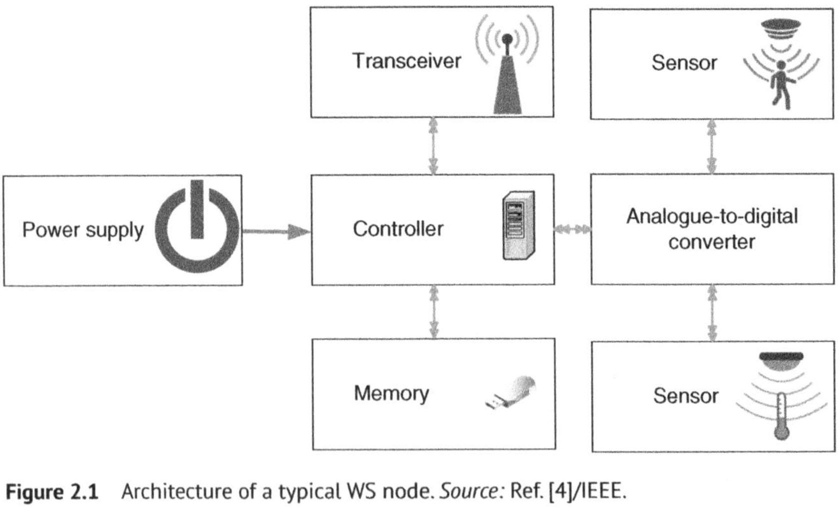

Figure 2.1 Architecture of a typical WS node. *Source:* Ref. [4]/IEEE.

the energy efficiency is one of the critical and open research statements for WS nodes. A WS node performs three main operations including sensing, communication, and data processing that consumes power.

Among these three operations, the processing of data transmission requires a significant amount of battery power. The cost of energy required in transmitting 1 kb of data at approximately a distance of 330 ft is similar to execution of three million instructions per second [6]. The batteries used by these sensors are classified based on the material used for manufacturing, for example, lithium-ion, nickel-cadmium, and nickel-metal hydride [7]. Some new WS nodes are also capable to harvest energy from various sources, such as radio frequency (RF), solar, and vibrations but require additional electronics [8].

- **Controller:** The controller is referred to as the brain of any WS node, as it controls the functionality of the wireless mote. Commonly, a microcontroller serves the purpose of a controller, but there are other alternatives as well, for example, field programmable gate arrays (FPGAs), digital signal processors (DSPs), and microprocessors, but generally, a microcontroller is preferred because of its lower cost and lower power consumption [9].
- **Memory:** The memory module present on a WS node serves two purposes, one is storing sensing data and another is for programming the device. Generally, the memory requirements are set based on the applications. Mostly, a microcontroller equipped with an onboard memory chip serves as the memory module. In particular, the flash memory is preferred because of its lower cost and more storage [10].
- **Transceiver:** WS nodes equipped with wireless transceiver (ability to both uplink and downlink data) possess the capability of transmitting the sensing information over the wireless channel. The WS nodes, usually transmit their sensing data in industrial, scientific, and medical (ISM) band. The transceiver associated with WS nodes usually operates in four states, that is, receive, transmit, idle, and sleep. The modern-day transceivers are generally equipped with state machines that are able to perform some of these operations automatically. It is noted that the power consumption in idle mode is equal to power consumption in receiving mode and requires some innovative techniques to have a sheer difference between the two states. Similarly, a significant amount of power is consumed while switching between different modes [11].
- **Analogue-to-digital converter:** The sensors produce analogue signals, which need to be converted into a digital form so that it can be readable by a controller. Analogue-to-digital converter usually acts as an interpreter between the controller and the sensor(s) [12].
- **Sensors:** The sensor module is basically the front-end of a WS node and directly interacts with the environment. Based on the intended applications, the sensors can be of various types as discussed in Chapter 1.

2.1.2 Operating Systems

Similar to a general purpose computer, an operating system with less complexity is usually required for WS nodes. The operating system for WS nodes closely resembles an embedded system and the operating systems, such as embedded configurable operating system (eCos) and microcontroller operating systems, can serve the application-specific needs of WSN [13, 14]. The first ever WSN-specific operating system, TinyOS, was developed collaboratively by the University of California, Intel Research, and Crossbow Technology [15]. Instead of multithreading, TinyOS uses event-driven programming model, where a signal is triggered based on the occurrence of an external event [16]. The lightweight operating system (LiteOS), developed by Huawei, is fairly a new operating system for WSNs and supports C programming language [17]. RIOT, Contiki, and PreonVM are some other examples of operating systems developed specifically for IoT-based WS nodes [18–20]. A comparison of various attributes of well-known operating systems for WS nodes is shown in Table 2.1.

2.1.3 Classification of Wireless Sensors

WSs can be classified in a variety of ways. The most commonly used method is application-specific classification giving consideration to factors such as deployment, reporting, and monitoring [21–23] or operational environment, type of sensor node, and network [24]. They can also be classified based on different protocols that each WS node strictly follows to perform intended sensing and data transmission operations [25, 26]. A simple yet comprehensive classification of WSs is shown in Figure 2.2. This simple approach classifies the WSs into four main types listed below:

- Reporting method
- Detection
- Conversion
- Output

The WSs based on their reporting method can be classified into two groups, that is, active and passive. Active sensors are periodic, as they continuously transmit the data after meeting a certain threshold, whereas, passive sensors generally transmit the data upon receiving an acknowledgment signal from the sink node/gateway. The sonar and radar are excellent examples of active sensors that require a continuous amount of power supply [27]. The passive sensors can be further subdivided into omnidirectional sensors (no well-defined direction for measurement) and narrow-beam sensors (having a well-defined direction for measurement) [28, 29]. The detection-based classification of WSs depends on the means

Table 2.1 Comparison of different operating systems used in wireless sensors.

Attribute	TinyOS	LiteOS	RIOT	Contiki	PreonVM
Software language	nesC	C/LiteC++	C/C++/Rust	nesC	C, Java
Execution model	Event driven	Multithreading	Multithreading	Protothread	Multithreading
Kernel	Monolithic	Modular	Microkernel	Modular	—
Real-time support	✗	Partial	✓	Partial	—
Platforms	3ARM7, PXA271	ARM,RISC-V	ARM Cortex	MSP430	ARM Cortex-M3
Low power support	✓	✓	✓	✓	✓
Scheduling	Preemptive	Priority	Multithreading	Preemptive	Multithreading
Network stack	BLIP	—	✓	✓	✓
Communication security	✓	✓	✓	✓	✓
Simulation support	TinySEC	✓	✓	Cooja	✓

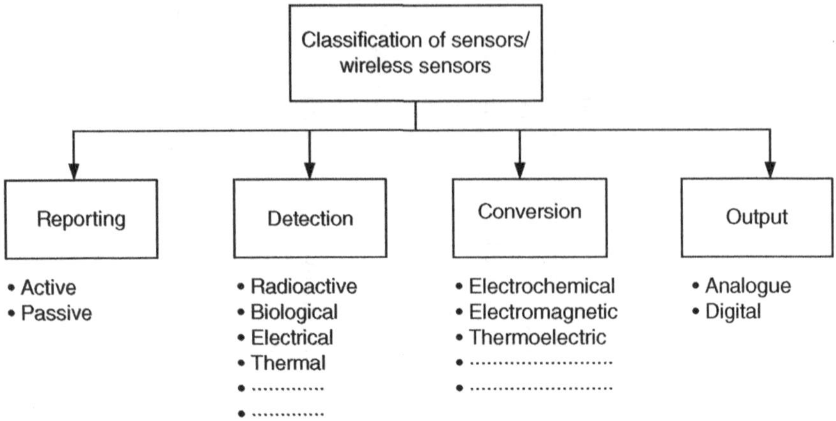

Figure 2.2 Classification of wireless sensors. *Source:* Ref. [4]/IEEE.

by which a sensor detects a change in its sensing area. Some classical examples of detection are radioactive, biological, electrical, etc. The classification based on the conversion phenomenon requires a different output in comparison to an input [30]. Electrochemical, electromagnetic (EM), thermoelectric, etc., are some of the examples of this type of classification [30]. The classification based on the output is divided into two types, that is, analogue and digital. The analogue sensors produce a continuous signal as an output, whereas the digital signal is produced as output by a digital sensor.

2.2 Multi-modal Wireless Sensing

Wireless sensing is an essential technology that leverages EM signal fluctuations to detect environmental changes caused by the specific motion of interest. RF sensing encompasses diverse application areas such as indoor and outdoor environmental monitoring, security surveillance, healthcare, and transportation. In the age of the Internet of Things (IoT), the prevalent use of mobile devices has made it easier to use wireless communication signals for environmental sensing. In smart home environments, RF-sensing applications are used to detect different degrees of human activity, including daily routine as well as detailed monitoring of vital signs and detection of falls in elderly people [23, 31]. Existing solutions tackle this problem by utilizing different kinds of wireless techniques such as Wi-Fi-based systems at 2.4 and 5 GHz ISM band, or radar system to sense EM signatures.

In Wi-Fi-based systems, channel state information (CSI) measured by Wi-Fi networks is widely used for different sensing purposes [31]. Several Wi-Fi-enabled IoT devices are strategically placed to capture human motion, which see diverse fluctuations in the local wireless channels as a result of identical human movement.

The radar-based technology is an alternative to Wi-Fi-based systems. Micro-Doppler signatures have been used for sensing and localization. The most pervasive example in this area is frequency-modulated carrier wave (FMCW) radars. Obtaining a high-dimensional, uncorrelated input that gives enough information about the monitoring target is the primary challenge to achieving high accuracy in RF sensing. Having sufficient spatial or frequency diversity allows wireless systems to accomplish this. For spatial diversity, we can increase the number of transmit and receive antenna pairs at several locations, each of which will have a unique impact on the local wireless signals caused by the tracked motion [32]. A few examples of this kind of sensing technology are antenna arrays, and multiple-input–multiple-output (MIMO) technology. Other than this, the other way is to leverage frequency diversity, which includes Wi-Fi-based sensing, and ultra-wideband (UWB) radars.

2.3 Antennas: The Sensory Gateway for Wireless Sensors

In all multi-modal wireless sensing applications, the antennas serve as a key interface between EM waves and the physical world. In order to facilitate the interchange of data between sensors and control units without the need for physical connections, the antennas play a crucial role in enabling these systems to send and receive EM waves that are essential for monitoring environmental conditions, recording vital signs, and providing remote control features. It is important to mention here that complete RF sensing involves end-to-end signal processing from the transmitter module to the receiver module and data processing [30]. Therefore, the design and performance of antennas play a significant role in driving the performance of contemporary technology by monitoring environmental conditions, recording vital signs, and allowing remote control.

In the context of recent technological breakthroughs, it is crucial to possess a fundamental understanding of antennas for wireless sensing. The importance of comprehending basic antenna characteristics and different antenna types in the field of wireless sensing is crucial to effectively build and implement wireless sensing systems. Furthermore, an analysis of antennas in wireless sensing at a system level reveals the complex relationship among sensor nodes, communication protocols, and antenna configurations [32]. This analysis offers valuable insights that are vital for enhancing system performance and guaranteeing reliable data transmission in various dynamic applications. In the context of wireless sensing, this chapter elucidates the need to comprehend basic

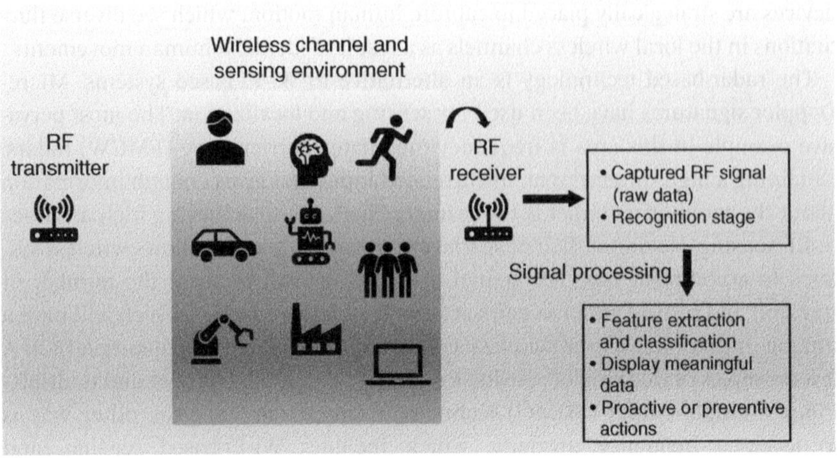

Figure 2.3 A generic wireless sensing environemnt.

antenna characteristics, RF front-end modules, and measurement schemes for RF sensing (Figure 2.3).

2.4 Fundamental Antenna Parameters

Passive antennas have the ability to both transmit and receive energy, making them reciprocal in nature. Hence, the terms "radiation" and "transmission" are equally applicable to the process of receiving radio waves. The description and evaluation of the physical factors that govern the behavior of each antenna are likely to have broader applicability. However, in this section, a brief introductory view of some important antenna parameters adhered to wireless sensing is discussed.

2.4.1 Bandwidth and Operating Frequency

Antenna bandwidth is not an inherent property of the antenna itself. It is defined as the range of frequencies in which the antenna functions based on predetermined parameters [33]. Depending on the useable frequencies, the bandwidth of an antenna is the difference between the highest frequency (f_H) and the lowest frequency (f_L):

$$Bandwidth = f_H - f_L \tag{2.1}$$

Note that the term useable frequency is relative. Often, this is relative to the input impedance matching of the antenna where the input reflection coefficient is equal to or below −10 dB. In other words, it is the range of frequencies over which 90% of the input power is delivered to the antenna, whereas 10% of power is reflected due to a mismatch between the antenna and the input circuit. Therefore, such antenna bandwidth is also known as −10 dB impedance bandwidth [34].

In the context of linear media, it is possible to characterize any wave pattern by considering the independent propagation of sinusoidal components. The expression representing the wavelength λ of a sinusoidal waveform propagating at a constant speed c is provided by:

$$\lambda = \frac{c}{f} \tag{2.2}$$

It is worth mentioning that the wavelength (unit in meters) is inversely related to the operating frequency and it directly translates to the physical size of the antenna. Hence, antennas at higher frequency bands will have a small physical size. For millimeter-wave (mmWave) antennas, the operating frequency is high and the antenna size is quite small which allows to use several antenna elements in an array configuration to enhance the effective radiation area and gain.

2.4.2 Gain

The gain of an antenna is the ratio of the radiation intensity in a given direction to the radiation intensity that would be produced if the power delivered to the antenna were radiated isotropically [33]. An isotropic radiator is the only kind of radiator that produces the same radiation in all spatial directions. However, the isotropic antenna cannot be realized for any specific polarization. Therefore, it is mostly used as a model and a standard for comparison, and the gain is usually measured in dBi. The directivity of an antenna is a similar concept; however, the reference is radiated power. Hence, the difference between gain and directivity is only the reference power by which the radiation intensity is normalized. As the factor between radiated power and accepted power is the radiation efficiency, therefore gain equals directivity and efficiency.

$$Gain = efficiency \times directivity \tag{2.3}$$

Note that the efficiency of a practical antenna is always less than 100%, therefore gain is always less than the directivity.

2.4.3 Radiation Pattern

The radiation pattern illustrates the spatial arrangement of a parameter that describes the EM field produced by an antenna. The radiation characteristics of

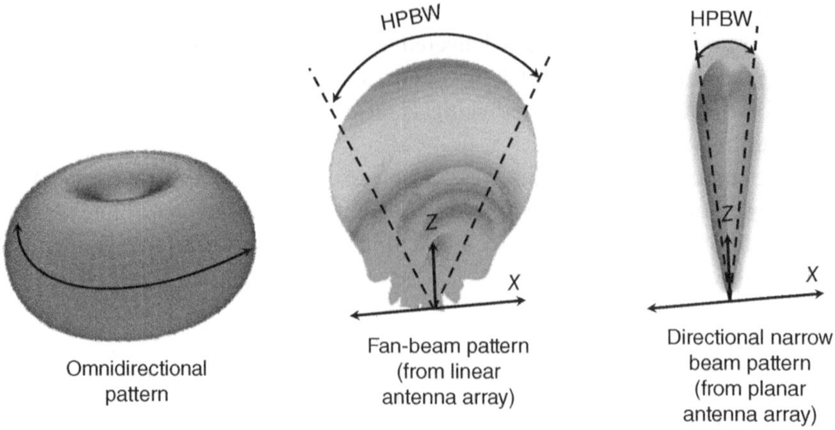

Figure 2.4 Different kinds of antenna radiation patterns.

antennas in three dimensions are characterized by their radiation pattern, often seen in the far field. A radiation pattern refers to a visual depiction that illustrates the comparative intensity of the EM field produced by or detected by an antenna, with respect to its direction. From a wireless sensing perspective, three different types of radiation patterns can be used depending on the application [33]. Omnidirectional radiation pattern covers a complete plane (either azimuth or elevation) and utilizes low gain antennas. For directional sensing, a fan-shaped radiation pattern can be used for relatively wider coverage in one plane. Linear antenna arrays produce fan-beam patterns. On the other hand, for mmWave high-resolution point-to-point sensing, highly directional beam from planar array antennas is used. Different types of radiation patterns are depicted in Figure 2.4.

Another important parameter related to antenna coverage and radiation pattern is its half-power beamwidth (HPBW). HPBW is the angular region where antenna power drops by 3 dB on both sides of the peak power. It therefore depicts that the fan beam pattern has a wider HPBW in elevation (or Azimuth) plane and narrow HPBW in the azimuth (or elevation plane). Similarly, directional pencil beams show narrow HPWB in both elevation and azimuth planes. The HPBW is often related to array antennas where a directional radiation pattern is observed. Therefore, it becomes trivial to discuss HPBW in omnidirectional antennas such as commonly used monopole and dipole antennas.

2.4.4 Polarization

The polarization of an antenna is an important parameter which is defined as the direction of the electric field vector. Antennas mainly exhibit three types of polarizations, namely, linear, circular, and elliptical. In linear polarization (LP),

Figure 2.5 Antenna polarization matching scheme.

	➡	⬆	↷	↶
➡ H	0 dB	–∞	–3 dB	–3 dB
⬆ V	–∞	0 dB	–3 dB	–3 dB
↷ RHCP	–3 dB	–3 dB	0 dB	–∞
↶ LHCP	–3 dB	–3 dB	–∞	0 dB

the E-field (electric field) vector has only one component and changes in magnitude only. It can be vertically oriented or horizontally oriented. In circular polarization (CP), two components of the E-field coexist (both horizontal and vertical) and the magnitude of the E-field vector is constant, but the direction (sense of polarization) changes and rotates around the direction of propagation. It can be right-hand CP (RHCP) or left-hand CP (LHCP) [35]. The elliptical polarization occurs when the magnitude and the direction of the E-field vector change and its peak position can be described by an elliptical equation.

From the RF sensing aspect, the knowledge of polarization of transmit and receive antennas is very important because if they are not oriented in the same polarization (i.e., if not co-polarized), then RF signal is not received. Polarization mismatch arises when there is a disparity between the polarization of the sensor antenna and the polarization of the incident wave. For instance, a vertically linear polarized antenna will not receive any power from horizontally polarized antenna (and vice versa) even if both antennas face each other. However, from LP to CP, the power is still received but half of the power is lost [33]. Similarly, LHCP and RHCP antennas offer cross-polarized orientation between each other and no communication or sensing can be observed in that case. The polarization of an antenna depends on various factors such as antenna geometry and feed mechanism. A conceptual view of antenna polarization mismatches and related ideal losses is presented in Figure 2.5.

2.5 Key Operating Frequency Bands for Sensing Antennas

The knowledge of potential frequency bands for wireless sensing unveils a multitude of possibilities that encompass a wide range of wireless sensing applications. The complete RF spectrum, which spans from few hundred megahertz to tens and hundreds of gigahertz, offers an extensive range of wireless sensing capabilities and related antenna designs. From antenna design and utilization perspective, the choice of correct frequency band is immensely important to achieve the desired functionality of any wireless sensing system.

The lower microwave region spans from a few MHz to 24 GHz. The most utilized frequency regions in this band include low-frequency RFID bands including 13.56 MHz and 860–960 MHz. Besides, this spectral region includes commonly utilized 2.4 and 5 GHz Wi-Fi bands and Bluetooth for wireless sensing. However, they are known to degrade in performance due to multipath and spectral interference issues at the 2.4 and 5 GHz bands. Modern Wi-Fi standards, such as Wi-Fi-6/6E (IEEE 802.11ax) [36–38], Wi-Fi-7 (IEEE 802.11be) [39–42], and Wi-Fi-8 (IEEE 802.11bn) [43, 44], have been developed to operate primarily between 5.17 and 7.125 GHz. These Wi-Fi generations aim to offer high efficiency, high data rate and reliability, low latency and multiple nodes connectivity through efficient and multi-beam antenna designs [45–49]. Moreover, various radar bands are in use such as ultra-wideband radar ranging from 1 to 8 GHz and X-band radar covering 8 to 12 GHz.

At higher frequencies, from 30 GHz and beyond, the spatial resolution of a propagated signal becomes considerably more precise, allowing for fine-grained spatial differentiation in high definition. In this view, mmWave and Terahertz (THz) spectra are highly envisaged for next-generation wireless sensing capabilities for accurate positioning, imaging, and spectroscopy. These bands permit the monitoring of environmental conditions, vital signs, and human activity in residences, offices, and healthcare facilities. They demonstrate promising potential for precise localization and high-resolution imaging in densely populated urban settings. However, in order to optimize the potential of mmWave and THz sensing, it is imperative to mitigate challenges such as efficient antenna designs, hardware intricacies, and signal attenuation.

2.6 Fabrication Methods for Sensing Antennas

Antenna fabrication involves various methods and materials that meet specific antenna performance criteria. Below are some key techniques used in the fabrication of antennas.

2.6.1 Printed Circuit Board (PCB) Antennas

In PCB-based fabrication, often different copper-clad laminates and dielectric substrates such as Rogers and FR-4 are used. First, antenna layout, the shape, and the size of the metal patch are designed using computer-aided design (CAD) software. Then the design file is extracted and transferred to a copper-clad laminate on a milling machine. The unwanted copper is removed through milling bits or chemical etching, leaving back the desired antenna pattern. The holes for vias and connectors are drilled using drill bits. For plating and soldering, plated through holes are created for electrical connections, and components are soldered onto the board. PCB fabrication is the most commonly used for planar antennas and offers a cost-effective and mass-production antenna fabrication capability [50, 51].

2.6.2 On-Chip and Integrated Antenna Fabrication

For on-chip antennas, semiconductor substrates (e.g. GaAs and SiGe) with metal interconnects are used [52]. Antenna elements are integrated into the integrated circuit (IC) design using electronic design automation (EDA) tools. Semiconductor fabrication processes including deposition, photolithography, and etching are used for fabrication. For the packaging of these antennas, ICs are packaged with connectors and other components [53, 54].

2.6.3 Stitching and Embroidery for Flexible Textile Antennas

The sewing and embroidering methods are innovative techniques for fabricating flexible and wearable sensor antennas, leveraging textile materials to create antennas that are adaptable, durable, and suitable for various applications [55, 56]. Typically, nonconductive fabrics like cotton, polyester, or other textile substrates are used for fabric selection, while materials such as silver-coated nylon or stainless steel are used for conductive threads. For sewing technique, a sewing machine or manual sewing process is used to stitch the conductive thread onto the fabric according to the antenna design. Care should be taken to maintain consistent stitch density and tension to ensure uniform conductivity [57]. The stitched antenna is inspected for uniformity and continuity of the conductive thread. Electrical testing is conducted to verify the antenna's performance, including impedance, radiation pattern, and gain.

Besides sewing fabrication method, the embroidered fabrication method is also used based on nonconductive fabrics [55, 58, 59]. High-conductivity threads, often silver-coated or gold-plated, are used for better performance. The embroidery machine stitches the conductive thread onto the fabric following the digital pattern with high precision. The process ensures consistent stitch quality and spacing, crucial for maintaining uniform electrical antenna properties. Sewing fabrication methods are well suited for low-cost, rapid prototyping, and simple designs [57, 60]. It is ideal for applications where slight variations in the antenna characteristics are acceptable. The embroidering technique is usually suitable for high-precision, durable, and stretchable antenna designs. It is preferred for applications requiring consistent performance and complex geometries.

2.7 Antenna Types for Wireless Sensing Networks

Antenna classification is subjective and does not follow strict categorizations. The categorization of antennas lacks clear criteria and is typically impacted by various aspects such as coverage (radiation pattern), geometry, and physical construction. It varies based on the different demands and applications in the field of WSN. The necessity of comprehending the complexities of antenna design and functioning is emphasized by its dynamic character, which guarantees optimum performance in various applications.

Antennas can be categorized into two main types based on their physical construction: flexible antennas and non-flexible antennas. Flexible antennas typically exhibit a degree of bendability or deformability, allowing them to adapt to different form factors and environments. Conversely, non-flexible antennas maintain a rigid structure and are often characterized by their fixed geometric design.

On the other hand, antennas can also be classified based on their radiation pattern into two broad categories: omnidirectional antennas and directional antennas. Omnidirectional antennas radiate or receive EM waves uniformly in all directions, making them suitable for applications where coverage in all directions is desired, such as broadcasting antennas in Wi-Fi sensing [34]. In contrast, directional antennas concentrate their radiation pattern in a specific direction, offering increased gain and longer distance in that direction while sacrificing coverage in other directions. These antennas are commonly used in point-to-point communication links or scenarios where targeted signal transmission or reception is required.

A single microstrip patch antenna is a directional antenna, offering relatively wide coverage within a specific sector. These single-patch antennas are frequently used for various sensing applications. Some of the major applications include monitoring soil and environmental moisture levels for agricultural and environmental studies, measuring temperature variations in different environments, monitoring physiological parameters like heart rate and body temperature, detecting and characterizing tissues for medical diagnostics, identifying cracks and defects in buildings, bridges, and other structures, monitoring stress and strain in materials and structures, as well as measuring moisture content in plants, which is crucial for agricultural research and crop management. Similarly, the array of patch antennas produces highly directional beam which is used in radar systems for remote sensing and imaging applications.

Consider a design use case of a patch antenna where if the moisture content in a material or a plant sample changes, the resonance frequency of the patch antenna system will change accordingly. This shift can be measured and translated into meaningful data, which is then displayed to the user. The system can notify the user of any significant changes, allowing for timely and appropriate actions to be taken. This ability to detect and respond to environmental variations makes patch antennas highly valuable toward WSN in various fields, including agriculture, environmental monitoring, and material science.

A typical patch antenna designed on FR-4 substrate around 5.2 GHz, along with its corresponding resonance behavior is shown in Figure 2.6. Note that the resonance dip shifts according to the change in the moisture content of the sensing environment (either a plant sample or the moisture content in a dielectric material). The flexible sensor is meant to provide the status of water content (or any other desired physical parameter) efficiently. The information provided by the sensor will help farmers, planters, and agriculturists to water every plant at the right time or make any proactive management. The communication between the end-user and the sensor node can be established via Wi-Fi, ZigBee, Bluetooth, or Long Range (LoRa) protocols to establish a smart IoT WSN. Thus, this type of system controlled through a WSN helps automate processes, improve efficiency and remote monitoring, and capture the meaningful data. It is important to note that the antenna can be made flexible and conformable to the desired shape by using flexible substrates, as discussed in the next section.

2.7.1 Flexible Antennas

One of the numerous benefits of current electronic gadgets would be the incorporation of flexible electronics onto a textile substrate or flexible materials that have mechanical qualities that allow them to bend and twist. Flexible antennas can be wearable or non-wearable [31]. Usually, wearable antennas are designed on flexible textile substrates and utilized to detect surroundings, send data, receive energy, and operate on the body premises. The simple designs, sensing capabilities, adaptability, and low cost of antenna sensors have contributed to their dramatic rise in popularity among wearable devices [34]. A number of parameters, including the frequency range, operational conditions, and required transmission strength, dictate the specific design of these antennas. When compared to traditional antennas, textile antennas have several benefits, including being lightweight, comfortable, and washable, and they may be easily incorporated into clothing.

Wearable flexible antennas can adopt the geometry and design principles of various conventional antennas such as microstrip patch antennas, monopole antennas, loop antennas, and planar inverted-F antennas. Flexible antennas can conform to the irregular shapes of wearable devices and the human body, ensuring comfortable and unobtrusive integration into clothing or accessories. In these antennas, the choice of substrate makes them flexible and the performance of antennas in terms of impedance bandwidth, gain, and radiation efficiency depends on the substrate as well as the geometry and construction of the antenna. Various flexible substrate materials are used for wearable antennas

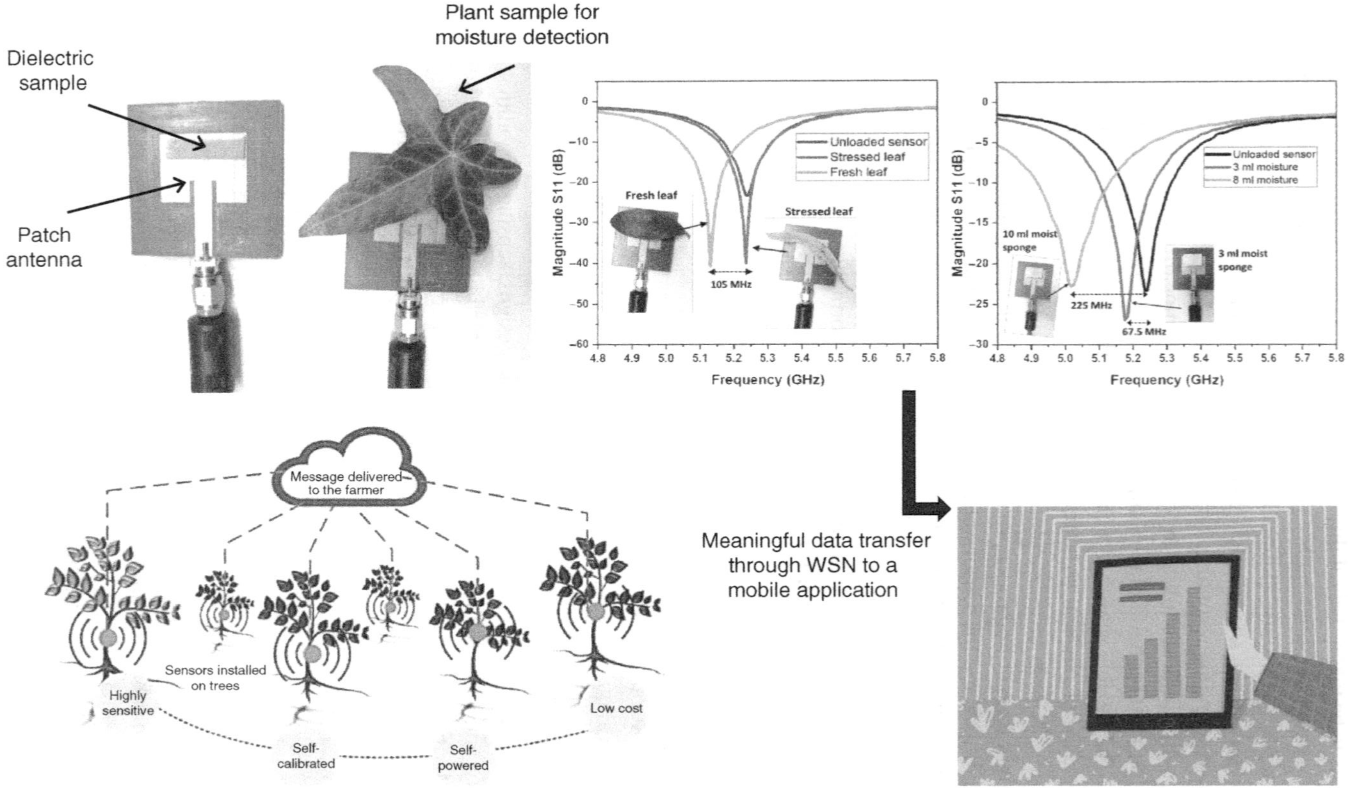

Figure 2.6 Patch antennas for environmental sensing. *Source:* Gemma can fly/Stocksy/Adobe Stock Photos.

such as Felt fabric, polyethylene terephthalate (PET), Jeans, polyurethane fabric, and polyimide fabric to name but a few. These substrates have low dielectric constant below 4. The optimal conditions for the flexible substrate materials are low dielectric primitivity, high thermal conductivity, low coefficient of thermal expansion, and low dielectric loss. For conduction, different conductive materials such as gold, silver, and copper can be used for radiation mechanism due to their high conductivity.

The fabric-based flexible wearable antennas integrate directly into clothing or textile substrates, offering seamless integration and comfortable wearability. Thin, planar antennas fabricated on flexible substrates such as polyimide or polyester offer compact size and high performance for sensing applications. Flexible wires or conductive threads are integrated into wearable textiles to form simple yet effective antennas for sensing applications. Antennas fabricated using inkjet printing techniques onto flexible substrates provide cost-effective and customizable solutions for wearable sensing.

2.7.2 Omnidirectional Antennas

Omnidirectional antennas transmit and receive EM waves in all directions, in a specific plane, usually around the axis of the antenna. These antennas provide large coverage in all directions, making them well suited for applications that need communication or sensing in every direction as shown in Figure 2.7. They are often used in situations when the whereabouts of the target or transmitter are

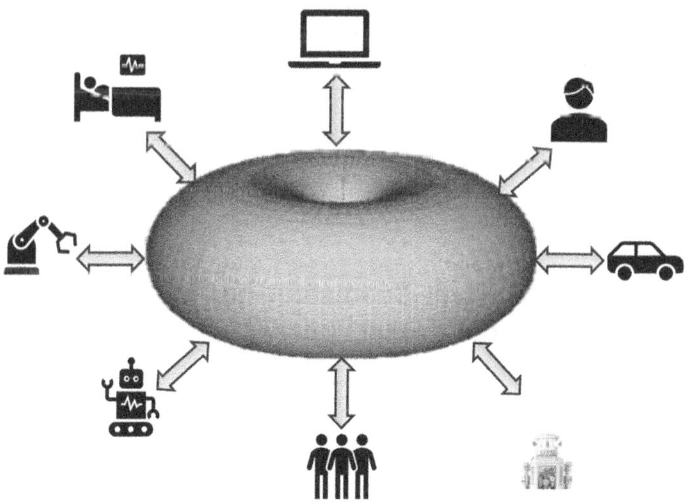

Figure 2.7 Conceptual view of omni-directional antennas and wide area coverage.

uncertain or constantly shifting. Omnidirectional antennas are suitable for wireless sensing applications, including environmental monitoring, asset tracking, and IoT networks. They perform very well in situations where the sensor nodes are spread out over a large region or when the sensing needs necessitate collecting data from numerous directions at the same time.

Omnidirectional antennas are often designed with a focus on simplicity and convenience of deployment. Typical designs include of monopole antennas, dipole antennas, and helical antennas. Wi-Fi sensing is typically employed using omnidirectional antennas. Antenna size, gain, and impedance matching are crucial factors that must be taken into account to provide the best possible performance within the intended frequency range: 2.4 GHz ISM band as well as 60 GHz mmWave ISM band antenna [61].

2.7.3 Directional Antennas

The directional antennas can be categorized into two main types, namely, traditional directional antennas using mechanical rotation and electronic beamsteering directional antennas. The beam produced by conventional directed antennas can be mechanically rotated to face a certain direction. Among the conventionally oriented antennas are parabolic reflectors, helixes, patch antenna grids, Yagi-Uda aperture horn-like structures, and pyramidal horns. However, mechanical wear and tear, as well as latency in beamswitching, are major drawbacks of such antennas.

Electronic beamsteering antennas are often regarded as smart antennas because they include reconfigurable electronic circuitry that allows the main beam to be directed through an external software program as depicted in Figure 2.8. These antennas constitute a comprehensive system, integrating antenna configuration, RF front end, and signal processing techniques to achieve specific beamforming patterns for enhanced radiation pattern diversity. Beamsteering antennas are an advanced type of directional antenna that can dynamically adjust the direction of their main lobe. This capability makes them particularly valuable in WSNs for improving security and energy efficiency [62].

2.8 Advantages of Electronic Beamsteering Antennas in Sensing Systems

In WSNs, antennas offering spatial diversity and frequency diversity are widely used techniques [63, 64]. However, through the use of spatial diversity, extra transceiver pairs are intentionally placed at different locations and this is usually achieved using MIMO antennas or multiple array antennas. Furthermore, frequency diversity is utilized by delivering the sensor signal across a wide range of

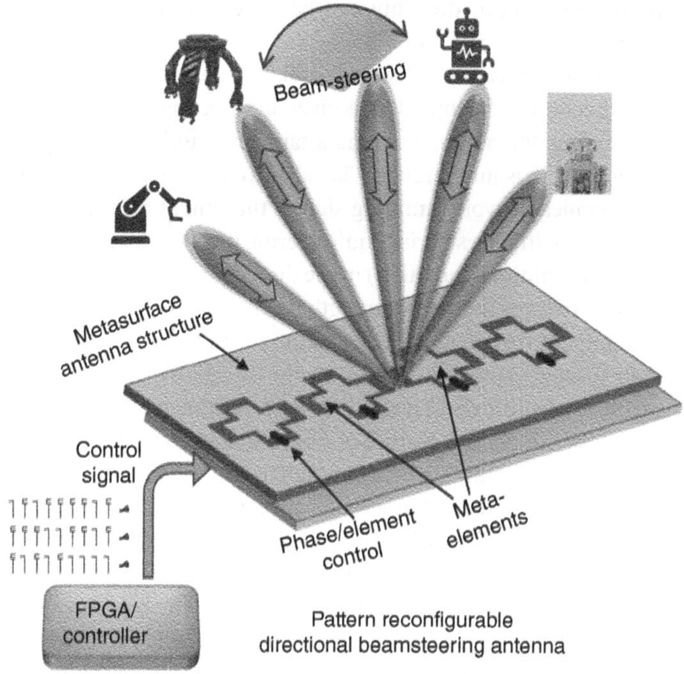

Figure 2.8 Conceptual depiction of a pattern reconfigurable directional antenna for IoT and WSN.

frequencies. Dependence on either frequency diversity or spatial variety poses challenges and incurs high expenses. For instance, in spatial diversity, Doppler radars and antenna arrays are expensive in terms of both the physical components and the computational resources required for signal processing. Augmenting the quantity of antennas not only results in a bulky network but also amplifies the intricacy of digital signal processing. Moreover, frequency diversity requires a wide frequency spectrum in RF components, such as amplifiers and oscillators, and increases their complexity and cost compared to narrowband devices.

In order to overcome the above-mentioned challenges in wireless sensing, pattern diversity is another technique in WSN [65]. By concentrating the radiation pattern in one direction, directional antennas may boost gain, eliminate interference and collisions, and enhance security against malicious assaults like eavesdropping and jamming. Usually, arrays of basic antenna components, such as dipoles or microstrip patches, are combined to form directional antennas. The total radiation patterns of the directional antennas are affected by the geometrical configuration, the number and type of elements, and the properties of the applied signals in each element to steer the main beam. A new class of beamsteerable metasurface antennas with pattern diversity can be effectively utilized for RF sensing and WSNs [65].

Adaptive beamsteering antennas can adjust their radiation pattern in real time to focus the signal in the direction of the intended receiver. By directing the beam away from the source of jamming, the antenna can maintain communication integrity even in the presence of malicious interference. The adaptive nature of these antennas allows them to dynamically avoid jamming signals that might be detected in certain directions. Besides main lobe steering, null steering involves creating radiation pattern nulls (points of minimal radiation) in the direction of known jammers. This technique effectively reduces the signal strength received by the jammer, making it difficult for the jammer to interfere with the communication. Such beamsteering antennas use sophisticated algorithms to identify the direction of jammers and adjust the antenna pattern accordingly to place nulls in those directions.

Spatial filtering is another important feature of directional beamsteering antennas, It refers to the ability to discriminate between signals based on their direction of arrival. Beamsteering antennas can differentiate between the desired signal and jamming signals based on their direction. This selective reception reduces the impact of jamming by only accepting signals from the intended direction while ignoring others and enhances resolution. Moreover, by focusing the beam in a specific direction, the required transmission power to reach the intended receiver can be significantly reduced. By concentrating the energy toward the receiver, beamsteering antennas ensure that less power is wasted in other directions. This efficient use of power leads to reduced energy consumption for transmitting the same amount of data. By reducing the radiation in undesired directions, beamsteering antennas decrease the likelihood of interfering with other nodes. This reduction in interference and collisions leads to fewer retransmissions, thereby conserving energy. Furthermore, directional beamsteering antennas can support energy-efficient routing by selecting routes that require fewer transmissions and lower power levels. By steering the beam to optimize the path, the network can balance load and avoid overburdening specific nodes, leading to longer network lifetime. Another important benefit of electronic beamsteering antennas is the improved stability of WSN. The targeted communication enabled by beamsteering antennas allows for denser network deployments without a proportional increase in interference. Efficient management of energy and spectrum resources ensures that the network can scale while maintaining performance and energy efficiency.

Another type of beamsteering antenna for WSNs is the reconfigurable metasurface antenna that can realize large-scale adaptive antenna arrays in a smaller form factor with a simple and low-cost design [65–70]. It does this without requiring complicated, expensive, and power-hungry RF components in its design. In order to achieve more antenna diversity, designs based on metasurfaces and metamaterials have proven to be highly effective in removing the mutual coupling effects between the various antenna elements. Therefore, to create a more precise and reliable RF sensing system, we may leverage the dynamic pattern diversity of metasurface antenna to measure sensing channels with a high degree of uncorrelation [66, 71]. Future low-cost, low-complexity RF sensing systems, where transceivers and bandwidth are scarce or accessible, can be made possible by such dynamic metasurface antenna solutions.

2.9 Summary

WSs are widely used in multi-modal sensing applications. Antennas act as the eyes and the ears of these sensors and there is a wide variety of antennas for WSNs encompassing a range of types and fabrication techniques tailored to meet specific requirements and challenges. Flexible antennas provide adaptable and durable solutions for wearable and conformable sensor applications. Omnidirectional antennas offer broad, uniform coverage, ideal for evenly distributed networks. However, they are not fully able to mitigate interference and jamming attacks. On the other hand, directional beamsteering antennas enhance performance by focusing signals, mitigating interference, and improving energy efficiency. Together, these antennas form the backbone of efficient, reliable, and secure WSNs. By offering adaptive and directed control over the signals, beamsteering antennas greatly improve the energy efficiency and security of WSNs. Furthermore, they enhance network scalability, decrease interference, promote energy-efficient routing, increase communication range, and minimize transmission power, all while optimizing energy utilization. With these features, beamsteering antennas are a great help for building efficient and reliable WSNs.

References

1 Wilson, J.S. (2004). *Sensor Technology Handbook*. Elsevier.

2 Malasri, K. and Wang, L. (2009). Design and implementation of a secure wireless mote-based medical sensor network. *Sensors* 9 (8): 6273–6297.

3 Fahmy, H.M.A. (2020). *Concepts, Applications, Experimentation and Analysis of Wireless Sensor Networks*. Springer Nature.

4 Jamshed, M.A., Ur Rehman, M., Abbasi, Q.H. et al. (2022). Challenges, applications and future of wireless sensors in internet of things: a review. *IEEE Sensors Journal* 22 (6): 5482–5494.

5 Wang, X., Wang, X., Xing, G. et al. (2012). Intelligent sensor placement for hot server detection in data centers. *IEEE Transactions on Parallel and Distributed Systems* 24 (8): 1577–1588.

6 Rice, A. and Hay, S. (2010). Measuring mobile phone energy consumption for 802.11 wireless networking. *Pervasive and Mobile Computing* 6 (6): 593–606.

7 Yasin, S., Ali, T., Draz, U. et al. (2019). A parametric performance evaluation of batteries in wireless sensor networks. In: *Recent Trends and Advances in Wireless and IoT-Enabled Networks* (ed. M.A. Jan, F. Khan, and M. Alam), 187–196. Springer.

8 Whitaker, M. (2010). Energy harvester produces power from local environment, eliminating batteries in wireless sensors. *Journal of Analog Innovation* 20: 1–36.

9 Di Nisio, A., Di Noia, T., Carducci, C.G.C., and Spadavecchia, M. (2016). High dynamic range power consumption measurement in microcontroller-based applications. *IEEE Transactions on Instrumentation and Measurement* 65 (9): 1968–1976.

10 Pruteanu, E. and Gabriel, P. (2019). Intelligent measuring system using network wireless sensors for structural diagnostics. *2019 22nd International Conference on Control Systems and Computer Science (CSCS)*, Bucharest, Romania (28–30 May 2019), 492–495. IEEE.

11 Matin, M.A. and Islam, M. (2012). Overview of wireless sensor network. *Wireless Sensor Networks-Technology and Protocols* 1–3.

12 Akyildiz, I.F., Su, W., Sankarasubramaniam, Y., and Cayirci, E. (2002). A survey on sensor networks. *IEEE Communications Magazine* 40 (8): 102–114.

13 Thomas, G. (2000). Ecos: an operating system for embedded systems. *Dr. Dobb's Journal: Software Tools for the Professional Programmer* 25 (1): 66–72.

14 Anh, T.N.B. and Tan, S.-L. (2009). Real-time operating systems for small microcontrollers. *IEEE Micro* 29 (5): 30–45.

15 Levis, P. and Gay, D. (2009). *TinyOS Programming*. Cambridge University Press.

16 Levis, P., Madden, S., Polastre, J. et al. (2005). Tinyos: An operating system for sensor networks. In: *Ambient Intelligence* (ed. W. Weber, J.M. Rabaey, and E. Aarts), 115–148. Springer.

17 Wang, X. (2021). Analysis of thread schedulability in huawei liteos. *MATEC Web of Conferences*, vol. 336. EDP Sciences, 05031.

18 Baccelli, E., Gündoğan, C., Hahm, O. et al. (2018). RIoT: an open source operating system for low-end embedded devices in the IoT. *IEEE Internet of Things Journal* 5 (6): 4428–4440.

19 Dunkels, A., Gronvall, B., and Voigt, T. (2004). Contiki-a lightweight and flexible operating system for tiny networked sensors. *29th Annual IEEE International Conference on Local Computer Networks*, Tampa, FL, USA (16– 18 November 2004), 455–462. IEEE.

20 Operating software "PreonVM" for Preon32 series. https://web.archive.org/web/20171111094628/; https://www.virtenio.com/en/preonvm-virtual-maschine.html (accessed 15 May 2024).

21 Gilbert, E.P.K., Kaliaperumal, B., and Rajsingh, E.B. (2012). Research issues in wireless sensor network applications: a survey. *International Journal of Information and Electronics Engineering* 2 (5): 702.

22 Rashid, B. and Rehmani, M.H. (2016). Applications of wireless sensor networks for urban areas: a survey. *Journal of Network and Computer Applications* 60: 192–219.

23 Li, Y. and Thai, M.T. (2008). *Wireless Sensor Networks and Applications*. Springer Science & Business Media.

24 Borges, L.M., Velez, F.J., and Lebres, A.S. (2014). Survey on the characterization and classification of wireless sensor network applications. *IEEE Communications Surveys & Tutorials* 16 (4): 1860–1890.

25 Singh, S.K., Singh, M., Singh, D.K. et al. (2010). Routing protocols in wireless sensor networks – a survey. *International Journal of Computer Science & Engineering Survey (IJCSES)* 1 (2): 63–83.

26 Huang, P., Xiao, L., Soltani, S. et al. (2012). The evolution of mac protocols in wireless sensor networks: a survey. *IEEE Communications Surveys & Tutorials* 15 (1): 101–120.

27 Minkoff, J. (1992). *Signals, Noise, and Active Sensors Radar, Sonar, Laser Radar*, 256. New York: Wiley.

28 Yagi, Y. (1999). Omnidirectional sensing and its applications. *IEICE Transactions on Information and Systems* 82 (3): 568–579.

29 Gianchandani, Y.B. and Najafi, K. (1996). Bent-beam strain sensors. *Journal of Microelectromechanical Systems* 5 (1): 52–58.

30 Fraden, J. and Fraden, J. (2004). *Handbook of Modern Sensors: Physics, Designs, and Applications*, vol. 3. Springer.

31 Abbasi, Q.H., Rehman, M.U., Alomainy, A., and Qaraqe, K. (ed.) (2016). *Advances in Body-Centric Wireless Communication: Applications and State-of-The-Art*. The IET.

32 Yang, S.-H. (2013). *Wireless Sensor Networks Principles, Design and Applications Signals and Communication Technology*. Springer.

33 Ur Rehman, M. and Safdar, G.A. (2018). *LTE Communications and Networks: Femtocells and Antenna Considerations*. Wiley.

34 Chen, Z.N. (2007). *Antennas for Portable Devices*. Wiley.

35 Matin, M.A. (2022). *Wideband, Multiband, and Smart Antenna Systems*. Springer Cham.

36 Oughton, E.J., Lehr, W., Katsaros, K. et al. (2021). Revisiting wireless internet connectivity: 5G vs Wi-Fi 6. *Telecommunications Policy* 45 (5): 102127.

37 Mozaffariahrar, E., Theoleyre, F., and Menth, M. (2022). A survey of Wi-Fi 6: technologies, advances, and challenges. *Future Internet* 14 (10): 293.

38 Wi-Fi Alliance (2020). Wi-Fi Alliance Introduces Wi-Fi 6. October 2018. https://www.wi-fi.org/news-events/newsroom/wi-fi-alliance-introduces-wi-fi-6 (accessed 15 October 2024).

39 Frascolla, V., Cavalcanti, D., and Shah, R. (2023). Wi-Fi evolution: the path towards Wi-Fi 7 and its impact on IIoT. *Journal of Mobile Multimedia* 263–276.

40 Naik, G., Ogbe, D., and Park, J.-M.J. (2021). Can Wi-Fi 7 support real-time applications? On the impact of multi-link aggregation on latency. *ICC 2021-IEEE*

International Conference on Communications, Montreal, QC, Canada (14–23 June 2021), pp. 1–6.

41 Garcia-Rodriguez, A., Lopez-Perez, D., Galati-Giordano, L., and Geraci, G. (2021). IEEE 802.11 be: Wi-Fi 7 strikes back. *IEEE Communications Magazine* 59 (4): 102–108.

42 Deng, C., Fang, X., Han, X. et al. (2020). IEEE 802.11 be Wi-Fi 7: new challenges and opportunities. *IEEE Communications Surveys & Tutorials* 22 (4): 2136–2166.

43 Giordano, L.G., Geraci, G., Carrascosa, M., and Bellalta, B. (2023). What will Wi-Fi 8 be? A primer on IEEE 802.11 bn ultra-high reliability. *arXiv preprint arXiv:2303.10442*.

44 Reshef, E. and Cordeiro, C. (2022). Future directions for Wi-Fi 8 and beyond. *IEEE Communications Magazine* 60 (10): 50–55.

45 Ku, C.C., Wu, M.T., and Chuang, M.L. (2022). Multiband antenna design for Wi-Fi 6E applications. *2022 IEEE International Conference on Consumer Electronics*, Taiwan, ICCE-TW (6–8 July 2022), Taipei, Taiwan, pp. 383–384.

46 Zhang, X., Zhou, D., Li, Y. et al. (2022). A simple dual-polarized patch antenna array for Wi-Fi 6/6E application. *IEEE Transactions on Antennas and Propagation* 70 (11): 11143–11148.

47 Ischenko, E.A., Pasternak, Y.G., Pendyurin, V.A., and Fedorov, S.M. (2021). Active patch antenna for Wi-Fi 5, 6 and Wi-Fi 6E applications. *Journal of Physics: Conference Series* 2096 (1).

48 Zhang, W., Li, Y., Wei, K., and Zhang, Z. (2022). A two-port microstrip antenna with high isolation for Wi-Fi 6 and Wi-Fi 6E applications. *IEEE Transactions on Antennas and Propagation* 70 (7): 5227–5234.

49 Jiang, H., Yan, N., Ma, K., and Wang, Y. (2022). A wideband circularly polarized dielectric patch antenna with a modified air cavity for Wi-Fi 6 and Wi-Fi 6E applications. *IEEE Antennas and Wireless Propagation Letters* 15–19.

50 Reig, C. and Avila-Navarro, E. (2013). Printed antennas for sensor applications: a review. *IEEE Sensors Journal* 14 (8): 2406–2418.

51 Pires, N., Parra, T., Skrivervik, A.K., and Moreira, A.A. (2017). Design and measurement of a differential printed antenna for a wireless sensor network node. *IEEE Antennas and Wireless Propagation Letters* 16: 2228–2231.

52 Karim, M.R., Yang, X., and Shafique, M.F. (2018). On chip antenna measurement: a survey of challenges and recent trends. *IEEE Access* 6: 20320–20333.

53 Mendes, P.M., Polyakov, A., Bartek, M. et al. (2006). Integrated chip-size antennas for wireless microsystems: Fabrication and design considerations. *Sensors and Actuators A: Physical* 125 (2): 217–222.

54 Rida, A., Yang, L., Vyas, R., and Tentzeris, M.M. (2009). Conductive inkjet-printed antennas on flexible low-cost paper-based substrates for RFID and WSN applications. *IEEE Antennas and Propagation Magazine* 51 (3): 13–23.

55 Mohamadzade, B., Hashmi, R.M., Simorangkir, R.B.V.B. et al. (2019). Recent advances in fabrication methods for flexible antennas in wearable devices: state of the art. *Sensors* 19 (10): 2312.

56 El Gharbi, M., Fernández-García, R., Ahyoud, S., and Gil, I. (2020). A review of flexible wearable antenna sensors: design, fabrication methods, and applications. *Materials* 13 (17): 3781.

57 Al-Haddad, M., Jamel, N., and Nordin, A.N. (2021). Flexible antenna: a review of design, materials, fabrication, and applications. *Journal of Physics: Conference Series* 12068.

58 Yu, M., Shang, X., Wang, M. et al. (2020). Exploiting embroidered UHF RFID antennas as deformation sensors. *IEEE Journal of Radio Frequency Identification* 4 (4): 406–413.

59 Hasani, M., Vena, A., Sydänheimo, L. et al. (2015). A novel enhanced-performance flexible RFID-enabled embroidered wireless integrated module for sensing applications. *IEEE Transactions on Components, Packaging and Manufacturing Technology* 5 (9): 1244–1252.

60 Zhang, L., Wang, Z., and Volakis, J.L. (2012). Textile antennas and sensors for body-worn applications. *IEEE Antennas and Wireless Propagation Letters* 11: 1690–1693.

61 Ding, Y.R. and Cheng, Y.J. (2019). A tri-band shared-aperture antenna for (2.4, 5.2) GHz Wi-Fi application with MIMO function and 60 GHz Wi-Gig application with beam-scanning function. *IEEE Transactions on Antennas and Propagation* 68 (3): 1973–1981.

62 George, R. and Mary, T.A.J. (2020). Review on directional antenna for wireless sensor network applications. *IET Communications* 14 (5): 715–722.

63 Jabbar, A., Rehman Ur, M., Pang, Z. et al. (2024). A wideband frequency beam-scanning antenna array for millimeter-wave industrial wireless sensing applications. *IEEE Sensors Journal* 24 (8): 13315–13325.

64 Mietzner, J., Schober, R., Lampe, L. et al. (2009). Multiple-antenna techniques for wireless communications – a comprehensive literature survey. *IEEE Communications Surveys and Tutorials* 11 (2): 87–105.

65 Lan, G., Imani, M.F., Manjarres, J. et al. (2021). MetaSense: boosting RF sensing accuracy using dynamic metasurface antenna. *IEEE Internet of Things Journal* 8 (18): 14110–14126.

66 Lan, G., Imani, M.F., Del Hougne, P. et al. (2020). Wireless sensing using dynamic metasurface antennas: challenges and opportunities. *IEEE Communications Magazine* 58 (6): 66–71.

67 Jabbar, A., Hamed, M., Rehman Ur, M. et al. (2024). 60 GHz programmable dynamic metasurface antenna (DMA) for next-generation communication, sensing and imaging applications: from concept to prototype. *IEEE Open Journal of Antennas and Propagation* 5 (3): 705–726.

68 Oesterling, A.X., Imani, M.F., Mizrahi, O.S. et al. (2020). Detecting motion in a room using a dynamic metasurface antenna. *IEEE Access* 8: 222496–222505.

69 Jabbar, A., Jamshed, M.A., Abbasi, Q. et al. (2023). Leveraging the role of dynamic reconfigurable antennas in viewpoint of Industry 4.0 and beyond. *Research* 6: 0110.

70 Del Hougne, P., Imani, F., Sleasman, M. et al. (2018). Dynamic metasurface aperture as smart around-the-corner motion detector. *Scientific Reports* 8 (1): 1–10.

71 Rezvani, M. and Adve, R. (2023). Channel estimation for dynamic metasurface antennas. *IEEE Transactions on Wireless Communications* https://doi.org/10.1109/TWC.2023.3328496.

3

Sensor Design for Multimodal Environmental Monitoring

Muhammad Ali Jamshed[1], *Bushra Haq*[2], *Syed Ahmed Shah*[2], *Kamran Ali*[3], *Qammer H. Abbasi*[1], *Mumraiz Khan Kasi*[2], *and Masood Ur Rehman*[1]

[1] *James Watt School of Engineering, University of Glasgow, Glasgow, UK*
[2] *Department of Computer Science, Balochistan University of Information Technology, Engineering and Management Sciences, Quetta, Pakistan*
[3] *Department of Computer Science, Middlesex University, London, UK*

The world is currently experiencing increasingly recurrent climate variations, such as droughts, floods, heatwaves, wildfires, cyclones, hurricanes, heavy rains, and unexpected snowfalls. There are several causes for this phenomenon, predominantly of human origin. Human activities such as rising urbanization, increasing industrialization, and mounting deforestation are leading to a much warmer planet due to the release of greenhouse gases. These climate changes pose an existential threat and have placed us in a state of emergency. Efficient, targeted, and effective technology-led smart solutions are the need of the hour to combat this catastrophic situation. Multi-modal wireless sensor networks (WSNs) can play a pivotal role in addressing climate change through sensing, monitoring, reporting, and intervening in real time. This chapter discusses the key environmental sensing variables and associated sensor node designs to effectively tackle this pressing issue by taking deforestation as a case study. A list of three major operations to monitor and preserve forests – unauthorized tree cutting, fire, and tree health – has been identified. Subsequently, a key list of sensors and other electronic components has been proposed. A comprehensive system architecture is outlined by identifying key elements, their interfaces, energy requirements, coverage, and deployment issues.

Multimodal Intelligent Sensing in Modern Applications, First Edition.
Edited by Masood Ur Rehman, Ahmed Zoha, Muhammad Ali Jamshed, and Naeem Ramzan.

Published 2025 by John Wiley & Sons, Inc.

3.1 Environment and Forests

Forests are an integral part of maintaining the balance between different environmental entities and ecosystems. It is estimated that the livelihood of approximately 1.6 billion people depends on forests [1]. They provide habitat for most animals and plants, are still undiscovered, supply oxygen (fundamental for humans and animals to survive), protect our watersheds, and supply timber for everyday use. Forests act as a carbon sink and soak up different greenhouse gases, contributing adversely to the environment, hence, playing a fundamental role in combating global warming. Forest ecosystems store about 80% of all aboveground and 40% of all subterranean earthbound natural carbon [2]. Despite their role in maintaining life balance, forests are being degraded and destroyed. There are numerous forms of deforestation, such as climate changes, illegal tree cutting, ranching and development, and fires, that directly impact humans, animals, and plants. It was estimated in 2020 that the tropics lose almost trees equal to 30 soccer fields, every minute. Similarly, in the last century, the Amazon rainforest has lost 17% of its forests due to human activities. Moreover, the island of Sumatra in Indonesia lost 85% of its forest due to the conversion to oil palm and pulp plantations [3]. Deforestation has numerous environmental, social, and monetary results, one of which is the deficiency of organic variety. It has also increased greenhouse gas emissions by 15%. Soil degradation and water pollution are some of the other impacts on the environment due to deforestation.

Forest decline can effectively be reduced by restricting human activities (cutting down trees, mining, etc.) and improving tree health. The manual real-time monitoring of the forest to prevent unauthorized activities and observe plant health/soil condition is practically unattainable.

3.2 Methods to Combat Deforestation

A plethora of work is available in the literature to combat deforestation by using mostly satellite imagery. Recently the use of machine learning has proven to be an effective approach to accurately map the deforested areas [4–6], which in return helps to set up effective deforestation containment policies. A neural network-based approach is found to be helpful in enhancing the resolution of maps and increasing the accuracy in predicting the deforested area in the Amazon rainforest [7]. Supervised maximum likelihood classification based on Bayes' theorem can also refine the satellite images and accurately identify the deforestation areas [8]. Satellite images refined using a neural network are also used to predict the deforestation area and other illegal tree-cutting activities [9]. The use of deep learning to identify deforested areas and then employing unmanned aerial vehicles to plot seeds to overcome the identified deforestation are also investigated [10].

These techniques purely rely on offline image processing techniques and lack reliability in making efforts toward combating deforestation in real time. The use of WSNs to combat deforestation offers the capability to address this shortcoming.

3.2.1 Combating Deforestation Using Wireless Sensor Networks

WSNs have emerged as a powerful tool in combating environmental degradation across the globe, particularly deforestation. These networks consist of small, autonomous sensors that can monitor and collect data on environmental conditions in real time. By strategically deploying these sensors in forested areas, researchers and conservationists can gather valuable multi-modal information to better understand and address deforestation. In addition, these networks can also monitor wildlife activity and biodiversity in and around deforested areas by tracking the movements of animals and identifying species present, sensors can help conservationists assess the impact of deforestation on local ecosystems and inform conservation strategies accordingly.

WSNs can help detect forest fires early on. By detecting changes in temperature, smoke, and humidity, sensors such as the RHT03 can alert authorities to the presence of a fire, enabling faster response times and minimizing damage to the forest ecosystem [11]. Forest fire detection is a complicated and crucial procedure. The system must be precise, and its reaction time must be quick. Several studies have recently been done to detect forest fires through sensor networks. A hierarchical WSN aimed at early fire detection in wildlands using fire simulations is discussed in [12]. A theoretical model of a ZigBee-based environmental sensing module containing temperature, humidity, oxygen, carbon monoxide, and carbon dioxide sensors for early detection of forest fires is proposed in [13]. A similar setup can be extended toward the tree health monitoring in real time. The network presented in [14] consists of a single node based on temperature, humidity, and carbon dioxide sensors that keep track of allowable thresholds on monitored parameters and dynamically adjusts the frequency of readings for forest fire detection. Josue et al. have discussed a fire controller based on fuzzy logic and decision-making methods and the analytic hierarchy process method having an Arduino platform and seven environmental sensors [15]. The work in [16] proposes a fuzzy system based on overlap indices to improve forest fire detection by implementing a WSN and analyzing different variables such as the lightness and the distance to the fire.

Another important area in which WSNs are used in the fight against deforestation is through monitoring forest cover. By capturing data on tree density, tree health, canopy cover, soil condition, and forest fragmentation, sensors can provide insights into the extent and pace of deforestation. This information is crucial for identifying areas at high risk of illegal logging, tree health degradation, or land conversion and for targeting conservation efforts effectively. An Internet of Things (IoT)-based smart solution to combat deforestation by using NodeMCU (Esp8266) as a wireless fidelity (Wi-Fi) module incorporating a three-axis accelerometer MEMS (ADXL335) sensor for post-processing is discussed in [17] and for tree cutting and fire detection. The monitoring system proposed is suitable for covering an area of 50 m. A wireless sensor-based monitoring system is proposed in [18] to combat deforestation by actively identifying illegal tree cutting, while relying on a global system for mobile communication (GSM) module for communicating with a gateway and by using a series of vibration and sound detection sensors. This sound monitoring system is capable of identifying the noise of a chainsaw within a 10-m range. Another study uses acoustic sensors for the same purpose, enhancing noise analysis with deep neural network techniques to improve the precision of detecting illegal activities [19].

An energy-efficient WSN architecture is proposed to combat deforestation by using a series of sound sensors in [19]. The authors in [20] proposed an energy-efficient acoustic sensor-based WSN to monitor and specify illegal tree-cutting activities. A monitoring system designed to enhance fire detection and humidity monitoring within forests is discussed in [21]. This system is based on AT Mega328 microcontroller, specifically an Arduino Nano, which is equipped with sensors to measure levels of carbon monoxide and carbon dioxide. These measurements are critical for monitoring forest fires and assessing tree health, thereby contributing to the overall safety and health of trees. Additionally, the system incorporates infrared cameras to combat deforestation by preventing illegal tree-cutting activities, further emphasizing its dual focus on forest preservation and environmental protection. Moreover, a light detection and ranging (LiDAR)-based strategy is used in [22] to monitor the tree health, while relying on unmanned aerial vehicles. In [23], a LiDAR-based unmanned aerial vehicles monitoring system is proposed to first identify the key deforested sites and then the unmanned aerial vehicles are used to reforest that area.

The use of WSNs in combating deforestation has the potential to revolutionize conservation efforts by providing real-time, high-resolution data on forest ecosystems. By harnessing the power of these technologies, we can work toward a more sustainable and greener future for our planet.

3.2.2 Sensor Types for Combating Deforestation

WSNs can broadly perform the following three tasks in their assignment to monitor and conserve the forests:

- Detect unauthorized tree cutting.
- Sense the wildfires.
- Perceive the tree health.

To monitor unauthorized tree cutting, a sensor that detects the sound and the movement is required. The fire can be monitored using carbon monoxide sensors, whereas a different series of sensors, such as sensors used to measure soil humidity and acidity, sensors that measure the environmental parameters around the trees, etc., are required to measure the health of a tree. A detailed comparison of the cited literature that uses different configurations of WSNs to combat deforestation is given in Table 3.1.

3.3 Design of a WSN to Combat Deforestation

WSNs offer the power and capability for real-time monitoring and intervention to environmental degradation. Deforestation is one of the best examples of it. Designing an efficient and effective WSN to combat deforestation is very challenging due to the real-time nature of the job, deployment in harsh forest environment, error-prone acquisition of multi-modal data and limitations around the energy usage, signal processing, and analysis and decision stages. In this section, this complex task is being dealt with by adopting a three-stage approach to study various verticals of the required sensing system. In stage 1, a thorough list of user requirements has been identified. In stage 2, an end-to-end system architecture, based on user requirements, is established. Finally, in stage 3, a detailed analysis of system implementation based on key performance indicators is discussed. Moreover, a comparison of each of the selected modules, as shown in system implementation, is also provided to justify the selection of components. The high-level end-to-end three-stage implementation of system architecture is shown in Figure 3.1.

3.3.1 Stage 1: System Requirements

The success of any efficient system depends hugely on putting together a targeted and comprehensive list of requirements. Following is the list of requirements from a user perspective to develop a complete, fully functional, low-cost, and real-time IoT-based WSN.

- Each wireless sensor node deployed in a forest and attached to a tree is required to perform the following three main operations.
  1) **Monitor Tree Health:** Each wireless sensor node will be able to measure tree health, which includes soil quality, water content, light intensity, pH level, air quality (including temperature and humidity), and the amount of oxygen it produces.
  2) **Monitor Illegal Tree Cutting:** Each wireless sensor node is required to measure the noise of loggers' tools, such as chainsaws and trucks to haul away the logs. Moreover, the noise and vibrations need to be used to predict human or animal movements.

Table 3.1 Comparison of wireless sensor network-based systems for combating deforestation.

Citations	Combating Deforestation			Area covered/ detection range (meters)	Sensors		
	Cutting	Fire	Health		Tree cutting	Fire	Tree health
Sivasankari et al. [17]	✓	✓	✗	50	MEMS (accelerometer)	RKI 3100	N/A
Prasetyo et al. [18]	✓	✗	✗	10	Sound (FC-04), vibration (SW420)	N/A	N/A
Kalhara et al. [24]	✓	✗	✗	N/A	Acoustic sensor	N/A	N/A
Petrica and Stefan [19]	✓	✗	✗	50	(MP3DT01)	N/A	N/A
Darlis et al. [25]	✗	✓	✓	N/A	N/A	(MQ4)	DHT11
Silva et al. [20]	✓	✗	✗	N/A	Acoustic sensor	N/A	N/A
Ghazali et al. [21]	✓	✗	✗	10	IR cameras	N/A	N/A
Temuulen et al. [22]	✗	✗	✓	100	N/A	N/A	LIDAR sensor
Liu et al. [26]	✗	✓	✓	N/A	N/A	Smoke sensor	Temperature sensor
Almedia et al. [23]	✗	✗	✓	N/A	N/A	N/A	LIDAR sensor

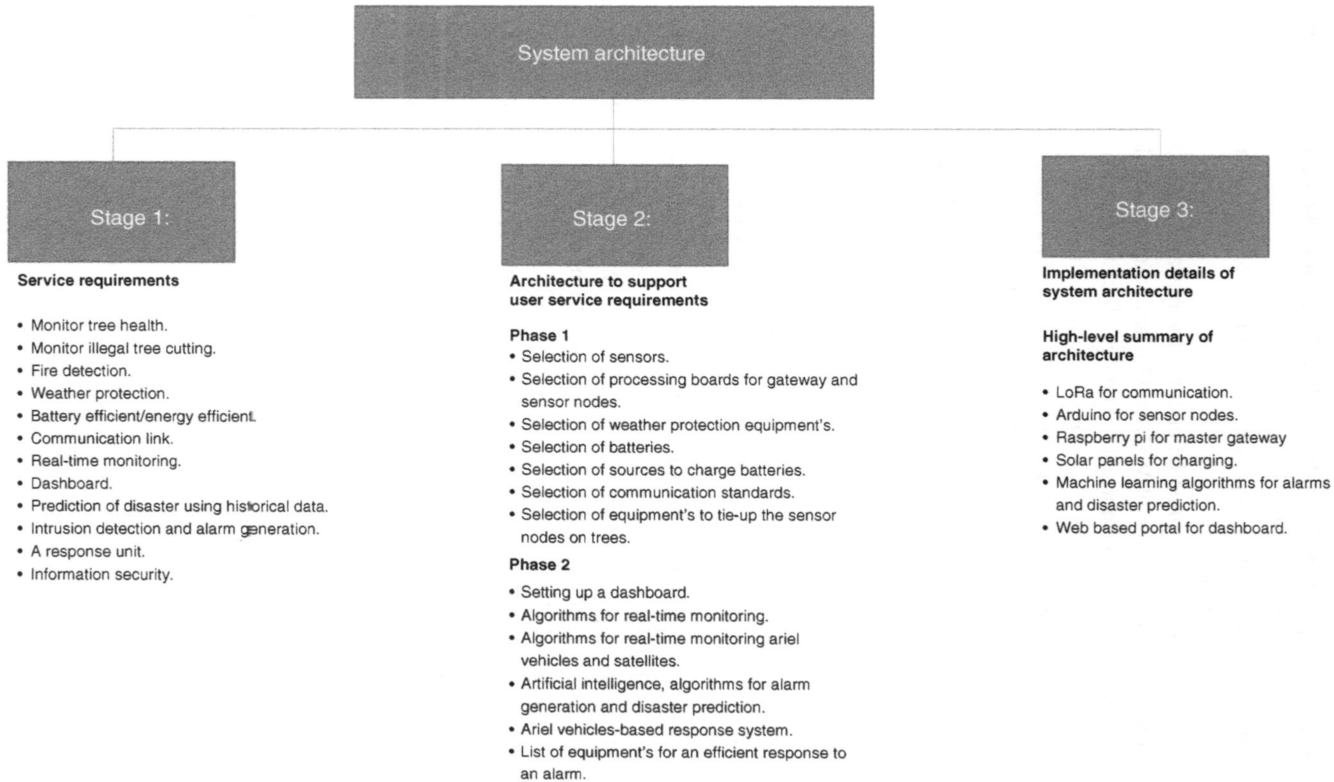

Figure 3.1 High-level end-to-end three-stage WSN system architecture for combating deforestation.

3) **Detect Wildfire:** Each wireless sensor node is required to measure the change in temperature, humidity, and evidence of carbon monoxide in the surrounding environment of a tree.

- Each wireless sensor node should be protected from forest environmental conditions as well as wildlife.
- Each wireless sensor node must be energy efficient and able to operate for a longer duration. The lifetime of WSN will be set, based on available batteries, their size, and cost. Furthermore, some energy-efficient algorithms can be utilized or developed to find the best optimal lifetime value.
- Each wireless sensor node will be able to send sensing data (related to tree health, gases, sound, etc.) to a gateway (usually connected to the internet or the server, available in the monitoring center) or other wireless sensor node to closely monitor the tree health, tree cutting, and fire detection. The hybrid connectivity of the wireless sensor nodes can be altered based on the available resources.
- The central monitoring station is equipped with a dashboard (visualization of results) to measure (using artificial intelligence and machine learning tools) and show various trends based on sensing data received from each of the wireless sensor node in real time. This dashboard is required to be able to be accessible remotely from anywhere around the world by using the internet cloud.
- The central monitoring station is required to generate an alarm at the surveillance (a central monitoring station) if the presence of illegal activities (tree cutting based on noise), or prediction of fire, is detected. It is also required to predict the occurrence of a disaster by using artificial intelligence or other appropriate algorithms.
- The central monitoring station is equipped with a response unit based on remotely operated drones, water sprinklers, and CO_2 cylinders to counter the anomaly. The drones are also required to provide coverage to sensor nodes in case of a hardware failure.
- Central monitoring is required to ensure end-to-end information security.
- Central monitoring is required to provide remote access to each of the deployed wireless sensor nodes, to perform algorithm updates and re-routing of data, in case of hardware or software failure of the wireless sensor node.

3.3.1.1 Key Performance Indicators

To measure the performance and justify the selection of different components, the key performance indicators of a real-time WSN-based deforestation combating system are discussed below.

Energy Efficiency: It is one of the critical key performance indicators in any WSN application. The lifetime of a WSN is defined based on the numerical values of energy efficiency. The identification of appropriate batteries, use of energy harvesting circuits, the choice of appropriate routing and scheduling schemes, limiting data processing at wireless sensor nodes, choice of appropriate communication standards, sleeping mechanisms, etc. are some of the factors that need to be considered before the designing of energy-efficient WSN system.

Coverage: Coverage of a WSN is defined by the ability of each wireless sensor node to accurately perform the following three main operations.

- Sensing.
- Processing.
- Transmitting.

The processing of sensing information can be reduced as much as possible to increase energy efficiency. Whereas the number of retransmissions can be reduced effectively by choosing energy-efficient routing and scheduling schemes.

Scalability: It is used to measure the flexibility by which the proposed WSN can be extended for a relatively larger area and increased number of devices/users.

Delay: Delay can be defined in three ways. First, it is the time required for a packet to be received and displayed on the dashboard. Second, it is the time difference of renewal of data, that is, the interval after which the new set of values is being sent by each wireless sensor node. Third, it is the time taken by the monitoring unit to respond to the generation of an alarm.

Resilience: It is the ability to cope with node failures. Resilience is defined in three ways. One, if any node undergoes software failure. Second, if any node undergoes hardware failure. Finally, how to cope with weather and harsh conditions.

Security: It is defined by how securely the information is shared in the end-to-end IoT-based WSN system.

Cost: It is used as a key performance indicator to make the product usable by industry and governmental organizations.

3.3.2 Stage 2: Architecture

To effectively support the set of user requirements gathered in stage 1, a sustainable and meticulously planned system is architecture. The end-to-end system architecture is divided into two phases. Phase 1 is related to the deployment of wireless sensor nodes and to setting up an end-to-end reliable communication link. Phase 2 deals with the intervention/response schemes to facilitate the necessary actions that must be taken manually/automatically by the system in case an alarm is raised by the central monitoring unit. Figure 3.2 illustrates a high-level end-to-end system architecture for phase 1.

To fulfill the user requirements of tree health monitoring, illegal tree-cutting detection, and wildfire sensing, a number of sensing parameters need to be observed by the multi-modal WSN. To measure the health of a tree, it is necessary to estimate the air quality in its vicinity, the quality of soil beneath it, and the

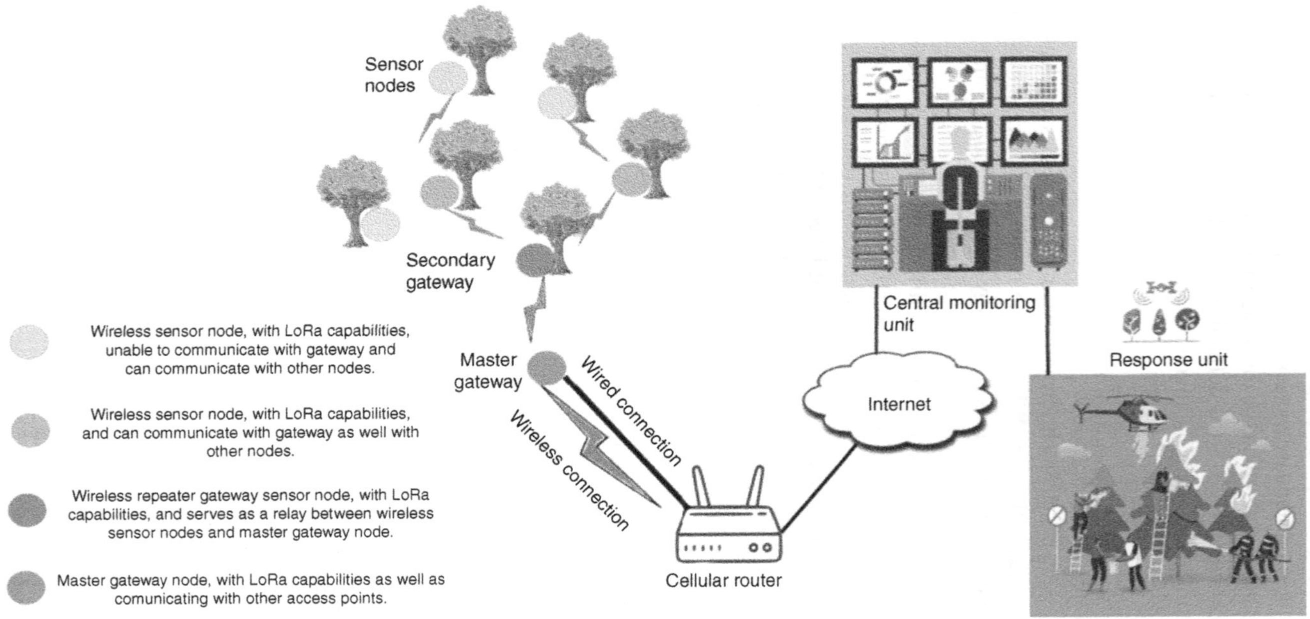

Figure 3.2 An end-to-end system architecture to set up a reliable communication link. *Source:* (bottom) Sergii Pavlovskyi/Adobe Stock Photos; (top) K3Star/Adobe Stock Photos.

amount of light it is exposed to. To monitor illegal tree-cutting activities, a vibration/motion sensor and an acoustic sensor are required. To detect the wildfire, carbon dioxide/carbon monoxide levels and temperature should be noted. Moreover, to pinpoint the adversarial activity and facilitate an informed, quick, and precise containment response, accurate location is also needed. Thus, the required sensing parameters are listed as the following:

- Motion
- Acoustic
- Air quality
- pH levels
- Soil humidity and acidity
- Temperature
- Carbon dioxide levels
- Illumination levels
- Positioning and localization
- Orientation
- Carbon monoxide levels

The system architecture uses the long-range (LoRa) communication standard to provide communication between sensor nodes and the gateway (as illustrated in Figure 3.2). Two separate gateways are suggested, that is, the secondary gateway gathers the sensing data received from multiple wireless sensor nodes deployed on trees, and then transmits it to the master gateway, present in the monitoring unit. The selection of LoRa is based on its inherent features of low power and long range enabling it to cover large areas, making it ideal for monitoring vast forest areas. Based on the key performance indicators, Arduino as a board/minicomputer for sensors and a secondary gateway is selected, whereas a Raspberry Pi could be a suitable option for the master gateway. The use of small solar panels is proposed to increase the lifetime of wireless sensor nodes, whereas two different battery-based solutions have been recommended to make the sensing platform capable of coping with different weather conditions around the world. The use of a number of machine learning algorithms is suggested to increase the accuracy of event detection and reduce the delay in generating an alarm. Aerial-based intervention strategies are recommended to cope with disaster situations with minimum latency.

3.3.3 Stage 3: System Implementation

To implement the system architecture discussed in Section 3.3.2, efficient selection of sensors, communication standards, deployment strategy, and intervention schemes is required.

3.3.3.1 Type of Sensors

Based on the required sensing parameters, different types of wireless sensors are needed to implement the system. Selection of an appropriate sensor is a

Table 3.2 Comparison of different commercially available temperature sensors.

Sensor	Temperature	Humidity	Current	Sampling rate	Accuracy
DHT11	(0 to +50 °C)	20–80%	2.5 mA	1 Hz	1 °C
Grove temperature Sensor	(−40 to +125 °C)	N/A	3 mA	0.5 Hz	1.5 °C
DHT22	(−40 to +80 °C)	5–99%	1.5 mA	20 Hz	0.5 °C
STH31	(−40 to 125 °C)	0–100%	4.8 μW	1000 kHz	0.3 °C
BME280	(−40 to 80 °C)	0–100%	0.4 mA	—	1 °C
ATH21	(−40 to 85 °C)	0–100%	—	1000 kHz	0.3 °C

systematic process led by strategic adherence to various technical, geometrical, performance, and economic factors such as size, power consumption, operating temperatures, accuracy, sampling rate, and cost. There is a wide variety of sensors available in the market and can be chosen to fulfill the prioritized system requirement. For example, Table 3.2 provides a comparison of different temperature sensors available commercially.

A similar approach should be adopted in the selection of other sensor types required for the proposed multi-modal wireless sensing system architecture.

3.3.3.2 Processing Boards

A wide variety of processing boards are available to facilitate the proposed system architecture. Some of the major names include Arduino Uno, Raspberry PI 3 B+, Raspberry Pi 4, Raspberry Pi Pico, Portenta H7 lite, and LoRa-E5 STM32WLE5JC.

The Arduino Uno, developed by Arduino, is a single microcontroller-based board and supports C and C++ programming languages. A relatively huge amount of open-source data is available to support the technical needs of Arduino Uno. The price of Arduino Uno is relatively low with the ease of availability in any local market. The Raspberry Pi is a series of small single-board computers supported by Raspberry Pi Foundation and Broadcom and developed in the United Kingdom. Raspberry PI 3 B+, Raspberry Pi 4, Raspberry Pi Pico, etc., are some of the variants of Raspberry Pi available in the market. The Raspberry PI 3 B+ contains a mini operating system known as Rasberian and Noobs. It supports Python programming language by default, but it can also support C, C++, Rust, etc. A relatively huge amount of data is available to support the technical needs of Raspberry PI 3 B+. It can easily be connected to any Wi-Fi-enabled device by using ethernet or wireless channel but possesses a slow processing speed.

Table 3.3 Comparison of various processing boards for wireless sensor networks.

Name	Clock rate	Power consumption	Processor	Bluetooth	Wi-Fi
Arduino UNO	16 MHz	200 mW	Atmega 328	—	—
Raspberry PI 3 B+	1.4 GHz	1.4–2.1 W	ARM 11	4.2	2.4 and 5 GHz
Raspberry Pi 4	1.5 GHz	3.8–5.5 W	ARM V8	5	2.4 and 5 GHz
Raspberry Pi Pico	133 MHz	500 mW	ARM RP2040	—	—
Portenta H7 lite	480 MHz	<1 mW	STM32H747XI dual Cortex®-M7+M4	—	—

Similarly, Raspberry Pi 4 possesses the same features as that of Raspberry PI 3 B+ but has a relatively high processing speed and low-power consumption on standby. The Raspberry Pi Pico possesses a similar capability as that of Raspberry Pi 4 but with an extremely low form factor. Portenta H7 Lite is a powerful and extremely low-energy consumption platform developed by Arduino, with a relatively higher cost but lower than Raspberry Pi 4. A detailed comparison of the commonly used boards is shown in Table 3.3. Again, selection would be as per system requirements, form factor, and processing capabilities.

3.3.3.3 Communication Modules

In the high-level system architecture, LoRa is proposed to be the suitable option to facilitate communication between the different entities of the multi-modal forest sensing system. To support the LoRa communication, both gateways and each of the wireless sensor nodes need to be equipped with LoRa communication modules. A detailed comparison of different commercial LoRa communication modules is shown in Table 3.4.

Based on this, a wireless sensor node can be set up by using Arduino with a grove LoRa E5 communication module and creating the edge device/master gateway by using the Raspberry Pi 3/4 as shown in Figure 3.3.

3.3.3.4 Batteries

Batteries play a crucial role in the functionality and effectiveness of wireless sensor nodes. As the proposed forest sensing system is meant to be deployed in remote or inaccessible areas where continuous power supply is not available, the importance of an efficient battery source is further prevalent.

Table 3.4 Comparison of various LoRa communication modules used for wireless sensor networks.

Name	Type	Technology	Compatibility	Frequency
LoRa-E5 mini (STM32WLE5JC)	LoRa and LoRaWAN	LoRA-E5	Stm32	EU868/US915/AU915/AS923/KR920/IN865
LoRa-E5 Development Kit	LoRa and LoRaWAN	LoRA-E5	Stm32	EU868/US915/AU915/AS923/KR920/IN865
Dragino LoRaST IoT Module	LoRa and LoRaWAN	LoRaWAN 1.0.2	Stm32	915 MHz
Grove – LoRa-E5 (STM32WLE5JC)	LoRa and LoRaWAN	LoRa-E5	Arduino, Raspberry pi and Stm32	EU868/US915/AU915/AS923/KR920/IN865
Arduino MKR WAN 1300	LoRa and LoRaWAN	—	Arduino	—
RHF0M301 8 channels LoRaWAN	LoRa and LoRaWAN	—	Raspberry Pi 3B/3B+/4	434, 470–510, 780, 868, and 915 MHz
Adafruit Feather 32u4 RFM96 LoRa Radio	LoRA	—	ATmega32u4	433 MHz

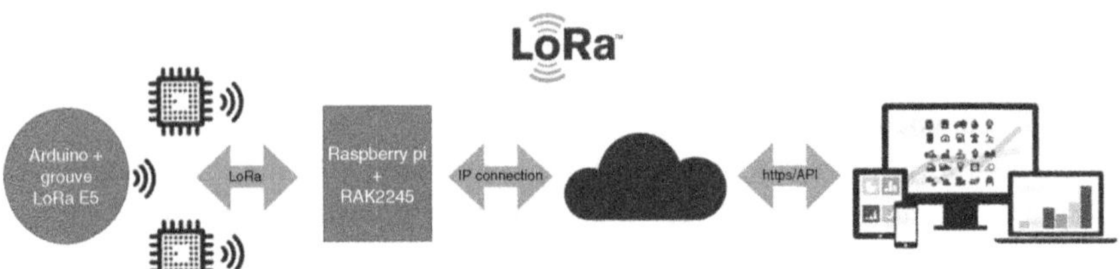

Figure 3.3 A high-level representation of proposed communication links.

Batteries serve as the primary power source for wireless sensor nodes, providing the necessary energy to collect, process, and transmit data. Without a reliable and long-lasting battery, the sensor node would not be able to perform its intended functions. The efficiency of the battery directly impacts the overall energy consumption of the sensor node. A high-quality battery can ensure that the node operates efficiently, prolonging its lifespan and reducing the frequency of battery replacements. A dependable battery is also essential for ensuring the continuous operation of the sensor node. Unreliable or

Table 3.5 Comparison of batteries used for wireless sensor networks.

Types	Cell voltage (V)	Temperature (°C)	Cell size (mAh)	Battery life (days)
NiCd	1.2	−20 to 60	1100	4.5
NiMH	1.2	−20 to 70	2300	3.9
Low self-discharge NiMH	1.2	−20 to 70	1900	6.5
Li-ion (LCO)	3.6	−40 to 70	1300	5
LiFePo4	3.6	−40 to 80	1100	6.5
LiPo (LCO)	3.7	−40 to 80	2200	5.3
Li-Ti (LTO)	2.4	−40 to 75	3500	6.5
AGM	2	−15 to 50	7500	4

low-quality batteries can lead to frequent power failures, resulting in data loss and interruptions in data transmission.

Selecting the right type of battery is important for minimizing the environmental impact of wireless sensor nodes. Rechargeable batteries, for example, are more sustainable than single-use batteries as they can be reused multiple times, reducing waste and pollution. The performance of a wireless sensor node is closely tied to the performance of its battery. A high-capacity battery with fast charging capabilities can enhance the overall performance of the node, enabling it to collect and transmit data more effectively.

Batteries are indispensable components of wireless sensor nodes, playing a critical role in ensuring their functionality, efficiency, reliability, and overall performance. Selecting the right type of battery and managing its usage effectively are key considerations in maximizing the effectiveness of sensor nodes in monitoring and data collection applications. Table 3.5 presents a comparison of different commercially available rechargeable batteries for the proposed sensing system.

For cold temperatures below 0 °C, the batteries which are chargeable at such a temperature are preferred. Lithium-titanate (LTO)-based batteries are the most suitable option, due to their temperature tolerance, containing huge cycle life. An alternative option is an Absorbent Glass Mat (AGM) battery, but the capacity of these batteries decreases with a decrease in temperature. For normal temperatures, the suitable option is Lithium-ion cells as they are cheaper as well as possess a decent lifetime.

3.3.3.5 Energy Harvesting Circuit and Equipment

Energy harvesting is a technology that enables wireless sensor nodes to generate power from their surrounding environment, reducing their dependency on batteries or external power sources. It allows sensing nodes to tap into renewable energy sources such as solar. It provides a sustainable power source for the nodes, reducing the need for frequent battery replacements and minimizing the environmental impact. It also enhances the reliability of sensing nodes by providing a continuous power supply, very useful in environments such as forests where traditional batteries may not be practical or reliable. This ensures uninterrupted data collection and transmission, making the nodes more dependable. It can help reduce the overall operating costs of the system by minimizing the expenditure on batteries and the logistics associated with battery replacement.

There are different energy harvesting techniques available for WSN-based applications. Some of these techniques include pulse width modulation (PWM) and maximum power point tracking (MPPT). In a dense forest environment, the use of advanced power harvesting techniques like MPPT is preferred, which can obtain maximum power through continuous tracking of sunlight. Energy harvesting using a solar panel can be done by using MPPT and a battery management system. An illustration of such a scenario is shown in Figure 3.4. The power output will be obtained through the DC–DC converter (buck converter) for the conversion of 12-volt DC battery input into 3.3 and 5-V output, respectively. A detailed comparison of different solar panels is given in Table 3.6.

3.3.3.6 Weather Protection

Weather protection boxes for forest monitoring sensors are essential components in ensuring the longevity and reliability of sensor systems deployed in outdoor environments. These specialized enclosures are designed to shield sensors from harsh weather conditions, such as extreme temperatures, moisture, and debris, while allowing for accurate data collection and transmission.

Weather protection boxes are typically constructed from durable and weather-resistant materials, such as UV-stabilized plastics, fiberglass, tin, and corrosion-resistant metals like aluminum or stainless steel. An alternative option is a custom-made box by 3D printing. The design of the enclosure should be compact,

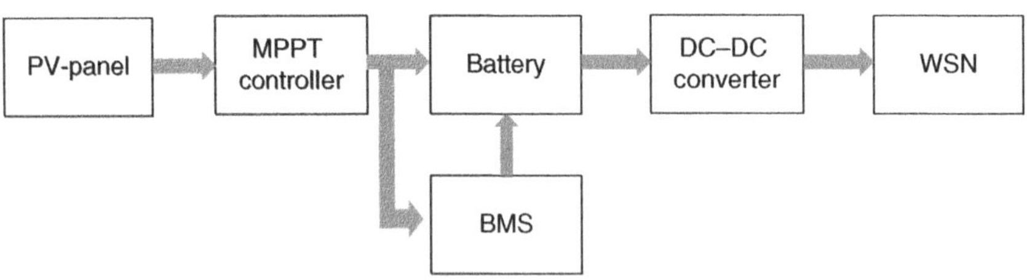

Figure 3.4 A high-level block diagram of an energy harvesting and battery management system.

Table 3.6 Pros and cons of different solar panels for wireless sensor networks.

Panel type	Disadvantages	Advantages
PERC	High initial cost.	Requires least space, energy efficiency, and higher power capacity.
Monocrystalline	High initial cost.	Low cost, alternative to PERC panels, and without the passivating layer.
Polycrystalline	Low heat tolerance, not suitable in hot environments	Moderated in terms of cost, efficiency, and power capacity.
Copper indium gallium selenide (CIGS)	Shorter lifespan than crystalline panels	Lowest cost, easier to install, flexible, and unbreakable.
Cadmium telluride (CdTe)	Requires more space	Lowest cost, easier to install, flexible, and unbreakable.
Amorphous silicon (a-Si)	Least efficient	Lowest cost, easier to install, flexible, and unbreakable.

yet spacious enough to accommodate the sensors and associated electronics, while providing protection from rain, snow, wind, and sunlight. These boxes should also be designed for easy mounting on trees or other structures, with secure mounting hardware to withstand strong winds or inclement weather.

3.3.3.7 Wireless Communication Link

Wireless sensor nodes, deployed in the area of interest, require some protocols to communicate and transmit their sensing data to the gateway or other nodes by using a wireless channel. The following are some of the commonly used WSN communication protocols.

• ZigBee	• Thread	• LoRa
• Z-Wave	• Digimesh	• SigFox
• Bluetooth 4.0+	• MiWi	• LTE Cat-M1
• Bluetooth 5	• EnOcean	• NarrowBand-IoT (Cat M2)
• Bluetooth low energy	• 6LoWPAN	• Cellular 3G/4G
• Wi-Fi	• Weightless (W, N, P)	
• Wi-Fi-ah (HaLow)	• mcThings	

The selection of a suitable communication protocol depends on numerous factors. For instance, the deployment region, energy consumption, licensing, required data rate, availability of hardware, transmission range, interoperability, cost, intended usage, etc. are some of the key parameters that need to be considered while selecting an appropriate communication standard.

ZigBee is one of the popular wireless communication standards for IoT devices, which is maintained and published by Zigbee Alliance. Zigbee is usually suited for applications requiring a low range, data rate, and power. ZigBee faces interoperability issues, due to the presence of multiple ZigBee standards [27]. On the other hand, the Z-wave (an open-source standard by Zigbee Alliance) overcomes the flaws of ZigBee but possesses relatively high cost, reliability, and latency issues. Bluetooth 4.0+, Bluetooth 5, and Bluetooth low energy are used for proximity-based applications and have a low cost and high data rate, except for Bluetooth low energy, which possesses a very low data rate [28]. Wi-Fi and Wi-Fi-ah (HaLow) are generally popular for IoT applications requiring very high data. Since Wi-Fi-ah (HaLow) works at a lower-frequency band, it possesses a better wall penetration and range than Wi-Fi, but at the expense of a relatively lower data rate [29]. Thread, developed by Google, uses a similar radio as ZigBee and works for a low-range application [30]. Digimesh and MiWi are modified versions of ZigBee. Digimesh works for long-range communication at a relatively high cost, whereas MiWi is suitable for short-range communication and the cost of the chipset is low [31]. EnOcean is a proprietary standard for short-range communication and relies on energy harvesting [32].

IPv6 over low-power wireless personal area networks (6LoWPAN) by IEEE possesses similar features as that of Wi-Fi but it is suitable for applications requiring low data rate [33]. LoRa and Sigfox are usually suitable for long-range applications (up to 20 mi), but LoRa possesses relatively high data and is open-source [34]. The weightless (W, N, P) is a controversial IoT communication standard, as weightless-W was rejected by governing bodies. The weightless N and P versions are wide area standards but possess a relatively lower range than LoRA and SigFox [31]. The mcThings is a short-range communication standard with a high-cost and low-power consumption and can increase the battery life of wireless sensor nodes [31]. LTE Cat-M1 and NarrowBand-IoT (Cat M2) are IoT standards added to the LTE cellular communication technology and require a cellular backbone to operate [35]. Finally, cellular technology could be another option to provide a communication link to IoT devices. A detailed comparison of the commonly used standards is shown in Table 3.7.

The way the node communicates with each other or with the gateway is determined by their deployment within the network. This deployment of wireless sensor nodes is termed the topology of WSNs. Following are some of the popular topologies of WSNs [36]:

- Star Topology.
- Mesh Topology.
- Hybrid Topology.

Table 3.7 Comparison of various wireless protocols.

Wireless protocols	Security	Frequency	Range	Data rate
ZigBee	Encrypted	2.4 GHz, 915 MHz (US), 868 MHz (EU)	100–325 ft	250 kbps (2.4) 40 kbps (915) 20 kbps (868)
Z-Wave	Encrypted	915 MHz (US) 868 MHz (EU)	100–325 ft	40 kbps (915) 20 kbps (868)
Bluetooth 4.0+	Encrypted	2.4 GHz	200 ft	25 Mbps
Bluetooth 5	Encrypted	2.4 GHz	800 ft	50 Mbps
Wi-Fi	Optional	2.4 GHz/5 GHz	115–230	7 Gbps
Wi-Fi-ah (HaLow)	—	900 MHz	3000 ft	347 Mbps
Thread	Encrypted	2.4 GHz	100 ft	250 kbps
Digimesh	Encrypted	2.4 GHz/900 MHz (US)/868 MHz (EU)	~20 mi	250 kbps (2.4) 40 kbps (915) 20 kbps (868)
MiWi	Encrypted	2.4 GHz or subGHz	800 ft	250 kbps
EnOcean	Encrypted	900 MHz (US) 868 MHz (EU) 315 MHz	30–100 ft	125 kbps
6LoWPAN	Optional	2.4 GHz	380 ft	250 kbps
Weightless (W, N, P)	Encrypted	white-spaces, 915 MHz, 868 MHz, 780 MHz, 470 MHz, 433 MHz, 169 MHz	1.2 mi (P), 3 mi (W, N)	200 bps–100 kbps
mcThings	—	2.4 GHz	650 ft	50 kbps
LoRa	Encrypted	150 MHz–1 GHz (lots of options)	up to 20 mi	50 kbps
SigFox	Encrypted	900 MHz (US) 868 MHz (EU)	~20 mi	100 bps
LTE Cat-M1	—	1.4 MHz	~20 mi	1 Mbps
NarrowBand-IoT (Cat M2)	Encrypted	Below 1 GHz	~20 mi	100 kbps
Cellular	Encrypted	700, 800, 850, 1700, 1900, 2100, 2300, and 2500	~20 mi	200 kbps (3G) 10 Mbps (4G)

In start topology, all the sensor nodes directly communicate with the gateway and are not allowed to communicate with each other. In mesh topology, the nodes can communicate with each other as well as with the gateway. Whereas hybrid topology is the combination of both start topology and mesh topology. In comparison, the sensors consume less power in the start topology due to a single point of contact but are highly unreliable due to the restrictions imposed on the start topology. The mesh topology is more scalable and is highly reliable as well but consumes more power. The hybrid possesses the advantages and the disadvantages of both mesh and star topologies, but interoperability is one of the key challenges in hybrid topology. The following list of parameters must be kept in mind while choosing a suitable topology.

- Reduced packet loss probability.
- Less energy consumption.
- Reduced radio interference.
- Less retransmissions.
- High reliability.

Although it is difficult to achieve all these goals in one topology, a tradeoff can be set. Overall, the choice of the topology is entirely application and environment dependent. Keeping in mind the user requirements, key performance indicators, and to perform real-time monitoring, initially, a star topology will be adopted to avoid any interoperability issues. To scale the proposed real-time WSN-based monitoring system, a hybrid topology will be adopted, by merging two or more topologies. Hence, dividing the load of large-scale WSNs and providing high reliability, scalability, and lower-power consumption. An illustration of such a strategy is shown in Figure 3.5.

3.3.3.8 Data Processing Algorithms

Machine learning and data analytics techniques are fast and accurate ways to predict the anomalies of any system(s), which in return helps to devise strategies to intervene/respond in real time. In the context of predicting and combating deforestation, different data acquisition and analysis approaches play a significant role in identifying the key areas being affected due to deforestation. In [7], a neural network-based approach is used to enhance the resolution of maps to increase the accuracy in predicting the deforested area in the Amazon rainforest. The satellite images can be further refined using supervised maximum likelihood classification based on Bayes' theorem to accurately identify the deforestation areas as proposed in [37]. Moreover, the image processing approach is also utilized in [38] to improve the prediction of deforestation areas in the Amazon rainforest. The use of satellite images is further refined using a neural network to

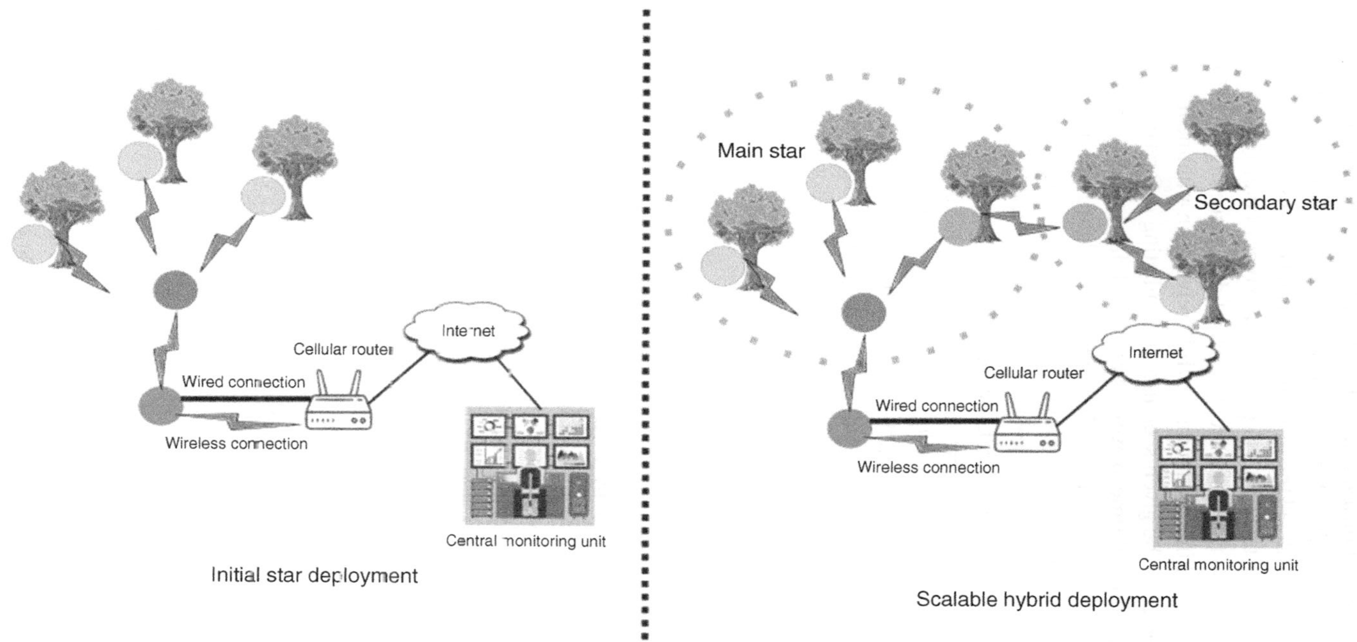

Figure 3.5 Illustration of sensor deployment scenarios. *Source:* K3Star/Adobe Stock Photos.

predict the deforestation area and other illegal tree-cutting activities [39]. In [40], the authors first used deep learning to identify deforested areas, and then to overcome it, they have utilized the unmanned aerial vehicles to plot seeds. The choice of machine learning technique depends on the type of data being fed to a learning algorithm. Based on the literature, the use of deep learning algorithms is proven to be useful in making predictions, while relying on nonlinear and time-series data [41]. Since, in the given scenario of combating deforestation, the end user will be relying on real time as well as historical data, to make predictions and to detect anomalies, the use of deep learning-based algorithms will certainly help to make fast and accurate predictions. Although there are various types of deep learning algorithms, the selection of an appropriate deep learning algorithm will entirely be dependent on the data being received.

Appropriate communication modules and protocols should be selected by emphasizing low-cost, energy-efficient solutions suitable for extensive outdoor deployment. The choices should be made to align with the objectives of maximizing range, security, and compatibility across diverse sensor nodes, ensuring the scalability and effectiveness of the multi-modal forest monitoring system.

3.4 Summary

In this chapter, complete architecture of a real-time, multi-modal forest monitoring system is discussed by identifying key elements, their interfaces, energy requirements, coverage, and deployment strategies. A set of key performance parameter definitions is also provided. A three-stage approach has been utilized to study various verticals of the required system to effectively combat deforestation. In stage 1, a complete list of user requirements has been identified. In stage 2, a high-level system architecture is explained by highlighting the required sensors, boards, batteries, energy harvesting equipment, communication standards, and establishing a central monitoring unit. In stage 3, a detailed explanation of each of the selected technology, module, and algorithm is provided. The discussion provides guidelines to develop a complete end-to-end low-cost, energy-efficient IoT-based WSN to combat deforestation.

References

1 United Nations (2011). *Forests for People*. London: Tudor Rose.

2 Food and Agriculture Organization of the United Nations (FAO). "Forests and climate change". http://www.fao.org/3/ac836e/AC836E03.htm (accessed 22 October 2024).

3 World Wildlife Fund (WWF), "Endangered species threatened by unsustainable palm oil production". https://www.worldwildlife.org/stories/endangered-species-threatened-by-unsustainable-palm-oil-production (accessed 21 October 2024).

4 Arai, E., Shimabukuro, Y.E., Pereira, G., and Vijaykumar, N.L. (2011). A multi-resolution multi-temporal technique for detecting and mapping deforestation in the Brazilian Amazon Rainforest. *Remote Sensing* 3 (9): 1943–1956.

5 Karimi, N., Golian, S., and Karimi, D. (2016). Monitoring deforestation in Iran, Jangal-Abr Forest using multi-temporal satellite images and spectral mixture analysis method. *Arabian Journal of Geosciences* 9: 1–16.

6 Candra, D.S. (2020). Deforestation detection using multitemporal satellite images. *IOP Conference Series: Earth and Environmental Science* 500: 012037.

7 Maretto, R.V., Fonseca, L.M.G., Jacobs, N. et al. (2021). Spatio-temporal deep learning approach to map deforestation in Amazon rainforest. *IEEE Geoscience and Remote Sensing Letters* 18 (5): 771–775.

8 Rehman, A., Shahid, H., Sultan, A. et al. (2019). Deforestation analysis of Northern Areas (Pakistan) using image processing and maximum likelihood supervised classification. *2nd International Conference on Communication, Computing and Digital systems (C-CODE)*, Islamabad, Pakistan (6–7 March 2019).

9 Kehl, T.N., Todt, V., Veronez, M.R., and Cazella, S.C. (2012). Amazon rainforest deforestation daily detection tool using artificial neural networks and satellite images. *Sustainability* 4 (10): 2566–2573.

10 Lohit, G.V.S. and Bisht, D. (2021). Seed dispenser using drones and deep learning techniques for reforestation. *5th International Conference on Computing Methodologies and Communication*, Erode, India (8–10 April 2021).

11 Aksamovic, A., Hebibovic, M., and Boskovic, D. (2017). Forest fire early detection system design utilising the WSN simulator. 26th *International Conference on Information, Communication and Automation Technologies (ICAT)*, Sarajevo, Bosnia and Herzegovina (26–28 October 2017).

12 Molina-Pico, A., Cuesta-Frau, D., Araujo, A. et al. (2016). Forest monitoring and wildland early fire detection by a hierarchical wireless sensor network. *Journal of Sensors* 2016: 1–8.

13 Abdullah, S., Masar, S., Bertalan, S. et al. (2017). Wireless sensor network for early forest fire detection and monitoring as a decision factor in the context of a complex integrated emergency response system. *IEEE Workshop on Environmental, Energy, and Structural Monitoring Systems*, Salerno, Italy (21–22 June 2018).

14 Yanjun, L., Zhi, W., and Yeqiong, S. (2006). Wireless sensor network design for wildfire monitoring. *6th World Congress on Intelligent Control* and Automation (WCICA), Dalian, China (21–23 June 2006).

15 Toledo-Castro, J., Caballero-Gil, P., Rodríguez-Pérez, N. et al. (2018). Forest fire prevention, detection, and fighting based on fuzzy logic and wireless sensor networks. *Complexity* 2018 (1): 1639715.

16 Garcia-Jimenez, S., Jurio, M.P.A., Miguel, L.D. et al. (2017). Forest fire detection: a fuzzy system approach based on overlap indices. *Applied Soft Computing* 52: 834–842.

17 Sivasankari, N., Mounika, S.C., and PriyaDharshini, K. (2020). Implementation of wireless sensor networks to prevent deforestation using node MCU. In: *Intelligent Systems and Computer Technology* (ed. D.J. Hemanth, V.D.A. Kumar, and S. Malathi), 45–51. IOS Press.

18 Prasetyo, D.C., Mutiara, G.A., and Handayani, R. (2018). Chainsaw sound and vibration detector system for illegal logging. *International Conference on Control, Electronics, Renewable Energy and Communications (ICCEREC)*, Bandung, Indonesia (5–7 December 2018).

19 Petrica, L. and Stefan, G. (2015). Energy-efficient WSN architecture for illegal deforestation detection. *International Journal of Sensors and Sensor Networks* 3 (3): 24–30.

20 da Silva, B., Segers, L., Braeken, A. et al. (2018). A low-power FPGA-based architecture for microphone arrays in wireless sensor networks. In: *Applied Reconfigurable Computing. Architectures, Tools, and Applications* (ed. N. Voros, M. Huebner, G. Keramidas, et al.), 281–293. Springer.

21 Ghazali, M.A., Othman, K.A., Isa, M.A.H.M., and Baharuddin, M.A. (2017). Wireless sensor development for Malaysian forest monitoring and tracking system. *21st International Computer Science and Engineering Conference (ICSEC)*, Bangkok, Thailand (15–18 November 2017).

22 Sankey, T., JonathonDonager, J.M.V., and Sankey, J.B. (2017). UAV lidar and hyperspectral fusion for forest monitoring in the southwestern USA. *Remote Sensing of Environment* 195: 30–43.

23 Almeida, D., Broadbent, E., Zambrano, A. et al. (2019). Monitoring the structure of forest restoration plantations with a drone-lidar system. *International Journal of Applied Earth Observation and Geoinformation* 79: 192–198.

24 Kalhara, P.G., Jayasinghearachchd, V.D., Dias, A.H.A.T. et al. (2017). Treespirit: Illegal logging detection and alerting system using audio identification over an IoT network. *11th International Conference on Software, Knowledge, Information Management and Applications (SKIMA)*, Malabe, Sri Lanka (6–8 December 2017).

25 Darlis, D., Sirait, D.S.P., and Maulana, D.B. (2018). Sustainable smart forest monitoring system for burning forest and deforestation detection. *The 3rd Annual Applied Science and Engineering Conference* (AASEC), Bandung, Indonesia (8 April 2018).

26 Liu, Y., Liu, Y., Xu, H., and Teo, K.L. (2018). Forest fire monitoring, detection and decision making systems by wireless sensor network. *Chinese Control And Decision Conference (CCDC)*, Shenyang, China (9–11 June 2018).

27 Connectivity Standards Alliance, "ZigBee: The Full-Stack Solution for All Smart Devices". https://csa-iot.org/all-solutions/zigbee/ (accessed 21 April 2024).

28 Texas Instruments (2010). "ANT and Bluetooth". https://www.ti.com/pdfs/wtbu/SWPT042.pdf (accessed 20 October 2024).

29 Colbach, G. (2019). *The WiFi Networking Book: WLAN Standards: IEEE 802.11 Bgn, 802.11n, 802.11ac and 802.11ax.* Independently Published.

30 Herrero, R. (2022). Thread architecture. In: *Fundamentals of IoT Communication Technologies* (ed. R. Herrero), 213–225. Springer.

31 Glow labs. Comparing wireless protocols for IoT Devices. https://glowlabs.co/wireless-protocols/ (accessed 20 May 2024).

32 EnOcean: Sustainable IoT. https://www.enocean.com/en/ (accessed 15 October 2024)

33 Shelby, Z. and Bormann, C. (2011). *6LoWPAN: The Wireless Embedded Internet.* Wiley.

34 Lora Alliance (2019). "RP002-1.0.0 LoRaWAN® Regional Parameters", November 2019. https://lora-alliance.org/resource_hub/rp002-1-0-0-lorawan-regional-parameters/ (accessed 30 April 2024).

35 Nauman, A., Jamshed, M.A., Ali, R. et al. (2021). Reinforcement learning-enabled Intelligent Device-to-Device (I-D2D) communication in Narrowband Internet of Things (NB-IoT). *Computer Communications* 176: 13–22.

36 Yick, J., Mukherjee, B., and Ghosal, D. (2008). Wireless sensor network survey. *Computer Networks* 52 (12): 2292–2330.

37 Rehman, A., Shahid, H., Sultan, A. et al. (2019). Deforestation analysis of Northern Areas (Pakistan) using image processing and maximum likelihood supervised classification. *2nd International Conference on Communication, Computing and Digital systems (C-CODE)*, Islamabad, Pakistan (6–7 March 2019), pp. 205–210.

38 Dominguez, D., del Villar, L.d.J., Pantoja, O., and González-Rodríguez, M. (2022). Forecasting amazon rain-forest deforestation using a hybrid machine learning model. *Sustainability* 14: 691.

39 Shoui, T., Saidi, S., and Lakti, A. (2023). A machine-learning-based approach to predict deforestation related to oil palm: conceptual framework and experimental evaluation. *Applied Science* 13: 1772.

40 Whiting, K. (2019). World Economic Forum. https://www.weforum.org/agenda/2019/12/technology-artificial-intelligence-ai-drone-trees-deforestation/ (accessed 13 January 2024).

41 Pouyanfar, S. and Sadiq, S. (2018). A survey on deep learning: algorithms, techniques, and applications. *ACM Computing Surveys* 51 (5): 1–36.

4

Wireless Sensors for Multi-modal Health Monitoring

Nadeem Ajum[1], Shagufta Iftikhar[1], Tahera Kalsoom[2], and Masood Ur Rehman[3]

[1] *Department of Software Engineering, Capital University of Science and Technology, Islamabad, Pakistan*
[2] *Manchester Fashion Institute, Manchester Metropolitan University, Manchester, UK*
[3] *James Watt School of Engineering, University of Glasgow, Glasgow, UK*

The healthcare sector has undergone a significant transformation due to the widespread adoption of wearable and implantable sensor devices. These innovative technologies have revolutionized healthcare monitoring by enabling the collection of huge amounts of multi-modal health and activity data. While this big collection of data offers immense potential for advancements in disease diagnosis, treatment, and timely prevention, traditional data analysis methods struggle to effectively utilize this data. Existing approaches have limitations in adaptability to rapidly evolving data, feature extraction, and lack processing efficiency required to handle such complex and high-dimensional datasets [1]. These methods face significant performance degradation as any change occurs in data or environments and hence struggle to effectively extract meaningful insights from such multi modal data [2]. This is where advancements in artificial intelligence (AI), particularly machine learning (ML)-based data analysis methods, offer a powerful solution. ML algorithms show potential to address limitations of traditional data analysis approaches and possess a unique ability to learn and adapt from data resulting in continuous improvements in their performance over time. By leveraging these algorithms, researchers can develop sophisticated models capable of handling the complexities of data collected using body-worn wearable/implantable sensors.

ML and deep learning (DL) models can help medical professionals to achieve improved disease detection and to create personalized treatment plans for their patients. This can be accomplished through the analysis of vital signals collected

Multimodal Intelligent Sensing in Modern Applications, First Edition.
Edited by Masood Ur Rehman, Ahmed Zoha, Muhammad Ali Jamshed, and Naeem Ramzan.

Published 2025 by John Wiley & Sons, Inc.

by sensors attached to the body of patients. This advanced data analysis facilitates the development of more accurate disease detection models. For example, ML models can be trained on historical patient data and sensor readings from wearable heart rate monitors to identify subtle patterns that might indicate the early stages of heart disease. This early detection allows for prompt intervention and potentially life-saving treatment options. Remote patient monitoring also becomes a reality by integrating wearable sensors with ML models, enabling doctors to track patients with chronic conditions remotely and adjust treatment plans based on real-time data. Medication adherence can also be improved by using ML models to analyze data from pill dispensers or activity levels, identifying patients who might need additional support in sticking to their medication schedules. Hence, there is a variety of application areas where ML could integrate with implantable and wearable wireless sensors for multi-modal health monitoring taking digital healthcare to a new level of advancement [3].

Integration of ML with body-worn sensors has recently garnered significant interest, leading to a new wave of studies. This chapter aims to discuss the power of ML-based approaches, specifically focusing on DL techniques within the context of implantable and wearable devices for healthcare. A robust framework is provided having sensor data from wearable devices, such as smartwatches and activity trackers, to identify subtle deviations from an individual's baseline physiological patterns. This is critical to detect early signs of various diseases by analyzing the multi-model data collected from sensor devices, allowing for prompt medical intervention and potentially improved patient outcomes in order to improve the complete treatment process.

4.1 Wearable Sensors

Our relationship with healthcare is constantly evolving, and the way we monitor and manage it is no exception. In recent years, wearable sensor technology has emerged as a powerful tool in the healthcare sector. These small, comfortable devices, worn on the body like watches, bracelets, or even clothing, are quietly transforming how we track and understand our health. Unlike traditional medical equipment that often requires clinical settings, wearable sensors offer a convenient and continuous way to monitor various health indicators. This allows individuals to gain valuable insights into their well-being, going beyond simple checkups at the doctor's office. Wearable sensors offer a continuous stream of real-time data. This data encompasses a broad spectrum of physiological information, including:

- **Vital Signs:** Heart rate, blood pressure, respiratory rate.
- **Activity Levels:** Steps taken, distance covered, calories burned.
- **Sleep Patterns:** Sleep duration, sleep quality, sleep stages.

- **Biochemical Markers:** Electrolyte, metabolite, drug, dopamine, etc. (through integrated sensors in smart patches or ingestible capsules).

For medical analysis, reliance only on a single sensor can lead to a limited understanding of the physiological state and can potentially provide insufficient data required for an accurate diagnosis. In contrast, systems that use multiple sensors offer many advantages such as redundancy of data acquired from those sensors which results in improved reliability and complementary data through the fusion of sensors. By incorporating multiple sensors and leveraging sensor fusion techniques, medical analysis can achieve a new level of sophistication and accuracy, leading to better diagnosis and patient care. For example, heart rate can be measured more accurately by combining photoplethysmography (PPG) and electrocardiogram (ECG) sensors. Different sensor combinations also enable classifying tasks like sleep stages (ACC, ECG, temperature [TEMP]) [4].

Wearable devices have become powerful tools for health and wellness monitoring thanks to their integration of multi-sensor systems. These systems combine various sensors onboard the device to unlock a wide range of functionalities. For instance, smartwatches often utilize ECG sensors and motion sensors to simultaneously track an individual's heart rate and steps [5]. Similarly, smart glasses leverage microphones (Mics) for voice interaction and cameras for capturing pictures and videos [6]. Beyond these examples, wearable devices can also track various physiological parameters such as nutrition intake, sleep patterns, and activity levels. Sensors that are commonly used in healthcare sector are illustrated in Figure 4.1. The potential location of different wearable sensors on the human body is shown in Figure 4.2.

4.1.1 Electrocardiography (ECG) Sensors

ECG sensors are like tiny detectives, using electrodes to capture the electrical whispers of your heart. These electrical signals are then translated into a visual representation, known as an ECG, revealing the rhythm and activity of this vital organ [7]. This ability to detect subtle electrical changes within the heart makes ECG sensors a powerful tool in the fight against cardiovascular diseases (CVDs) [3]. ECG sensors, precisely due to their compact and comfortable design, empower continuous monitoring, allowing healthcare professionals to identify potential issues like heart failure, arrhythmias, and atrial fibrillation early on as it is the silent nature of many heart problems, often lacking symptoms until a critical event occurs.

4.1.2 Electroencephalography (EEG) Sensors

The human brain is a constantly buzzing hive of electrical activity, and electroencephalogram (EEG) sensors offer a window into this fascinating world. These sensors can detect subtle changes in brain waves, making them highly sensitive

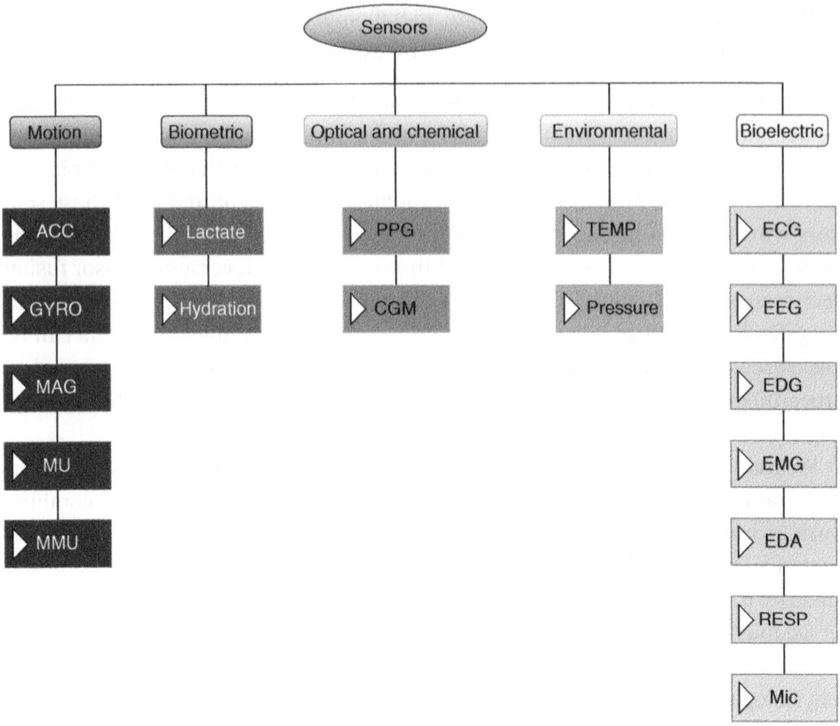

Figure 4.1 Wireless sensors for healthcare.

to fluctuations in emotional state and offering real-time insights. Traditionally, EEG sensors have relied on head-mounted interfaces, which, while effective, can be cumbersome and impractical for everyday wear. However, research is exploring alternative approaches, such as portable ear-mounted sensors, to overcome these limitations and enable long-term monitoring. Analysis of EEG can play a crucial role in the early diagnosis of neurological conditions like seizures and strokes by identifying characteristic patterns in brain waves. Studies suggest that interpreting EEG signals can provide clues about emotional states and potentially even aid in disease rehabilitation through attention control techniques [8]. This technology has the potential to revolutionize our understanding of the brain and nervous system, leading to breakthroughs in diagnosis, treatment, and overall well-being.

4.1.3 Electrooculography (EOG) Sensors

Beyond monitoring the heart and brain, wearable sensors can also shed light on our eye movements and muscle activity. Electrooculography (EOG) sensors, placed around the eyes, detect electrical changes generated by eyeball movements. This

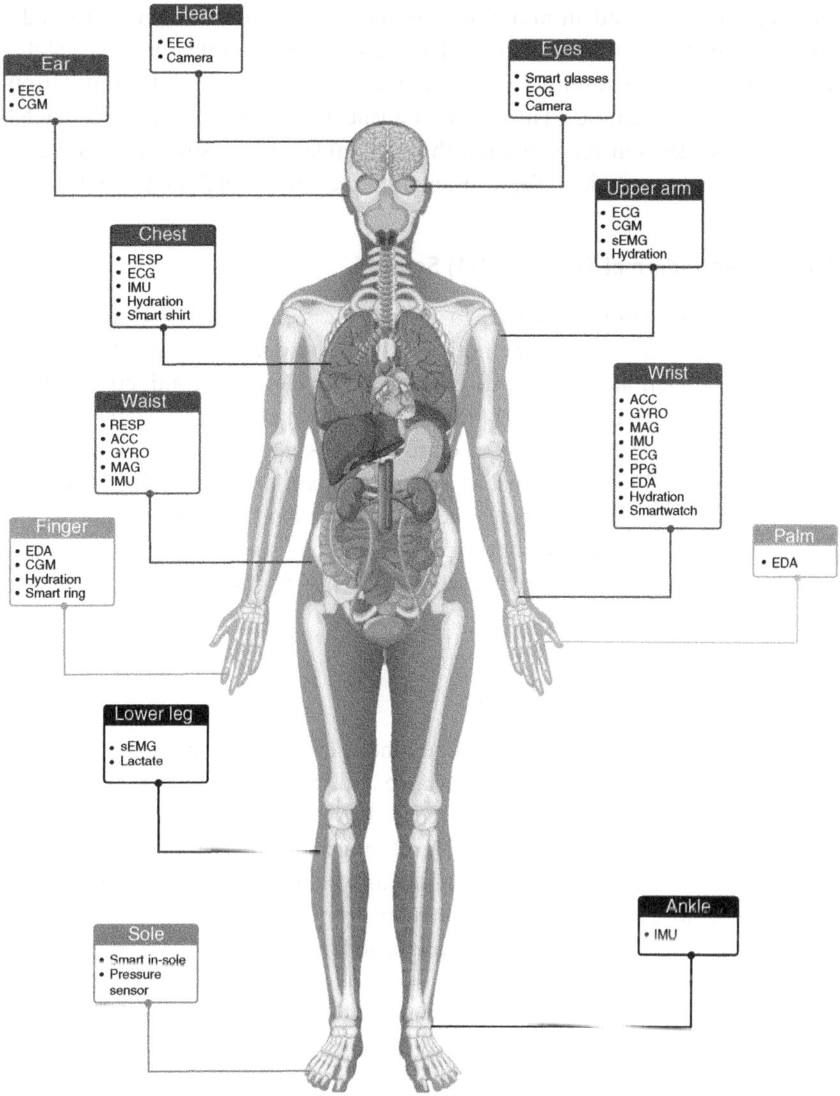

Figure 4.2 Wireless sensors positioned on the human body for remote healthcare. *Source:* GraphicsRF/Adobe Stock Photos.

allows for the determination of gaze direction, making EOG sensors valuable in wearable eye-tracking devices [9]. These sensors measure electrical signals produced by muscle contractions and relaxations. Electromyography (EMG) comes in two forms: surface EMG (sEMG: noninvasive, placed on the skin overlying the target muscle) and intramuscular EMG (iEMG: more invasive, electrodes inserted directly into the muscle). sEMG offers greater versatility due to its ability to measure activity in multiple directions and across various muscle groups, from upper to lower extremities. This makes it a valuable tool for applications like fall detection, exoskeleton rehabilitation through ankle torque estimation, and even medical disease prevention, diagnosis, monitoring, and rehabilitation efforts [10].

4.1.4 Electrodermal Activity (EDA) Sensors

EDA sensors, sometimes referred to as galvanic skin response (GSR) sensors, offer a unique window into our emotional state. These sensors, typically worn on the palm or finger, measure subtle changes in the electrical conductivity of the skin. This conductivity is directly linked to sweat gland activity, which in turn fluctuates with our emotions. The close relationship between EDA and emotional arousal makes these sensors a powerful tool for emotion identification, classification, anxiety monitoring, and stress detection [11]. Imagine a wearable device that can provide real-time feedback on your stress levels, allowing you to take proactive steps toward relaxation. This is the potential that EDA sensors hold.

4.1.5 Respiratory (RESP) Sensors

Traditional respiratory monitoring equipment often relies on bulky masks or mouthpieces, which can be uncomfortable and impractical for long-term wear. Fortunately, advancements in sensor technology have led to the development of noninvasive respiratory sensors (RESP), offering a more user-friendly alternative. One such example is the respiratory inductance plethysmography (RIP) belt. This belt, worn around the chest and abdomen, measures changes in circumference as you breathe. These changes are then translated into an estimate of your respiratory volume. Other research explores the use of low-cost breathing belts with strain gauges. These belts convert the physical strain caused by your breathing into electrical signals, allowing for respiration monitoring. Additionally, studies have investigated the potential of Fiber Bragg Grating (FBG) sensors for monitoring respiration. These sensors use changes in light caused by chest expansion and contraction to track your breathing patterns [12].

4.1.6 Motion Sensors

Inertial sensors, also known as motion sensors, possess the capability to convert inertial forces into electrical signals, facilitating the measurement of object motion parameters like acceleration, inclination, and vibration. The accelerometer (ACC) and gyroscope (GYRO) are pivotal in this regard, dedicated, respectively, to measuring inertial acceleration and angular rate. In everyday activities, the need often arises for multidimensional measurements. An established solution entails the utilization of an inertial measurement unit (IMU) that amalgamates multiple ACCs and GYROs. Additionally, certain IMUs incorporate magnetometers (MAGs) to gauge the ambient magnetic field, transforming them into magneto-inertial measurement units (MIMUs).

ACCs excel in gauging activity intensity and type, such as fall detection and gait monitoring. Conversely, GYROs are proficient in determining orientation and angular rate, facilitating tasks like ankle sprain detection and fall monitoring. Compared to standalone inertia sensors, IMUs or MIMUs enjoy broader acceptance due to their integration of ACC, GYRO, and MAG characteristics, enabling the capture of multidimensional data. Their applications span various domains of human activities, encompassing disease classification, gait analysis, rehabilitation monitoring, and training optimization, among others.

4.1.7 Temperature (TEMP) Sensors

Temperature plays a crucial role in various biological processes within the human body. To monitor these fluctuations, we rely on temperature sensors, often abbreviated as TEMP sensors. These sensors convert body temperature into an electrical signal, allowing for continuous monitoring and analysis. TEMP sensors often work hand-in-hand with bioelectric sensors. By combining the data from both, we gain a more comprehensive understanding of how the body is functioning. For instance, TEMP sensors can detect changes in body temperature associated with specific activities or physiological states. Beyond their role in medical applications, TEMP sensors have become an essential part of our daily lives. Embedded in wearable devices like smartwatches and smart rings, they allow us to continuously monitor our body temperature.

4.1.8 Pressure Sensors

Pressure sensors play a vital role in analyzing human movement. These electronic devices convert the pressure applied to them into electrical signals, providing valuable insights into how we move. Embedded in wearables like smart insoles and gloves, pressure sensors become powerful tools for understanding and optimizing human movement across various applications. For instance, pressure sensors in smart insoles can analyze the pressure distribution on your foot during gait activities. This data helps assess walking patterns, identify potential gait abnormalities, and even improve athletic performance. Similarly, pressure sensors integrated into wearable gloves can measure grip strength and muscle activity, offering valuable information for rehabilitation programs or training specific muscle groups.

4.1.9 Hydration Sensors

Maintaining proper hydration is crucial for overall health, while dehydration can lead to a variety of issues. This is where hydration monitoring sensors come into play. These innovative devices allow us to track our hydration levels conveniently and accurately, helping us stay on top of our fluid intake [13]. This can be particularly beneficial for athletes, those engaged in physical activity, or individuals with health conditions that require close monitoring of hydration levels.

There are three main types of hydration sensors:

- **Impedance-based** sensors estimate hydration by measuring the electrical resistance of your body tissues. Since well-hydrated tissues conduct electricity more readily, changes in resistance can indicate fluctuations in hydration levels.
- **Optical-based** sensors use light to assess hydration. They analyze how light interacts with your skin, with changes in light absorption potentially revealing variations in hydration.
- **Sweat-based** sensors directly measure the concentration of electrolytes in your sweat. As you sweat, electrolytes are lost along with fluids. By monitoring these levels, sweat-based sensors can provide insights into your hydration status.

4.1.10 Lactate Sensors

Lactate, a molecule produced during anaerobic metabolism, is a valuable indicator of physical exertion and metabolic stress [14]. Traditionally, monitoring lactate levels relied on invasive blood draws, which can be inconvenient and disruptive to training regimens. However, advancements in sensor technology have introduced noninvasive lactate sensors, offering a more comfortable and practical alternative for athletes and fitness enthusiasts.

One approach utilizes enzymatic reactions. These sensors incorporate enzymes that specifically react with lactate molecules in sweat, producing a measurable signal that can be translated into lactate concentration. In another method, biosensors combine biological elements like enzymes with electronic components, creating a highly specific lactate detection system. Finally, some noninvasive lactate sensors employ spectroscopy [14]. This technique analyzes how light interacts with sweat, with specific wavelengths potentially revealing the concentration of lactate present. By continuously monitoring lactate levels in sweat, these sensors can provide real-time insights into workout intensity and recovery status. This valuable information allows individuals to optimize their training programs, prevent overexertion, and ultimately achieve their fitness goals.

4.1.11 Photoplethysmography (PPG) Sensors

PPG sensors are a type of optical sensor that utilizes light to measure changes in blood volume within tiny blood vessels under the skin. Typically placed on the wrist or soles of the feet in wearables like wristbands or smart insoles [15], PPG sensors track how light interacts with these blood vessels. Variations in the light signal provide valuable insights into various physiological parameters, including heart rate, blood pressure, and even blood oxygen levels. While offering a more convenient alternative to ECG sensors, PPG sensors have a drawback: they are susceptible to noise caused by movement, which can lead to inaccurate measurements. Fortunately, these errors can be mitigated using appropriate techniques [16]. Despite these limitations, PPG sensors hold immense potential due to their ability to generate data that ML models can analyze and extract hidden, valuable information to aid in disease detection and monitoring. For example, PPG data has shown promise in automatically detecting seizures [17] and monitoring conditions like blood pressure, sleep apnea, and hypopnea [18].

4.1.12 Continuous Glucose Monitoring (CGM) Sensors

Traditionally, monitoring blood glucose levels (BGL) involves finger pricking, an invasive procedure that carries the risk of infection and can be inconvenient. Fortunately, research is exploring the potential of non-invasive BGL monitoring, offering a more comfortable and user-friendly alternative. While these sensors are not yet clinically approved, they hold significant promise for the future [19]. Current CGM sensors primarily rely on three main technologies: optics, microwaves, and electrochemistry [20]. Some examples include near-infrared sensors like the C8 MediSensors, radio wave technology used in GlucoWISE, and the Freestyle LibrePro which analyzes electrochemical signals without the need for finger pricking.

4.2 Flexible Sensors

Flexible sensors are not a distinct category by themselves; it is more about their physical form. Unlike rigid sensors, these are designed to be stretchable and adaptable, conforming to the shape of whatever they are attached to. This unique characteristic makes them ideal for integration into wearables like smart clothing. Imagine smart gloves with embedded flexible ECG sensors that can continuously monitor your heart rhythm for irregularities [21]. Or smart socks containing flexible pressure sensors woven right into the fabric. These sensors can track foot pressure distribution in real time, allowing for early detection of high-pressure areas that could lead to ulcers, particularly for diabetics [22].

4.3 Multi-modal Healthcare Sensing Devices

The Internet of Things (IoT) has made an impactful contribution to the medical field by facilitating remote healthcare systems. These multi-modal systems effectively integrate different sensing devices and use powerful ML models for extracting valuable insights from the sensed physiological data. Whether it be wearable devices monitoring, for example, cardiovascular parameters, environmental sensors tracking air quality, or motion sensors detecting gait abnormalities, each sensor contributes unique streams of data that offer valuable insights into various facets of patient health and well-being. By integrating these disparate data streams and subjecting them to ML or DL analyses, researchers can unlock deeper layers of understanding and derive actionable insights. ML algorithms can discern patterns, trends, and anomalies within complex datasets, facilitating disease diagnosis, treatment optimization, and personalized healthcare interventions. DL models, with their ability to process vast amounts of data and extract high-level features, hold promise for advancing precision medicine and predictive analytics in the medical domain. Thus, by leveraging the wealth of information captured by these multi-modal sensing devices and harnessing the power of ML and DL techniques, we stand poised to revolutionize healthcare delivery and enhance patient outcomes on a global scale. Figure 4.3 depicts a typical multi-modal remote healthcare system.

4.3.1 Wearable Sensing Devices for Healthcare

As the global population ages and the rapid development of digital technologies, the demand for digital medical care is increasing. Wearables are generating large amounts of information-rich data every day. ML has become a promising data analysis assistant for doctors, helping them efficiently and rapidly diagnose diseases and helping patients recover from diseases.

4.3.1.1 Wearable Devices in Detection

In the medical field, the early diagnosis of diseases holds paramount significance for both healthcare providers and patients, contributing to the preservation of medical resources and ultimately improving outcomes. With the proliferation of multiple sensors capable of capturing various physiological signals from the human body, there exists the substantial potential for disease detection across a spectrum of conditions including CVDs, blood circulation disorders, and mental health disorders. CVD detection and monitoring have garnered significant attention given the staggering annual mortality rate of 17.9 million reported by the World Health Organization [23]. Sensors such as ECG and PPG offer avenues for individuals to monitor their cardiovascular health and explore heart activity. ML techniques applied to data from these sensors have been instrumental in diagnosing and

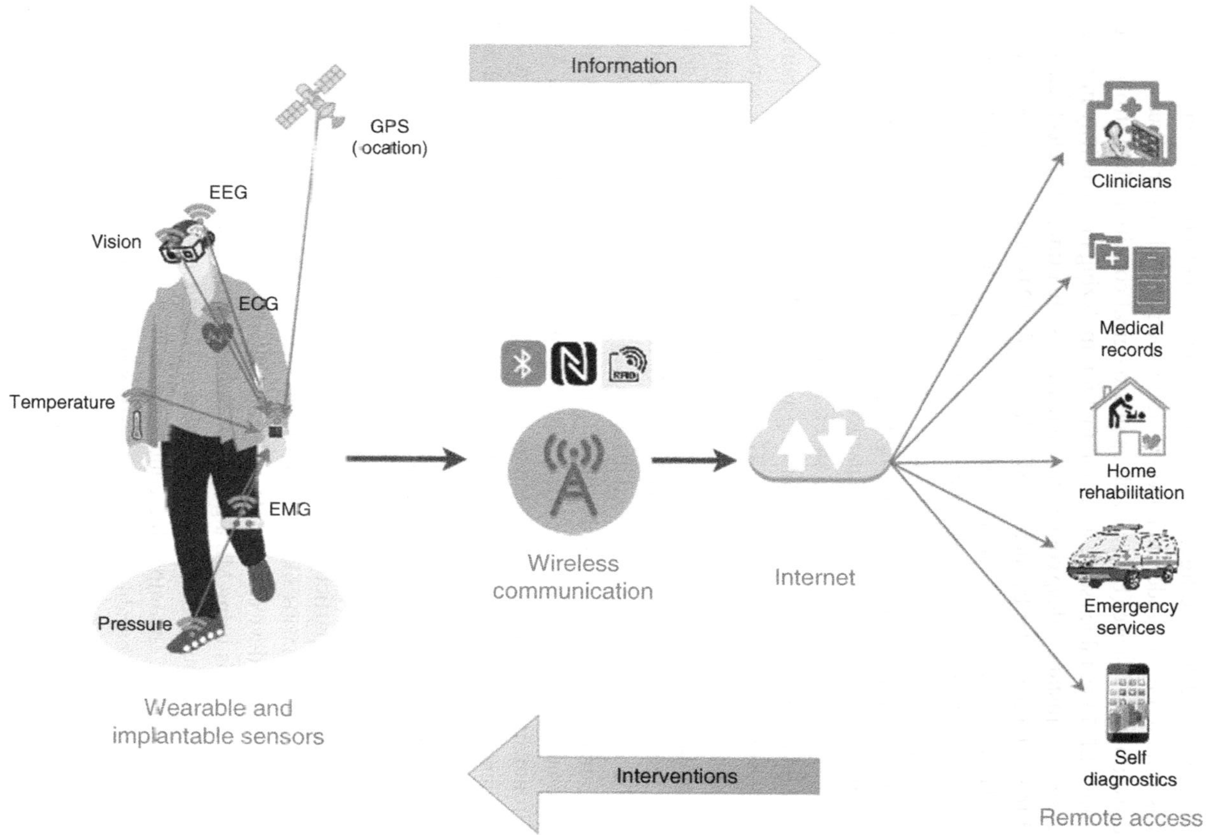

Figure 4.3 A typical multi-modal remote healthcare system.

monitoring CVDs, encompassing conditions like atrial fibrillation, arrhythmia, heart failure, and aortic stiffness.

Wearables and ML are frequently employed in the detection of neurological diseases, including neurodegenerative disorders such as Alzheimer's disease (AD) and Parkinson's disease (PD), as well as neurological conditions like stroke and seizures. EEG wearables serve as a vital tool for analyzing brain activity in these diseases. For instance, deep convolutional networks are utilized to detect seizures and their characteristic frequencies simultaneously [24]. Motion sensors play a crucial role in detecting limb and movement disorders, such as postural instability in patients with PD. Also, single waist-worn triaxial ACCs are used to capture motion signals, aiding in fall detection and tracking of motion [25].

Emotion detection presents a challenging yet significant aspect of understanding human behavior and well-being, particularly in the context of mental health. Mental states, including happiness, anger, sadness, stress, and depression, are closely linked to cognitive decision-making and overall health. ML models trained on data from sensors such as EEG, ECG, or electrodermal activity (EDA) sensors have been employed to identify and classify human mental states. For example, multimodal sensor data is utilized to recognize daily emotions [26], while some utilize multiple sensor data to detect stress [27]. In the realm of everyday healthcare, the concept of ambient assisted living (AAL) holds promise for saving medical resources and enhancing quality of life, particularly for the elderly and disabled individuals requiring daily care. ML applied to sensor data aids in building AAL environments, facilitating tasks such as cough detection and fall detection. An intelligent method to detect coughs based on ACCs has been developed and reported [28], while fall detection using motion sensors has emerged as a promising area of research aimed at reducing fall-related injuries among the elderly [29].

4.3.1.2 Wearable Devices in Monitoring

Health monitoring, particularly in remote settings, is of paramount importance across various scenarios like daily activities including walking, sleeping, and driving. Wearable devices facilitate real-time data transmission to cloud or edge servers for analysis using ML algorithms. Subsequently, users receive feedback for health assessment. Chronic diseases demand long-term monitoring for effective management, encompassing conditions such as CVDs, diabetes, and pulmonary diseases. Wearable devices like textiles and ear-mounted sensors provide real-time ECG data for monitoring CVDs. ML algorithms, notably convolutional neural networks (CNNs), exhibit prowess in detecting abnormalities within ECG signals [30].

Traditional methods for measuring blood pressure and blood glucose levels lack continuous monitoring capabilities and are often invasive. Wearables offer a promising avenue for continuous monitoring, with ML algorithms aiding in noise reduction and data extraction. For example, in [31], they demonstrated continuous blood glucose level monitoring by integrating CGM wearables with ML techniques. Precision enhancement in monitoring systems is a focal point of several studies, with various optimization techniques proposed for blood pressure monitoring using wearable data [32]. Moreover, wearables are leveraged for monitoring diabetes, postprandial hyperglycemia, and classifying blood pressure in ill patients [33].

Routine monitoring of fundamental physiological parameters like heart rate, respiratory rate, sleep quality, and gait analysis plays a crucial role in detecting underlying health conditions. ML algorithms aid in enhancing heart rate precision by addressing motion artifacts in PPG measurements [16]. Sleep monitoring, enabled by wearable sensors, assists in diagnosing sleep disorders and provides insights into cardiovascular and neurological diseases. ML models classify different sleep stages using ACC and ECG data [34]. Similarly, ML analysis of PPG data aids in detecting obstructive sleep apnea (OSA) [35]. Daily respiratory monitoring is vital for assessing overall health status and detecting conditions such as asthma and pulmonary disease. ML algorithms applied to wearable data facilitate monitoring of respiratory rate, pattern, and airflow, contributing to disease management and early detection [12]. Gait analysis, crucial for assessing balance and coordination, benefits from wearable devices such as smart insoles, enabling real-time monitoring and classification using ML algorithms [36]

4.3.1.3 Wearable Devices in Rehabilitation

Some disorders, such as PD, stroke, and arthritis, can severely impede the movement of both upper and lower limbs. Conversely, rehabilitation processes are often protracted and necessitate attentive care from medical professionals. In response to these challenges, the integration of ML and wearable technology into rehabilitation processes has emerged as a promising avenue for enhancing patient recovery.

Limb rehabilitation stands as one of the most common applications in the realm of recovery, particularly for patients requiring meticulous attention to restore limb function post-surgery. In [37], they developed a platform utilizing smartphones and motion sensors to monitor the rehabilitation progress of patients following knee arthroplasty. Lin et al. proposed an ML based approach to assist in the rehabilitation of stroke patients' activities of daily living [38]. Lee et al. employed ACCs to track upper-limb motor recovery in stroke or traumatic brain injury patients, leveraging ML analysis for intervention strategies [39].

Beyond limb recovery, wearable technology facilitates the monitoring of general postoperative recovery across various domains. For instance, Ghomrawi et al. utilized smart wristbands to monitor the conditions of pancreatic surgical patients, employing ML to predict postoperative complications based on wristband data [40]. Hunter et al. similarly utilized smartwatches to assess patients' recovery outcomes in hospital settings [41].

Intelligent wearable systems have also propelled the development of exoskeletons for limb rehabilitation. Some studies utilize wearables to translate patients' recovery intentions into signals, which are then analyzed using ML algorithms. Based on this analysis, exoskeleton robots receive instructions for continuous movements to aid in the patient's recovery process [42]. To generate continuous instructions, studies have explored methods for estimating continuous numerical values. Zhang et al. utilized an EMG-driven ML model to estimate ankle joint torque, facilitating patient recovery [43]. Zhang et al. designed a human–machine interaction system based on surface EMG (sEMG), where ML algorithms analyze signals to identify human motion intentions [44].

4.3.1.4 Wearable Devices in Personalized Medicine

Patients often encounter challenges in adhering to their prescribed medication regimen, influenced by various factors including comprehension barriers, skepticism toward healthcare providers, or simply forgetting to take medication as directed. The overarching goal of medication adherence is to optimize therapeutic outcomes by tailoring prescriptions to meet each patient's individual needs and circumstances. Consequently, several studies have delved into the realm of intelligent wearables aimed at enhancing medication adherence.

These studies predominantly fall into two categories: motion-based and policy-based approaches. Motion-based methodologies typically utilize motion sensors to discern whether patients have ingested their medication. For instance, Fozoonmayeh et al. [45] employed ML techniques to analyze data from smartwatches, enabling the detection of patients' medication adherence levels. Other research endeavors have also explored motion sensors as evidence of medication intake. Conversely, policy-based investigations leverage wearable-generated data to devise personalized medication management strategies for patients. This entails proposing tailored drug dosages and administration schedules through the application of deep reinforcement learning (RL) algorithms. Beyond medication adherence, ML can also facilitate comprehensive interventions based on wearable data, addressing various aspects of a patient's health status through end-to-end analysis. [46].

4.3.1.5 Wearable Devices in Skin Patches

The field of wearable health technology is witnessing a surge in popularity for attachable skin patches. These patches incorporate soft, flexible, and stretchable electronic sensors that adhere directly to the skin. This novel approach provides a valuable platform for monitoring physiological signals, offering advantages over traditional wearable devices. Skin patches have several key benefits. First, they can be discreetly hidden under clothing, improving aesthetics and user comfort. Second, their direct contact with the skin allows for more accurate data collection, minimizing disruptions caused by movement. As a result, skin patches are ideal for continuous health monitoring applications.

Researchers have explored various applications for skin patches, including monitoring temperature, strain, sweat, and cardiovascular activity. In the critical area of cardiovascular monitoring, significant advancements have been made. For instance, researchers have developed a thin, flexible patch-like sensor for continuous blood pressure monitoring. This innovative patch utilizes specially designed ferro-electric film, along with flexible electrodes and circuitry, to capture ECG and ballistocardiogram (BCG) readings simultaneously [47]. Another promising development is a cuffless blood pressure monitoring patch. This patch incorporates a flexible piezo-resistive sensor (FPS) and epidermal ECG sensors. By simultaneously monitoring ECG and pulse signals, the system can estimate blood pressure in real time using the pulse transit time (PTT) approach [48]. This ultra-low-power (3 nW) patch demonstrates exceptional sensitivity, making it suitable for tracking subtle physiological changes, including pre- and postexercise responses. The rise of wearable skin patches signifies a significant leap forward in wearable health technology. These innovative patches offer a comfortable, discreet, and highly accurate approach to continuous health monitoring, paving the way for improved healthcare management and personalized health insights.

4.3.1.6 Wearable Devices for Body Fluid Monitoring

Sweat, a readily accessible biofluid, has emerged as a valuable source of health information. It contains a rich tapestry of biomarkers, including electrolytes, small molecules, and proteins, offering a window into our physiological state. Recent advancements in wearable technology have led to the development of innovative sweat analysis sensors. These sensors often incorporate sophisticated features like inductive coils, planar capacitors, and absorbent substrates. The absorbent substrates utilize capillary forces to draw in sweat, enabling the analysis of various sweat components like electrolytes (OH^-, H^+, Cu^{2+}, Fe^{2+}) through colorimetric detection methods. Beyond colorimetric detection, researchers are exploring other sensing mechanisms. For instance, a potentiometric sodium ion sensor utilizes a layered design, with a polyvinyl chloride membrane embedded within a conductive polymer and a pH sensing layer based on high iridium oxide. Additionally, amperometric-based lactate sensors are being developed that employ selective doping enzymes within a copolymer membrane.

Another exciting development is the PDMS dermal patch. This patch, made from polydimethylsiloxane, utilizes microfluidic technology for noninvasive sampling of biomolecules present in interstitial fluids, the fluid between our cells. This technology holds immense promise for monitoring glucose levels in diabetic patients, eliminating the need for invasive blood draws. The ability to analyze sweat through wearable sensors opens a new frontier in health monitoring. These sensors offer a comfortable, noninvasive, and continuous approach to health data collection, providing valuable insights for personalized health management and disease monitoring.

4.3.1.7 Wearable Devices in Monitoring Body Temperature

Monitoring variations in skin temperature plays a crucial role in the early diagnosis and treatment of various conditions [49]. Researchers have addressed this need by developing innovative wearable patches for continuous skin temperature monitoring. These human stress monitoring patches are designed for comfort and wearability. They are flexible, have a small footprint (roughly the size of a postage stamp), and minimize skin contact area. This is achieved through a multilayered design and advanced micro-fabrication techniques, significantly reducing the contact area compared to traditional single-layer sensors.

The patches incorporate three key sensors: skin temperature, skin conductance, and pulse wave. The pulse wave sensor, a novel innovation, utilizes a flexible piezoelectric membrane and a perforated polyimide layer to achieve exceptional flexibility and chemical resistance. This sensor demonstrates high sensitivity within the human physiological temperature range and a rapid pulse response time for accurate skin temperature measurements [50]. Beyond temperature monitoring, researchers are developing transparent and stretchable (TS) sensors for wearable applications. These sensors can detect even subtle variations in skin temperature and deformation during physical activity. The fabrication process for TS sensors has been streamlined, making them easier to integrate into wearable patches. These sensors exhibit high sensitivity (around 1.34%/°C) and remarkable durability, maintaining their responsiveness even after repeated stretching cycles [51]. These comfortable, unobtrusive patches offer continuous monitoring of skin temperature and stress markers, providing valuable insights for early disease detection, stress management, and personalized healthcare.

4.3.1.8 Wearable Devices as Contact Lens

The world of contact lenses is undergoing a revolution with the development of smart contact lenses. These innovative lenses go beyond correcting vision and offer a noninvasive way to monitor various physiological details of the eye and tears. By employing both optical and electrical techniques, researchers are creating smart contact lenses capable of tracking a range of health markers: *Tear chemistry*: glucose and lactate levels in tears can provide valuable insights into blood sugar control and overall metabolic health; *Tear electrical conductivity*: this can offer information about tear production and potential dry eye issues; *Transcutaneous gases*: these are gases that pass through the mucous membrane of the eye. Monitoring these gases can offer additional insights into overall health.

Here are some examples of the exciting advancements in smart contact lens technology: *Photonic crystals*: Researchers have developed contact lenses embedded with photonic crystals made of hydrogel and colloidal particles. These crystals can detect glucose levels within the physiological range found in tears [52]. *Fluorescent contact lenses*: Another design utilizes fluorescent contact lenses that change fluorescence intensity based on glucose concentrations [53]. *Electronic enzyme-based sensors*: For lactate monitoring, researchers have developed miniature sensors integrated within the contact lens itself. These sensors can accurately detect lactate levels in tear fluids [54].

4.3.1.9 Wearable Devices in Daily Use Objects

The realm of wearable health tech is steadily expanding beyond the typical wrist-worn devices. Researchers are exploring innovative ways to incorporate health monitoring capabilities into everyday objects, creating a more holistic approach to wearable health. Here are a few fascinating examples.

Mouthguards equipped with integrated wireless electronics are emerging as a promising tool for monitoring health. One study demonstrated the ability of a mouthguard to analyze salivary uric acid, a potential indicator of gout and other conditions [55]. Another study explored the use of mouthguards for noninvasive glucose monitoring through saliva [56].

Tiny, tri-layer radio frequency sensors mounted directly on teeth hold immense potential for monitoring various health parameters. When coated with specific analyte-sensitive layers, these biocompatible sensors can detect a range of substances, including alcohol, pH, salinity, sugar, and even temperature [57]. This opens doors for real-time tracking of dietary habits and potential health issues.

Wearable technology is making inroads into the realm of musculoskeletal health as well. Specific wearables can monitor muscle activity, strain, and posture, aiding in injury prevention and rehabilitation. Additionally, wearable thermotherapy devices can deliver localized heat therapy for pain management and muscle relaxation [58]. These advancements showcase the expanding potential of wearable health tech. By incorporating health monitoring functionalities into everyday objects, researchers are paving the way for a more comprehensive and convenient approach to personal health management.

4.3.2 Implantable Sensing Devices for Healthcare

Since the introduction of cardiac pacemakers in the 1960s, the use of implantable electronic devices has grown steadily [59]. These devices, which improve or restore function in various organs and tissues, play a vital role in modern healthcare. Made from biocompatible materials, implantable devices typically house batteries and programmable circuitry. Beyond electrical stimulation, some advanced implants can now monitor physiological data and even deliver medication directly to the target site. We can expect to see closed-loop systems that can automatically adjust stimulation or medication delivery, and potentially even integrate with AI for even more personalized care. However, implantable devices also come with considerations. Surgical procedures are often required for implantation, battery life may necessitate replacements, and the cost of these devices can be a factor. Despite these challenges, the potential benefits of implantable medical devices for improving patient lives are undeniable.

4.3.2.1 Implantable Cardioverter Defibrillators

Many implantable devices rely on a combination of batteries, biocompatible materials, and programmable circuitry to function. Pacemakers, for example, play a crucial role in managing arrhythmias, or irregular heartbeats. These devices deliver low-energy electrical pulses to the heart muscle, nudging it back into a normal rhythm when abnormal heartbeats are detected. Implantable cardioverter defibrillators (ICDs) represent a more advanced form of pacemakers. They work in a similar way, but with a critical difference. While pacemakers use low-energy pulses for correction, ICDs can deliver high-energy electric shocks if a traditional pacemaker fails to restore the heart's normal rhythm. This is particularly important because sudden cardiac death (SCD) caused by ventricular arrhythmias is a significant risk factor for heart disease. Studies have shown that ICDs significantly reduce mortality rates in patients with a high risk of SCD [60]. In essence, both pacemakers and ICDs act as tiny electronic lifesavers, constantly monitoring and regulating the heartbeat to ensure its proper function.

4.3.2.2 Bioinks and 3D Print Implants

Researchers at Harvard and MIT have developed a revolutionary concept called bioinks [61]. These bioinks are essentially special inks that utilize the body's chemistry for data transmission. Imagine your skin acting as an interactive display! Bioinks achieve this feat through a clever chemical reaction, eliminating the need for external power. Traditional tattoo inks are being replaced with these color-changing biosensors. Why? Because the fluid between your cells, known as interstitial fluid, reflects the chemical composition of your blood. By analyzing this fluid, bioinks can provide real-time data on various health parameters.

Researchers have developed bioinks that change color based on glucose, pH, and sodium levels. For instance, a glucose-monitoring bioink can change from green to brown as your blood sugar rises. Similarly, another bioink can turn bright green if your sodium levels become too high, potentially indicating dehydration.

Beyond color-changing biosensors, 3D bio-printing inks hold immense promise. These inks are designed to print living cells at high resolutions. Imagine printing programmed bacterial cells onto a biomaterial with precise 3D structures! This could lead to exciting applications in tissue engineering and regenerative medicine. The future of bioinks is full of possibilities. From wearable health monitors to "living tattoos" that respond to chemical signals, these innovations blur the line between human and technology, paving the way for a new era of personalized healthcare and biocompatible interfaces.

4.3.2.3 Deep Brain Stimulation

Deep brain stimulation (DBS) is a surgical procedure that uses implanted electrodes to deliver electrical stimulation to specific areas of the brain. This technique offers a powerful tool for modulating brain circuits and treating various neurological disorders. Currently, DBS is the gold standard treatment for PD, dystonia, and essential tremor. Research is also exploring its potential for treating other conditions with abnormal brain circuit activity, such as Alzheimer's disease and major depressive disorders.

Modern DBS devices share similarities with pacemakers but are specifically designed for the brain. They consist of three main components: an intracranial electrode implanted in the brain, an extension wire connecting the electrode to a pulse generator, and the pulse generator itself, which is typically placed under the collarbone. Over the past two decades, advancements in technology, brain imaging, and our understanding of neurological conditions have led to significant improvements in DBS. These advancements are paving the way for a future where DBS can be even more precisely targeted and tailored to individual patients, offering a more personalized approach to treatment [62].

4.3.2.4 Biosensor Tattoos

The field of biosensor technology is undergoing a fascinating transformation with the emergence of biocompatible biosensor tattoos. These novel tattoos transcend the realm of mere aesthetics, offering a groundbreaking approach to real-time health monitoring. By leveraging advancements in nanotechnology and biocompatible materials, researchers are developing biosensor tattoos capable of detecting and quantifying various physiological parameters. These sensors can be strategically embedded within the dermal layers, utilizing biofluids such as sweat or saliva as the analyte source. One promising avenue of exploration involves the use of graphene-based nano-sensors. These sensors exhibit exceptional sensitivity, enabling the detection of specific microorganisms like bacteria within saliva or breath samples [63]. Furthermore, the integration of enzyme-based biosensors within tattoo designs holds immense potential. Recent studies have demonstrated the feasibility of tattoo biosensors for noninvasive sweat analysis. These sensors can monitor lactate levels, a valuable marker of physical exertion, and even glucose levels, offering significant benefits for diabetic patients [64].

The unique advantages of biosensor tattoos are manifold. Unlike traditional blood-based monitoring methods, these tattoos are noninvasive, eliminating the need for needles or blood draws. Additionally, they offer the distinct benefit of continuous monitoring, providing a more comprehensive picture of an individual's health compared to discrete, single-point-in-time readings. Moreover, some tattoo designs exhibit remarkable resilience, maintaining functionality even with the movement of the skin. While still in the early stages of development, biosensor tattoos represent a significant leap forward in wearable health technology. These innovative tattoos have the potential to revolutionize the way we monitor and manage our health, paving the way for a future of personalized and proactive healthcare. By harnessing the power of biocompatible materials and advanced sensor technology, biosensor tattoos hold the promise of transforming healthcare delivery and empowering individuals to take a more active role in their well-being.

4.4 AI Methods for Multi-modal Healthcare Systems

As wearable technology continues to advance, so too does the role of AI, particularly its branch known as ML. ML goes beyond traditional statistical methods by enabling wearables to learn and adapt from the data they collect. This allows the system to automatically optimize itself without the need for constant human intervention. The true power of ML in wearables lies in its ability to analyze data and uncover hidden patterns or regularities that might otherwise be missed. By employing ML models, researchers and developers can extract valuable insights from the vast amount of data generated by wearables. ML encompasses four main categories: supervised ML, unsupervised ML, semi-supervised ML, and RL.

4.4.1 Supervised Learning

While basic classification models like support vector machine (SVM) and K nearest neighbors (KNN) can be used to analyze electrical signals from wearables (like ECG or CGM data), more advanced techniques are unlocking deeper insights. These techniques consider the temporal nature of the data, meaning they take into account how the signal changes over time. Temporal models such as temporal convolutional networks (TCNs), causal networks, recurrent neural networks (RNNs), and transformers are specifically designed to handle this type of data and excel at classification tasks. Additionally, some approaches transform 1D electrical signals into 2D images (like spectrograms) to leverage the power of CNNs for analysis [65].

By combining the strengths of various models and mitigating their individual weaknesses through ensemble methods, researchers can achieve even more accurate results. This concept applies to regression problems as well, where the goal is to predict a continuous value (like blood sugar levels) rather than a category. Here too, researchers often employ single models or ensemble methods to optimize results. When dealing with massive datasets or highly complex data, DL techniques come into play. DL models, particularly those based on CNNs like ResNet and U-Net, have proven to be highly effective and versatile in these scenarios.

4.4.2 Unsupervised Learning

While supervised learning relies on labeled data, real-world scenarios often involve vast amounts of unlabeled data. This is where unsupervised learning comes into play. Unlike supervised learning models that require predefined categories, unsupervised learning allows the model to discover and classify patterns within this unlabeled data itself. Imagine a wearable sensor collecting a continuous stream of data, but the specific activities or physiological states corresponding to each data point are unknown. Unsupervised learning algorithms can analyze this data and identify hidden patterns, group similar data points together (clustering), and even reduce the complexity of the data while preserving key information (dimensionality reduction).

This ability to find meaning in unlabeled data makes unsupervised learning particularly valuable in the medical field. By analyzing large datasets of medical sensor data, researchers can uncover hidden patterns and relationships that might hold significant clues for disease diagnosis, treatment planning, and even drug discovery. For instance, research by Milanko et al. demonstrates the potential of unsupervised learning in wearables by creating a smart device that provides real-time feedback to weightlifters based on their workout data [66].

4.4.3 Semi-supervised Learning

Imagine having a vast amount of data from wearable sensors, but only a small portion of it is labeled with specific categories. This is where semi-supervised learning excels. It bridges the gap between supervised and unsupervised learning by leveraging both labeled and unlabeled data to train a model. The core idea is to use a smaller set of labeled data along with a much larger set of unlabeled data. By analyzing the unlabeled data alongside the labeled data, the model can learn a better overall representation of the underlying patterns. This approach proves particularly beneficial when collecting labeled data can be expensive or time consuming.

Semi-supervised learning employs various techniques like self-training and autoencoders. Autoencoders, for instance, utilize an encoder–decoder structure. The encoder learns a compressed representation of the labeled data, while the decoder attempts to generate new data instances based on this compressed form. This approach holds immense potential for wearables. For example, researchers have developed a CNN-based autoencoder model for ECG analysis. Here, the limited labeled data is used to train the encoder, and then the decoder leverages a larger set of unlabeled data to refine its ability to generate new data representations. This not only improves the model's overall performance but also allows for ECG analysis even in situations where labeled data is scarce [67].

4.4.4 Reinforcement Learning

Unlike supervised learning that relies on labeled data, RL takes a different approach. Imagine an agent (like a wearable) interacting with its environment (the wearer's body) and constantly learning through trial and error. The agent receives rewards for desired actions and modifies its behavior based on this feedback. This allows RL to find the optimal course of action in uncertain situations. There are three main categories of RL:

- **Value-Based RL:** The agent focuses on learning a value function. This function predicts the long-term reward for taking a specific action in a given state. Examples include Q-learning and deep Q-network (DQN), which are commonly used algorithms in this approach.
- **Model-Based RL:** The agent builds an internal model of its environment. By simulating different scenarios within this model, the agent can choose the action with the most desirable outcome. This can be more efficient but requires complex algorithms and computing power.
- **Policy-Based RL:** It involves directly learning a policy function. This function maps situations (states) to the most suitable actions. Researchers are exploring the potential of RL for tasks like customizing treatment plans for diseases based on individual responses [68] and analyzing a patient's musculoskeletal movements to optimize physical therapy [69].

4.5 Summary

The transformative impact of wearable and implantable sensor devices within the healthcare sector has been profound, initiating a new era of healthcare monitoring characterized by the collection of vast amounts of multi-modal health and patient activity data. However, the effective utilization of this wealth of data has posed significant challenges for traditional data analysis methods, which often struggle to adapt to the complexities of rapidly evolving datasets. The limitations of these approaches underscore the pressing need for advanced data analysis techniques, particularly those based on AI and ML. ML algorithms offer a promising solution to overcome the shortcomings of traditional data analysis methods. By leveraging the inherent learning capabilities of ML models, researchers can develop sophisticated algorithms capable of extracting meaningful insights from complex and high-dimensional datasets collected by wearable sensors. ML and DL models have the potential to revolutionize disease detection and treatment planning by analyzing vital signals obtained from wearable sensors.

The integration of ML with wearable and implantable sensor devices holds immense potential to advance digital healthcare to new heights. Through the development of robust ML-based frameworks, such as the one proposed in this chapter, healthcare professionals can identify early signs of various diseases and tailor personalized treatment plans accordingly. Remote patient monitoring has become a reality, enabling timely interventions and improved patient outcomes. While previous research has extensively explored various aspects of wearable sensor technology, there remains a notable gap in the literature concerning the integration of ML with these devices. This study aims to address this gap by developing a comprehensive framework that utilizes DL techniques to analyze sensor data from wearable devices. By bridging this gap, we aim to contribute to the advancement of healthcare monitoring and treatment processes, ultimately improving patient care and outcomes.

References

1 L'heureux, A., Grolinger, K., Elyamany, H.F., and Capretz, M.A. (2017). Machine learning with big data: challenges and approaches. *IEEE Access* 5: 7776–7797.

2 Priestley, M., O'donnell, F., and Simperl, E. (2023). A survey of data quality requirements that matter in ML development pipelines. *ACM Journal of Data and Information Quality* 15 (2): 1–39.

3 Abbasi, Q.H., Ur Rehman, M., Alomainy, A., and Qaraqe, K. (2016). *Advances in Body-Centric Wireless Communication: Applications and State-of-the-Art*. The IET.

4 Boe, A.J., McGee Koch, L.L., O'Brien, M.K. et al. (2019). Automating sleep stage classification using wireless, wearable sensors. *NPJ Digital Medicine* 2 (1): 131.

5 Prieto-Avalos, G., Cruz-Ramos, N.A., Alor-Hernández, G. et al. (2022). Wearable devices for physical monitoring of heart: a review. *Biosensors* 12 (5): 292.

6 Borah, X., Thangam, A., & Kumari, N. (2024). "Smart glasses with sensors" in Hemachandran K, Manjeet Rege, Zita Zoltay Paprika, K. V. Rajesh Kumar, Shahid Mohammad Ganie, *Handbook of Artificial Intelligence and Wearables* (pp. 236–245). CRC Press.

7 Javaid, S., Zeadally, S., Fahim, H., and He, B. (2022). Medical sensors and their integration in wireless body area networks for pervasive healthcare delivery: a review. *IEEE Sensors Journal* 22 (5): 3860–3877.

8 Al-Nafjan, A., Hosny, M., Al-Ohali, Y., and Al-Wabil, A. (2017). Review and classification of emotion recognition based on EEG brain-computer interface system research: a systematic review. *Applied Sciences* 7 (12): 1239.

9 Bulling, A., Roggen, D., and Tröster, G. (2009). Wearable EOG goggles: Seamless sensing and context-awareness in everyday environments. *Journal of Ambient Intelligence and Smart Environments* 1 (2): 157–171.

10 Bhardwaj, S., Khan, A.A., and Muzammil, M. (2021). Lower limb rehabilitation robotics: the current understanding and technology. *Work* 69 (3): 775–793.

11 Setz, C., Arnrich, B., Schumm, J. et al. (2009). Discriminating stress from cognitive load using a wearable EDA device. *IEEE Transactions on Information Technology in Biomedicine* 14 (2): 410–417.

12 Yuasa, Y. and Suzuki, K. (2019). Wearable device for monitoring respiratory phases based on breathing sound and chest movement. *Advanced Biomedical Engineering* 8: 85–91.

13 Garrett, D.C., Rae, N., Fletcher, J.R. et al. (2017). Engineering approaches to assessing hydration status. *IEEE Reviews in Biomedical Engineering* 11: 233–248.

14 Lafuente, J.L., González, S., Aibar, C. et al. (2024). Continuous and non-invasive lactate monitoring techniques in critical care patients. *Biosensors* 14 (3): 148.

15 Majumder, S., Mondal, T., and Deen, M.J. (2017). Wearable sensors for remote health monitoring. *Sensors* 17 (1): 130.

16 Naeini, E.K., Azimi, I., Rahmani, A.M. et al. (2019). A real-time PPG quality assessment approach for healthcare Internet-of-Things. *Procedia Computer Science* 151: 551–558.

17 Glasstetter, M., Böttcher, S., Zabler, N. et al. (2021). Identification of Ictal tachycardia in focal motor-and non-motor seizures by means of a wearable PPG sensor. *Sensors* 21 (18): 6017.

18 Lazazzera, R., Deviaene, M., Varon, C. et al. (2020). Detection and classification of sleep apnea and hypopnea using PPG and SpO $ _2 $ signals. *IEEE Transactions on Biomedical Engineering* 68 (5): 1496–1506.

19 Lee, I., Probst, D., Klonoff, D., and Sode, K. (2021). Continuous glucose monitoring systems: current status and future perspectives of the flagship technologies in biosensor research. *Biosensors and Bioelectronics* 181: 113054.

20 Tang, L., Chang, S.J., Chen, C.J., and Liu, J.T. (2020). Non-invasive blood glucose monitoring technology: a review. *Sensors* 20 (23): 6925.

21 Pant, M., Jadon, J.S., Agarwal, R., and Sinha, S.K. (2021). Smart monitoring system using smart glove. *9th International Conference on Reliability, Infocom Technologies and Optimization (Trends and Future Directions) (ICRITO)*, Noida, India (3–4 September 2021), pp. 1–4. IEEE.

22 Arteaga-Marrero, N., Hernández-Guedes, A., Ortega-Rodríguez, J., and Ruiz-Alzola, J. (2023). State-of-the-art features for early-stage detection of diabetic foot ulcers based on thermograms. *Biomedicines* 11: 3209.

23 Honi, D.G. and Szathmary, L. (2024). A one-dimensional convolutional neural network-based deep learning approach for predicting cardiovascular diseases. *Informatics in Medicine Unlocked* 49.

24 Zhao, W., Zhao, W., Wang, W. et al. (2020). A novel deep neural network for robust detection of seizures using EEG signals. *Computational and Mathematical Methods in Medicine* pp. 1-9, 2020.

25 Kumar, P. and Pandey, P.C. (2013). A wearable inertial sensing device for fall detection and motion tracking. *2013 Annual IEEE India Conference (INDICON)*, Mumbai, India (13–15 December 2013), 1–6. IEEE.

26 Park, C.Y., Cha, N., Kang, S. et al. (2020). K-EmoCon, a multimodal sensor dataset for continuous emotion recognition in naturalistic conversations. *Scientific Data* 7 (1): 293.

27 Gedam, S. and Paul, S. (2024). Multi-sensor data fusion and deep machine learning models-based mental stress detection system. In: *Advances in Data-Driven Computing and Intelligent Systems. ADCIS 2023*, Lecture Notes in Networks and Systems, vol. 891 (ed. S. Das, S. Saha, C.A. Coello Coello, and J.C. Bansal). Singapore: Springer.

28 Pahar, M., Miranda, I., Diacon, A., and Niesler, T. (2022). Automatic non-invasive cough detection based on accelerometer and audio signals. *Journal of Signal Processing Systems* 94 (8): 821–835.

29 Ren, L. and Peng, Y. (2019). Research of fall detection and fall prevention technologies: a systematic review. *IEEE Access* 7: 77702–77722.

30 Hammad, M., Kandala, R.N., Abdelatey, A. et al. (2021). Automated detection of shockable ECG signals: a review. *Information Sciences* 571: 580–604.

31 Herrero, P., Reddy, M., Georgiou, P., and Oliver, N.S. (2022). Identifying continuous glucose monitoring data using machine learning. *Diabetes Technology & Therapeutics* 24 (6): 403–408.

32 Fati, S.M., Muneer, A., Akbar, N.A., and Taib, S.M. (2021). A continuous cuffless blood pressure estimation using tree-based pipeline optimization tool. *Symmetry* 13 (4): 686.

33 Jacobs, P.G., Herrero, P., Facchinetti, A. et al. (2024). Artificial Intelligence and machine learning for improving glycemic control in diabetes: best practices, pitfalls and opportunities. *IEEE Reviews in Biomedical Engineering* 17: 19–41.

34 Boe, A.J., McGee Koch, L.L., O'Brien, M.K. et al. (2019). Automating sleep stage classification using wireless, wearable sensors. *NPJ Digital Medicine* 2 (1): 131.

35 Shi, Y., Zhang, Y., Cao, Z. et al. (2023). Application and interpretation of machine learning models in predicting the risk of severe obstructive sleep apnea in adults. *BMC Medical Informatics and Decision Making* 23: 230.

36 Khera, P. and Kumar, N. (2020). Role of machine learning in gait analysis: a review. *Journal of Medical Engineering & Technology* 44 (8): 441–467.

37 Huang, Y.P., Liu, Y.Y., Hsu, W.H. et al. (2020). Monitoring and assessment of rehabilitation progress on range of motion after total knee replacement by sensor-based system. *Sensors* 20 (6): 1703.

38 Lin, W.Y., Chen, C.H., Tseng, Y.J. et al. (2018). Predicting post-stroke activities of daily living through a machine learning-based approach on initiating rehabilitation. *International Journal of Medical Informatics* 111: 159–164.

39 Lee, S.I., Adans-Dester, C.P., O'Brien, A.T. et al. (2020). Predicting and monitoring upper-limb rehabilitation outcomes using clinical and wearable sensor data in brain injury survivors. *IEEE Transactions on Biomedical Engineering 68* (6): 1871–1881.

40 Ghomrawi, H.M., O'Brien, M.K., Carter, M. et al. (2023). Applying machine learning to consumer wearable data for the early detection of complications after pediatric appendectomy. *NPJ Digital Medicine* 6 (1): 148.

41 Hunter, A., Leckie, T., Coe, O. et al. (2022). Using smartwatches to observe changes in activity during recovery from critical illness following COVID-19 critical care admission: 1-year, multicenter observational study. *JMIR Rehabilitation and Assistive Technologies 9* (2): e25494.

42 Shi, D., Zhang, W., Zhang, W., and Ding, X. (2019). A review on lower limb rehabilitation exoskeleton robots. *Chinese Journal of Mechanical Engineering* 32 (1): 1–11.

43 Zhang, L., Li, Z., Hu, Y. et al. (2020). Ankle joint torque estimation using an EMG-driven neuromusculoskeletal model and an artificial neural network model. *IEEE Transactions on Automation Science and Engineering* 18 (2): 564–573.

44 Zhang, L., Liu, G., Han, B. et al. (2019). sEMG based human motion intention recognition. *Journal of Robotics* 2019: 1–12.

45 Fozoonmayeh, D., Le, H.V., Wittfoth, E. et al. (2020). A scalable smartwatch-based medication intake detection system using distributed machine learning. *Journal of Medical Systems* 44: 1–14.

46 Nurmi, J. and Lohan, E.S. (2021). Systematic review on machine-learning algorithms used in wearable-based eHealth data analysis. *IEEE Access* 9: 112221–112235.

47 Le, T., Ellington, F., Lee, T.Y. et al. (2020). Continuous non-invasive blood pressure monitoring: a methodological review on measurement techniques. *IEEE Access 8*: 212478–212498.

48 Li, S., Zhang, C., Xu, Z. et al. (2022). Cuffless blood pressure monitoring: academic insights and perspectives analysis. *Micromachines* 13 (8): 1225.

49 Huan, J., Bernstein, J.S., Difuntorum, P. et al. (2021). A wearable skin temperature monitoring system for early detection of infections. *IEEE Sensors Journal* 22 (2): 1670–1679.

50 Xu, C., Song, Y., Sempionatto, J.R. et al. (2024). A physicochemical-sensing electronic skin for stress response monitoring. *Nature Electronics* 7: 168–179.

51 Verma, R.P., Sahu, P.S., Rathod, M. et al. (2022). Ultra-sensitive and highly stretchable strain sensors for monitoring of human physiology. *Macromolecular Materials and Engineering* 307 (3): 2100666.

52 Tang, W. and Chen, C. (2020). Hydrogel-based colloidal photonic crystal devices for glucose sensing. *Polymers* 12 (3): 625.

53 Deng, M., Song, G., Zhong, K. et al. (2022). Wearable fluorescent contact lenses for monitoring glucose via a smartphone. *Sensors and Actuators B: Chemical* 352: 131067.

54 Kumar, V. and Arora, K. (2024). Nano-inspired point-of-care enzyme-based wearable biosensors for global health care. In: *Enzyme-based Biosensors: Recent Advances and Applications in Healthcare*, 293–322. Singapore: Springer Nature Singapore.

55 Liu, Z., Chen, Y., Zhang, M. et al. (2021). Novel portable sensing system with integrated multifunctionality for accurate detection of salivary uric acid. *Biosensors* 11 (7): 242.

56 Arakawa, T., Kuroki, Y., Nitta, H. et al. (2016). Mouthguard biosensor with telemetry system for monitoring of saliva glucose: a novel cavitas sensor. *Biosensors and Bioelectronics 84*: 106–111.

57 Mukherjee, S., Suleman, S., Pilloton, R. et al. (2022). State of the art in smart portable, wearable, ingestible and implantable devices for health status monitoring and disease management. *Sensors 22* (11): 4228.

58 Kasparbauer, A., Reisner, V., Schenk, C. et al. (2022). Sensor devices, the source of innovative therapy and prevention. In: *The Future Circle of Healthcare: AI, 3D Printing, Longevity, Ethics, and Uncertainty Mitigation*, 207–226. Cham: Springer International Publishing.

59 DeForge, W.F. (2019). Cardiac pacemakers: a basic review of the history and current technology. *Journal of Veterinary Cardiology* 22: 40–50.

60 Alba, A.C., Foroutan, F., Posada, J.D. et al. (2018). Implantable cardiac defibrillator and mortality in non-ischaemic cardiomyopathy: an updated meta-analysis. *Heart* 104 (3): 230–236.

61 Bajaj, P., Schweller, R.M., Khademhosseini, A. et al. (2014). 3D biofabrication strategies for tissue engineering and regenerative medicine. *Annual Review of Biomedical Engineering* 16: 247–276.

62 Lozano, A.M., Lipsman, N., Bergman, H. et al. (2019). Deep brain stimulation: current challenges and future directions. *Nature Reviews Neurology* 15 (3): 148–160.

63 Castillo-Henríquez, L., Brenes-Acuña, M., Castro-Rojas, A. et al. (2020). Biosensors for the detection of bacterial and viral clinical pathogens. *Sensors* 20 (23): 6926.

64 Heo, Y.J. and Takeuchi, S. (2013). Towards smart tattoos: implantable biosensors for continuous glucose monitoring. *Advanced Healthcare Materials* 2 (1): 43–56.

65 Kiranyaz, S., Avci, O., Abdeljaber, O. et al. (2021). 1D convolutional neural networks and applications: a survey. *Mechanical Systems and Signal Processing* 151: 107398.

66 Milanko, S. and Jain, S. (2020). LiftRight: quantifying strength training performance using a wearable sensor. *Smart Health* 16: 100115.

67 Lichaee, F.K., Salari, A., Jalili, J. et al. (2023). Advancements in artificial intelligence for ECG signal analysis and arrhythmia detection: a review. *International Journal of Cardiovascular Practice* 8 (2): 1–9.

68 Yu, C., Liu, J., Nemati, S., and Yin, G. (2021). Reinforcement learning in healthcare: a survey. *ACM Computing Surveys (CSUR)* 55 (1): 1–36.

69 Van Dieën, J.H., Flor, H., and Hodges, P.W. (2017). Low-back pain patients learn to adapt motor behavior with adverse secondary consequences. *Exercise and Sport Sciences Reviews* 45 (4): 223–229.

5

Sensor Design for Industrial Automation

Abdul Jabbar[1], *Tahera Kalsoom*[2], *and Masood Ur Rehman*[1]

[1] *James Watt School of Engineering, University of Glasgow, Glasgow, UK*
[2] *Manchester Fashion Institute, Manchester Metropolitan University, Manchester, UK*

The significance of sensor design for industrial automation plays a transformational dimension within the framework of Industry 4.0 and Industry 5.0. This chapter discusses the key areas to understand the intersection of conventional production methods and cutting-edge digital technology in industrial automation. It sheds light on the crucial significance of sensor design in facilitating this transformative process. The increasing adoption of smart factories and interconnected ecosystems by companies necessitates the use of multimodal sensors that can effectively integrate with cyber-physical systems (CPS), Industrial Internet of Things (IIoT) platforms, and analytics driven by artificial intelligence (AI) for sensor fusion in industrial automation and smart manufacturing environments.

5.1 Multimodal Sensing in Industrial Automation

In general, human beings utilize various sensory organs to perceive and infer objects. For instance, the tactile system can discern attributes such as surface roughness, softness, and temperature of objects. The visual system can discern characteristics like color, shape, and even a portion of the interior structure of objects. Similarly, the olfactory system can detect the odor of objects. By utilizing this multimodal information in conjunction, human beings are able to recognize objects and make decisions in natural environments with greater precision and accuracy. The analogy can be applied to multimodal sensing in industrial settings

Multimodal Intelligent Sensing in Modern Applications, First Edition.
Edited by Masood Ur Rehman, Ahmed Zoha, Muhammad Ali Jamshed, and Naeem Ramzan.

Published 2025 by John Wiley & Sons, Inc.

which pertains to the application of various sensing modalities or sensor types in order to capture and analyze diverse aspects of a certain environment or system [1, 2]. Multimodal sensing is a method that combines multiple sensors to detect numerous physical phenomena or qualities, including temperature, pressure, proximity, motion, light, sound, and chemical composition, among others, instead of depending on a single type of sensor. The integration of data from various sensors in multimodal sensing facilitates a more holistic comprehension of the surrounding environment, hence offering enhanced and precise information for the purposes of decision-making, control, and automation.

In view of Industry 4.0/5/0, multimodal sensing for industrial automation allows for a holistic approach to monitoring and controlling various industrial entities and manufacturing processes [3–8]. It facilitates predictive maintenance, improving resource efficiency, and promoting human-robot collaboration in future smart factories through the potential of wireless sensor networks [6]. For instance, a manufacturing robot that is equipped with various sensors, including vision sensors for identifying objects, pressure sensors for providing tactile feedback, and proximity sensors for avoiding collisions, can efficiently move through its surroundings, manipulate objects with accuracy, and adjust to dynamic conditions in real time.

The incorporation of intelligent devices has fundamentally transformed manufacturing procedures, facilitating the streamlining of operations and the adoption of novel protocols designed to improve overall production effectiveness and guarantee the well-being of operators and equipment [9, 10]. Crucial for ensuring safety is the use of specialized sensors that can identify possible dangers and activate suitable actions to address them. An efficient system should possess the capability to immediately detect potentially dangerous situations and respond by either stopping activities or providing alerts through audio/visual signals.

In the context of indoor industrial environments, multimodal sensing systems have the capability to integrate sensors that enable the monitoring of many factors such as occupancy, temperature, humidity, air quality, and energy consumption. This integration aims to enhance indoor safety, energy efficiency, and the overall performance of the industrial automation environment [11]. In general, multimodal sensing enhances the capabilities of sensing systems by leveraging the complementary strengths of different sensor modalities, thereby enabling more robust, versatile, and intelligent applications across industrial automation, smart manufacturing, and real-time CPS [4, 7, 12–14].

Prior to discussing industrial multimodal sensors, it is important to examine the key enabling technologies driving industrial automation and smart manufacturing, with a particular emphasis on advanced and robust multimodal sensing networks. Industrial automation and intelligent manufacturing are closely related, interdependent concepts that are frequently used interchangeably [15]. The use of control systems, machinery, and processes to automate processes and duties in manufacturing and other industrial settings is referred to as industrial automation. It entails the implementation of technological advancements, including robotics, programmable logic controllers (PLCs), sensors and actuators, and CPS to execute duties with minimal human involvement [16–18]. As a result, there is an enhancement in customized and personalized productivity, quality, and efficiency, alongside a reduction in labor expenses and errors.

Smart manufacturing expands upon the basis laid by industrial automation through the integration of cutting-edge technologies including but not limited to the IIoT, smart sensors and sensor fusion, software remote control, machine learning (ML) and artificial intelligence (AI), cloud computing, big data analytics, advanced robotics, augmented reality (AR), 3D printing, and smart adaptive communication in order to establish manufacturing systems that are both intelligent and interconnected [3, 14, 17–19]. Smart manufacturing optimizes and transforms the entire manufacturing value chain, including product design, production, supply chain management, and customer service, by leveraging data and connectivity. As a whole, such digitalized industrial paradigms are referred to as Industry 4.0 and Industry 5.0 [20–22].

The notion of Industry 4.0 is based on the amalgamation of sophisticated manufacturing technologies, data exchange, and automation within industrial operations as depicted in Figure 5.1. Robotic systems are of utmost importance in this context where machinery and production systems that are outfitted with sensors, actuators, and connectivity execute tasks independently and effectively. Industry 4.0 centers on the attainment of elevated degrees of interconnectivity, digitization, and automation throughout the manufacturing value chain. This progress is intended to result in enhanced capabilities for mass customization, flexibility, and productivity.

On the other hand, Industry 5.0 places significant emphasis on the collaboration between humans and robots, commonly referred to as "cobotics." In contrast to the fully autonomous robotic systems envisioned in Industry 4.0, cobots are purpose-built to operate in conjunction with human operators, augmenting their capabilities and aiding them in diverse duties. Cobotic environments place a premium on safety, adaptability, and flexibility, enabling direct collaboration and interaction between humans and robots in shared workspaces. This methodology for smart manufacturing integrates the merits of human ingenuity, problem-solving capabilities, and manual labor with the accuracy, and uniformity of autonomous automation to promote mass personalization.

5.1.1 IIoT and Multimodal Sensing

The notion of IIoT is essentially the same as that of IoT, but with a more specialized and specific focus on industrial machines. Instead of establishing connections

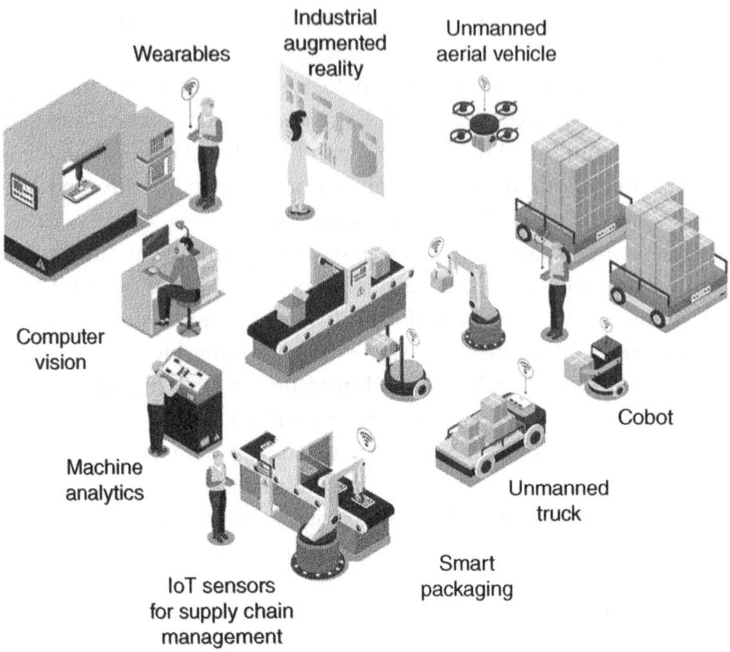

Figure 5.1 Industrial automation and smart manufacturing avenues. *Source:* Ref. [11].

between consumer devices, the interconnected sensors, instruments, and other networked devices collaborate with sophisticated industrial computing applications such as production and energy management in IIoT [13, 23]. A network of sensors gathers essential production data and transmits it to cloud software for analysis. The software then examines the data and provides useful insights regarding the quality and efficiency of industrial activities.

The interconnection of IIoT and multimodal sensors is complex, as they serve complementary functions in the transformation of conventional manufacturing settings into intelligent and networked systems. The core of this collaborative effort centers around the notion of sensor fusion, which involves the integration and analysis of data from several sensing modalities to offer a full understanding of industrial processes. The sensors provide substantial quantities of data, which, when integrated with IIoT systems, facilitate the continuous monitoring, analysis, and management of industrial activities in real time [24]. IIoT platforms play a crucial role in enabling connectivity by providing smooth communication across various components such as multimodal sensors, edge devices, cloud infrastructure, and enterprise systems. They facilitate bidirectional data exchange by utilizing wireless connectivity protocols like Wi-Fi, Bluetooth, Zigbee, and LoRaWAN. This enables the transmission of sensor data from the shop floor to the cloud for analysis, and the communication of actionable insights back to edge devices for prompt decision-making and control [25].

The incorporation of IIoT and multimodal sensors presents a wide range of advantages for industrial automation, including improved operational effectiveness, proactive maintenance, and quality control [24]. Manufacturers have the ability to monitor the health of equipment in real time and anticipate future problems by utilizing multimodal sensor networks enabled by the IIoT. Implementing this proactive maintenance strategy helps to minimize unexpected periods of inactivity, lower maintenance expenses, and extend the longevity of essential resources.

Furthermore, the state-of-the-art IIoT platforms facilitate the utilization of sophisticated analytics and ML algorithms to examine sensor data streams, identify patterns, and enhance personalized manufacturing processes [26]. By utilizing predictive analytics, manufacturers have the potential to forecast fluctuations in demand, enhance the allocation of resources, and reduce inefficiencies, hence enhancing overall productivity and profitability. Likewise, multimodal sensor networks in IIoT networks are crucial in enabling quality control and compliance monitoring through the continuous monitoring of production parameters and ambient conditions. By adhering to regulatory standards, product specifications, and quality benchmarks, product quality is enhanced and customer satisfaction is improved. Hence, leveraging sensor fusion and IIoT enables evolving smart industrial operations to achieve enhanced operational efficiency, resilience, and effectiveness.

5.1.2 Advanced Robotics

Smart sensor technology allows advanced robotics to interact with the physical environment. Advanced robotics is a combination of complex programming and powerful hardware and modern industrial automation strongly relies on sophisticated robotics. The widespread use of digital tools and cutting-edge computing technologies like AI in product development, production, logistics, and retail has led to a dramatic expansion in the breadth and depth of manufacturing processes [6]. Efforts to simplify and streamline processes are increasingly making use of applications of advanced robots. This involves the development of machines that can execute jobs similar to humans in complicated and unpredictable environments, such as ensuring safe teamwork or replacing humans in dangerous situations, it is crucial to gather a wide range of information.

Industry 4.0/5.0 makes use of sophisticated robots to streamline operations and boost output, which can adapt and correct themselves in response to changes in processes and procedures. Such a degree of flexibility is not available to traditional industrial robots [27–29]. Another benefit of using advanced robots instead of traditional ones is how much easier they are to configure and set up right from the start when they are used on an assembly line. Robots with more processing power can also use simulation software to learn new skills. On the assembly line, manufacturers use technology to fill talent gaps, which improves quality, reliability, and precision. This implies accurately identifying objects and making well-informed decisions in chaotic circumstances [28]. Therefore, it is necessary to use multi-modal sensors, which combine several detecting processes, in order to accurate movements and decisions taken by the robots.

5.1.3 Big Data Analytics

With the proliferation of IoT and numerous sensing nodes, a huge amount of data is generated, which gives birth to an entire single domain of "Big Data." Big data analytics employs sophisticated computational methods applied to massive data sets in search of meaningful relationships, trends, patterns, and customer preferences. Manufacturers can automate production management, optimize predictive maintenance, gain insight into real-time data with self-service solutions, and experience increased production efficiency through its use [30]. Various aspects of big data analytics such as self-organization, predictive maintenance, and production management automation help to improve the industrial automation processes. Self-organized analytics help in consolidating massive amounts of big data from manufacturing facilities. These systems provide visual representations for important decision-makers based on their analysis of real-time data, which helps to discover trends and possible errors [31, 32].

When it comes to predictive maintenance, the goal is to significantly reduce response time. The system's output from big data analytics is used for decision-making by CPS. In order to prevent unanticipated downtime or equipment malfunction, this information is used to prioritize modifications and actions. Production management can be automated with the use of big data analytics. This means that fewer actions and inputs from humans will be required in a smart factory. The system operates by examining past data of a manufacturing procedure gathered through a multi-modal sensing network, combining it with current information of the manufacturing process, and implementing physical modifications to machinery through actuators and sophisticated robots that are linked to control software [33].

5.1.4 Cloud Computing

Cloud computing refers to the utilization of both hardware and software to provide services across a network, typically over the internet. Smartphones exemplify sophisticated computational hardware that supports numerous cloud computing applications. Cloud computing applications optimize specific managerial and operational procedures and consolidate data storage, bandwidth, and processing, thereby eliminating the need for individual users to manually install software on their workstations [34, 35]. Users can easily retrieve them from the cloud, which facilitates real-time interchange of information.

One major issue for industries is that the cloud introduces increased security threats due to the fact that a third party has authority over the server where the data is housed. Contrary to this notion, the cloud offers distinct security benefits, such as the ability to initiate continuous updates that enhance infrastructure security [36]. Cloud computing offers a distinct approach to risk mitigation compared to local hosting, while also providing various operational advantages. It decreases the cost of hiring a person to oversee a local server, strengthens the ability to handle an increased workload, and improves the reliability of the system [37]. Data stored via cloud computing is highly resistant to loss [38, 39].

5.1.5 Artificial Intelligence

AI is crucial in advancing Industry 4.0/5.0 and transforming industrial automation. Advanced technologies enable production systems to analyze large volumes of data, extract practical insights, and autonomously make intelligent decisions [38, 40]. In the context of Industry 4.0, ML and deep learning (DL) algorithms (subsets of AI) are utilized to improve manufacturing processes, anticipate equipment malfunctions, and boost product quality by analyzing sensor data in real time [17]. AI-powered predictive maintenance systems utilize ML algorithms to detect trends that indicate upcoming equipment failures. This allows for preventative maintenance actions to be taken, reducing downtime and preventing expensive breakdowns [41–44]. Traditional ML algorithms are specifically developed to analyze and acquire knowledge from data that is organized or partially organized. These algorithms commonly depend on feature engineering, a process in which domain specialists manually choose or create significant characteristics from the data. DL employs artificial neural networks that consist of numerous layers [26]. These networks have the ability to learn hierarchical representations of data on their own through feature learning, which means there is no need for explicit feature engineering [45, 46].

Furthermore, robotic systems powered by AI have the ability to adjust to dynamic surroundings, acquire knowledge from past encounters, and enhance their performance gradually [47]. This results in increased adaptability and effectiveness in manufacturing operations. The use of ML/DL and AI technologies in Industry 4.0 and industrial automation enables the development of intelligent, flexible, and self-governing manufacturing systems that can effectively respond to the requirements of the current dynamic and competitive market environment [48].

Within the framework of industrial automation, the fusion AI technology with multimodal sensors facilitates data analysis and decision-making. ML/DL algorithms can analyze data from different sensing modalities, including vision sensors, proximity sensors, and temperature sensors, to extract important insights, identify patterns, and optimize manufacturing processes in real time [26, 43, 49]. AI-powered systems may utilize data streams from various sensors to forecast equipment breakdowns, optimize energy consumption, and improve product quality, hence increasing efficiency and productivity in manufacturing operations.

5.1.6 Augmented Reality

AR is a technology that superimposes digital items and information onto a screen, such as a tablet, phone, or headset, while simultaneously capturing the real-time physical world. It is used in enterprise and industrial environments to provide training for new workers on assembly lines in factories, as well as to instruct industrial workers in performing maintenance and repair tasks for various industrial objects.

AR is revolutionizing the product design lifecycle in the industrial automation paradigm and is also linked to digital twins [50]. Designers and engineers have the ability to create a product with precise specifications that accurately reflect real-world conditions [51]. AR enables virtual prototype testing for many items, such as equipment in a manufacturing plant or products designed to fit precise-sized shipping containers. In smart manufacturing, traditional paper manuals are replaced with digital instructions. AR headsets are available that superimpose digital instructions onto a workspace, eliminating the need for operators to switch between watching machines and instructions. Quality control applications enable industries to ensure the accuracy of component placement in assembly by validating the work instructions well in advance. AR significantly expedites the verification process compared to conventional approaches. Furthermore, AR and intelligent logistics have a symbiotic relationship, since AR applications enhance operational efficiency in crucial aspects of logistics such as warehousing, routing, and transportation of goods. Warehouse staff can wear AR glasses to efficiently discover the most direct route for locating and selecting items required for delivery [50, 51].

5.2 Sensors for Realizing Industrial Automation

A sensor is a device that employs a transducer to gather a particular category of data from either an external or internal physical environment. The information is processed by utilizing the integrated computing resources of the sensor to execute

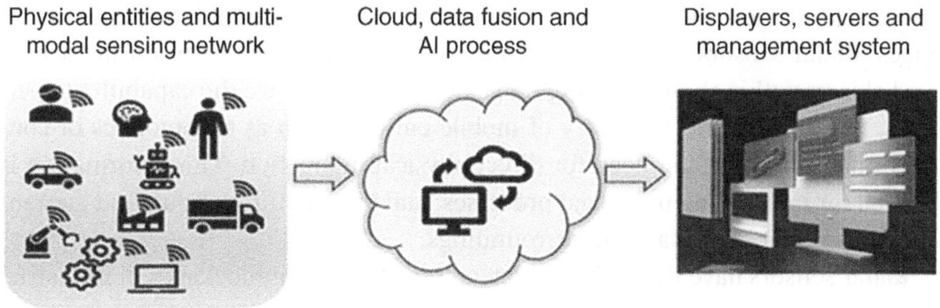

Figure 5.2 Conceptual depiction of a multi-modal sensing network in industrial automation. *Source:* (b) Nanang/Adobe Stock. (c) Allison Saeng/Unsplash.

a predetermined and planned operation on the specific data it is gathering. Subsequently, it transmits the data over the IoT network.

Smart sensors are integral to industrial automation, monitoring diverse processes, collecting data, conducting measurements, and transmitting this information to cloud computing platforms for analysis. Below, we illustrate examples of sensors in industrial automation and smart manufacturing which are usually combined in one way or the other to accomplish a dedicated task, as depicted in Figure 5.2.

5.2.1 RF Sensors

RF sensing plays a crucial role in industrial automation by providing noncontact detection, measurement, and monitoring capabilities across various applications. RF sensors have the capability to detect the presence or absence of items without making any physical touch [4, 12, 52, 53]. Conveyor systems frequently utilize them to oversee the motion of materials or products throughout the production line. RFID technology is a type of RF sensor, that is employed in industrial settings to monitor and locate assets, inventory, and work-in-progress objects [54, 55]. RFID tags affixed to things produce radio frequency (RF) signals that are detected by RFID readers, allowing for instantaneous visibility and efficient inventory management [56, 57]. RF sensors enable the wireless exchange of information between components in industrial automation systems. They provide the transfer of data between sensors, actuators, controllers, and other components without the necessity of physical wiring, hence, streamlining system installation and maintenance [58].

One of the major areas under RF sensing is radar technology [59]. Radars utilize radio waves to detect the presence, distance, speed, and direction of objects in their field of view. In industrial automation, radar sensors offer several advantages. Radar sensors precisely determine the exact distance between the sensor and objects within their detecting range. Radar sensors have the capability to perceive the velocity and trajectory of mobile entities, such as automobiles or conveyor belts, without the need for direct physical interaction. This information is crucial for monitoring industrial processes, maximizing throughput, and guaranteeing safety in ever-changing surroundings.

Radar sensors have the ability to detect impediments or blockages in industrial settings, offering proactive measures to mitigate dangers to machinery, equipment, or humans. They are frequently incorporated into automated guided vehicles (AGVs) and autonomous mobile robots (AMRs) to facilitate self-directed movement and the ability to avoid obstacles [60, 61]. By harnessing the power of radar waves, this technology offers unparalleled capabilities in detecting objects, measuring distances, and capturing movement with exceptional accuracy and reliability, even in challenging environments such as those with dust, smoke, or low visibility. Radar sensors provide real-time data on object location, speed, and trajectory, enabling precise control and coordination of robotic systems, conveyor belts, and other machinery on the factory floor. Furthermore, radar sensing enhances safety protocols by swiftly identifying potential hazards and initiating timely interventions to prevent accidents. As industries increasingly prioritize efficiency, flexibility, and safety, the integration of radar sensing technology in industrial automation continues to drive advancements toward smarter, more responsive manufacturing ecosystems [62–64].

5.2.2 Vision Sensors

Vision sensors play a crucial role in multimodal sensing in Industry 4.0. Vision sensors and Light Detection and Ranging (LiDAR) technology are closely interconnected in the field of perception and sensing, especially in sectors that are adopting sophisticated automation and robotics [5]. Vision sensors mainly use optical cameras to record and analyze visual data to offer exceptional ability in identifying objects, patterns, and textures in pictures. LiDAR system utilizes laser beams to measure distances and generate intricate 3D maps of the environment. It offers accurate spatial data and depth perception, even in difficult lighting situations or surroundings with barriers. The combination of two technologies is very significant and these sensors use sophisticated image processing techniques. By combining data from vision sensors with LiDAR, industrial systems may obtain a complete understanding of their surroundings, allowing them to perform tasks like autonomous navigation, geographical mapping, item recognition, and quality control with exceptional precision and efficiency [65]. By extracting valuable information from visual input and improving perception, significant progress in the realm of intelligent manufacturing and industrial automation is expected [66, 67]. This integration is also fundamental to the advancement of sophisticated robots, self-driving cars, and proactive maintenance.

5.2.3 Localization and Tracking Sensors

Location tracking encompasses a set of sophisticated computing systems that can detect, follow, and log the whereabouts of individuals and smart industrial entities. Using global positioning systems (GPSs), Global Navigation Satellite System (GNSS) and other location-tracking technologies, Industry 4.0 can ensure that mining and construction assets, including machinery, are never lost, while manufacturing productivity and worker safety are both enhanced by location tracking. GPS sensors are usually used for outdoor industrial settings. In the context of the broader Industry 4.0 network, indoor industrial automation often rely on real-time location systems (RTLSs) for crucial location tracking. RTLS delivers accurate real-time location tracking inside restricted locations like warehouses, factories, and production facilities.

Indoor positioning technologies like Wi-Fi, Bluetooth Low Energy (BLE), Ultra-Wideband (UWB), radio-frequency identification (RFID), or infrared (IR) beacons form the backbone of RTLS, as opposed to GPS system. These systems attach battery-operated or wearable tags to assets or objects, which interact with a network of fixed anchor nodes or receivers placed throughout the industrial floors. RTLS is able to pinpoint the precise whereabouts of assets or tagged objects in real time, usually within a few meters, by triangulating signals from several anchor nodes. Intelligent logistics and manufacturing industries make extensive use of RTLS for tasks such as process optimization, safety compliance monitoring, inventory management, workforce monitoring, and asset tracking.

5.2.4 Infrared Sensors

Infrared smart sensors are versatile and find applications in several industrial applications. They are employed in architecture, engineering, and construction to detect thermal energy losses in factory buildings and industrial automation facilities. Furthermore, they are employed in wearable devices designed for monitoring and improving the health and fitness of factory personnel.

5.2.5 Proximity Sensors

Proximity sensors are essential components in industrial automation, playing a crucial role in detecting the presence or absence of objects within a specific range without making physical touch. These sensors are essential for a wide range of automation applications such as several functions such as detecting objects, sensing positions, monitoring safety, and controlling machines. Proximity sensors function using different principles, including electromagnetic, capacitive, inductive, ultrasonic, and optical detection. Each principle is designed to meet unique application needs.

Proximity sensors are crucial components in industrial automation for manufacturing assembly lines to identify the existence of components or workpieces, guaranteeing precise positioning and alignment throughout assembly procedures. They are used in conveyor systems to identify when packages or products have arrived, which then initiates automated processes such as sorting, routing, or packing.

In addition, proximity sensors enhance workplace safety by detecting the presence of personnel in dangerous zones and triggering safety interlocks or alarms to prevent mishaps. Additionally, they play a crucial role in AGVs and AMRs, as they offer collision avoidance capabilities and facilitate safe interaction between humans and machines in shared workspaces.

5.2.6 IMU Sensors

Inertial measurement unit (IMU) sensors include accelerometers, gyroscopes, and sometimes magnetometers, which together quantify an object's orientation, velocity, and gravity forces. These sensors are used in several industries, such as in robotics and machinery to provide real-time tracking and monitoring of the motion and orientation of equipment. This is beneficial for activities such as predictive maintenance, remote monitoring, and streamlining processes. IMUs are included in autonomous cars, drones, and other mobile platforms to determine their exact location, direction, and alignment, even in situations when GPS signals are unavailable. Furthermore, IMUs are used to detect vibrations, structural deformations, and possible risks in industrial machinery.

5.2.7 Level Sensors

Level sensors are essential for obtaining real-time readings of containers, bins, and tanks. They provide key data to inventory management and process control systems. Their uses encompass various industries, such as waste management, irrigation, and diesel fuel gauging. These sensors are used to monitor liquid levels in storage tanks and control material flow in industrial processes. They help ensure effective operations and optimize resource management in a variety of applications.

5.2.8 Temperature Sensors

Temperature sensors are widely utilized in industrial environments. The thermal conductivity and temperature are inherent characteristics that determine the thermal properties of items. The temperature can be measured using temperature sensors, which consist of thermoelectric sensors, thermosensitive sensors, and thermistors. Thermal conductivity can be measured using thermosensitive metal resistors that have low resistance, as they operate depending on the heat transmission mechanism. For instance, temperature sensors are utilized to establish a connection with a specific piece of machinery or industrial equipment and to maintain the cooling effect of electronics machinery. This type of sensor is linked to the IIoT cloud computing platform and has the ability to identify instances when the machine or equipment is experiencing excessive heat and requires maintenance or shutdown.

5.2.9 Pressure Sensors

Pressure sensors are employed for the purpose of monitoring pipelines and notifying a centralized computing system about any leaks or abnormalities, which in turn notifies for preventive maintenance and repairs. These sensors are used to oversee and regulate variables such as fluid flow, pressure, and level in diverse industrial operations, including chemical processing, oil and gas production, and food processing. Pressure sensors are used in safety systems to identify leaks, excessive pressure situations, and irregular variations in pressure, so aiding in accident prevention and guaranteeing adherence to regulations. They are used in weather stations, water management systems, and heating, ventilation, and air conditioning (HVAC) systems to measure atmospheric pressure, water pressure, and air pressure for the purpose of environmental and operational monitoring.

5.3 Design Considerations for Effective Multimodal Industrial Automation

The successful implementation of multimodal industrial automation necessitates evaluation of several design aspects during the process of sensor selection and configuration, in order to effectively address the complex requirements of contemporary industrial automation and manufacturing settings. These design characteristics include aspects such as the range of sensing, the level of precision, the resolution, the time it takes to respond, the ability to withstand environmental conditions, and compatibility with communication protocols and data interfaces. For instance, design specifications for proximity sensors encompass various elements, including the range of detection, the type of sensing technology employed (inductive, capacitive, or ultrasonic), and the ability to withstand environmental influences such as dust, moisture, vibrations, and electromagnetic interference. Accurate object detection and analysis necessitate vision sensors that possess high-resolution imaging capabilities, employ advanced image processing algorithms, and demonstrate adaptability to varied lighting conditions. In order to accurately measure pressure exerted during manufacturing processes, pressure sensors must have high sensitivity, linearity, and dynamic range. On the other hand, temperature sensors must provide precise temperature measurements across a wide range of operating conditions with minimal drift and calibration requirements. To achieve smooth data exchange and interoperability across various automation workflows, it is crucial to prioritize sensor integration and interoperability with industrial control systems and automation platforms. This integration enables efficient decision-making and control in complicated manufacturing settings.

5.3.1 Design of AI-Assisted Multimodal Sensing

Multi-modal sensor fusion is a sophisticated undertaking that entails accurately managing data from multiple sources. Traditional signal processing and estimation methods, such as the Kalman filter, particle filter, and expectation maximization, form the foundation of data fusion technology [6]. Over time, various other fusion approaches have appeared, encompassing statistical inference (e.g., Bayesian reasoning, maximum-likelihood estimation), Dempster–Shafer theory, random finite set-based methods, Markov random field, information-theoretic fusion (e.g., minimum description length and entropy methods), and fuzzy logic [2, 26, 68]. The integration of these techniques has been instrumental in advancing information fusion technology. The rapid progress in processing technology, including computational resources and platforms such as graphic processing units (GPUs), cloud-based processing platforms, and associated big data processing approaches, has opened up possibilities for managing extensive datasets for multisensor data fusion. This domain pertains to big data analytics, which stands as one of the pivotal enabling technologies within modern industrial automation networks, as described before.

In the paradigm of modern sensing technologies, ML and DL play pivotal roles, in revolutionizing the capabilities and applications of industrial automation systems [43]. DL is a highly effective technique for extracting intricate features and patterns from multimodal data, thereby facilitating anomaly detection, predictive maintenance, and real-time decision-making. The integration of sensor data in this manner facilitates a more holistic comprehension of manufacturing procedures, resulting in improved quality assurance, decreased periods of inactivity, and escalated output. In addition, the perpetual advancement of DL methodologies guarantees consistent enhancements in the precision and robustness of multi-modal sensing fusion systems, establishing them as essential means for industrial automation.

DL has become a dominant technique for handling large datasets, demonstrating the following three distinct features. DL possesses exceptional nonlinear mapping capabilities, allowing it to autonomously construct intricate models that are challenging to manually create. Furthermore, DL-based models possess multiple layers that enable them to produce high-dimensional representations, beyond the capabilities of prior feature extraction methods. The versatility of DL models allows them to be applied across various areas, including computer vision, autonomous driving, robotics, medical diagnosis, and industrial manufacturing. Consequently, there is great anticipation that DL would significantly improve the overall performance of algorithms used for fusing data from several sensors [3, 4, 7–9].

The automation and control systems of Industry 4.0/5.0, which are used in machinery and plants, along with the expanding range of services linked to IoT devices, are based on sensing networks which have resulted in a noteworthy enhancement in data security and workplace safety. Hence, it is crucial to ensure that during the design and production stages of these devices, IoT security frameworks and best practices have been incorporated [69, 70]. Safety monitoring in a smart factory is of great significance as it seeks to safeguard human workers, equipment, and IT systems.

In revolutionizing manufacturing processes through automation and intelligence, robotics is a crucial cornerstone. The sophisticated robotic systems improve efficiency, accuracy, and adaptability, resulting in unparalleled levels of output and flexibility in contemporary manufacturing.

5.3.2 Design of Radar Sensing Networks

Radar sensing technology finds versatile applications in industrial automation, offering robust solutions for monitoring, control, and safety enhancement [62, 71] (Figure 5.3). Key design components of a radar sensing system in industrial automation include the following.

5.3.2.1 Transmitter and Receiver Antennas

Transmitting and receiving antennas are responsible for emitting and receiving radar signals, respectively. The transmitter antenna produces electromagnetic waves, usually in the microwave frequency range, while the reception antenna receives the reflected signals from objects within the detection range.

5.3.2.2 Data Collection and Interface

Radar sensors are fitted with interfaces that facilitate the transmission of processed data to either an industrial control system or cloud computing platforms. This allows for continuous monitoring, examination, and incorporation with other automated components in real time.

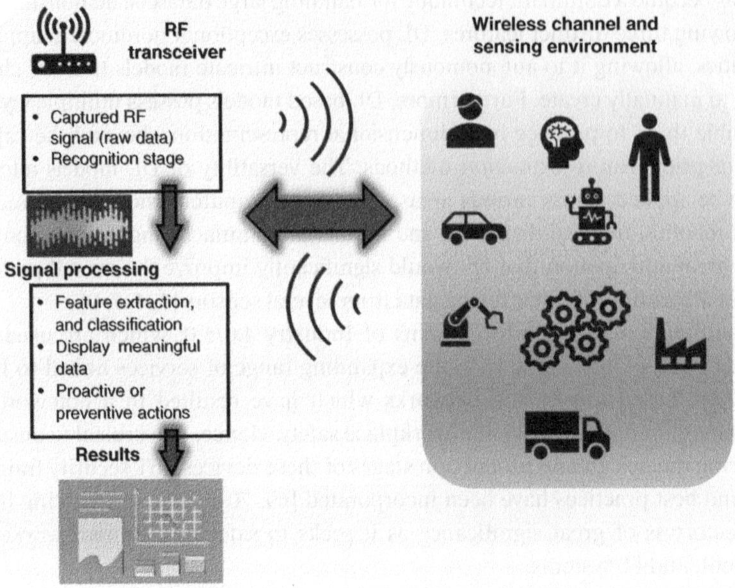

Figure 5.3 A high-level depiction of a typical radar sensing architecture.

5.3.2.3 Signal Processing

Signal processing is responsible for analyzing the radar signals that are received to extract important information, such as the position, velocity, and size of objects. The state-of-the-art signal processing algorithms are utilized to eliminate noise and interference, guaranteeing precise and trustworthy identification.

5.3.2.4 Housing and Enclosure

Radar sensors are placed in durable enclosures specifically engineered to endure the challenging conditions of industrial environments. These enclosures are meant to shield against dust, moisture, temperature fluctuations, and harsh industrial impacts. Enclosures can be constructed using materials such as metal or high-quality plastic.

5.4 Challenges and Opportunities of Multimodal Sensing in Industrial Automation

Multimodal sensing is an essential pillar in the ever-evolving realm of industrial automation and serves as a fundamental means to attain enhanced levels of effectiveness, security, and flexibility. Nevertheless, this auspicious technology is hindered in its seamless integration and optimization in industrial environments by an array of obstacles. A notable obstacle is an intricacy involved in integrating data from various sensing modalities, such as vision, LiDAR, radar, and acoustic sensors in real time, in order to derive actionable and coherent insights. Effectively managing the extensive volume and diverse variety of sensor data necessitates the implementation of sophisticated algorithms and computational resources for the integration of multiple sensing technologies. Additionally, a formidable challenge arises in guaranteeing the reliability and precision of multimodal sensor outputs in view of severe and dynamic industrial conditions. Electromagnetic interference, dust, fluctuating illumination conditions, and vibrations are all elements that can impair sensor performance and result in inaccurate interpretations. Furthermore, a significant obstacle to the extensive implementation of sophisticated multimodal sensing networks, especially among small and medium-sized businesses, is the substantial financial investment required for their deployment and upkeep.

The integration of AI (ML/DL) into multimodal sensing systems introduces substantial prospects as well as notable obstacles in the realm of industrial automation and Industry 4.0/5.0. A significant bottleneck is presented by the intricacy involved in instructing DL models to efficiently integrate data from various sensing modalities. The successful integration of various data streams, including auditory, visual, and tactile inputs necessitates the utilization of advanced neural network architectures that can comprehend and exploit intricate relationships across modalities. Moreover, a significant challenge that persists is the need to guarantee the resilience and consistency of AI-supported multimodal sensing algorithms when implemented in actual industrial settings. Dynamic work scenarios, unpredictable events, and fluctuating operating conditions necessitate AI models that are resilient and adaptable, with the capacity for ongoing learning and adjustment. In addition, a substantial obstacle arises from the requirement for extensive, varied, and annotated datasets to train AI models; this is especially true in industrial sectors where data acquisition may be difficult or expensive. In addition, it is critical to resolve privacy and security concerns that may arise from the implementation of AI-assisted multimodal sensing systems. This is necessary to safeguard sensitive industrial data and avert potential cyber threats. Addressing these obstacles necessitates the convergence of knowledge from various fields such as ML, sensor technology, industrial automation, and cybersecurity, to devise inventive resolutions that capitalize on the complete capabilities of AI to augment multimodal sensing capabilities within modern industrial settings.

Furthermore, the lack of universally accepted protocols and compatible platforms is a potential bottleneck for effective collaboration and communication between various sensor systems and components of industrial automation. To tackle these obstacles, it is imperative that policymakers, manufacturers, and researchers collaborate to create multimodal sensing solutions that are interoperable, durable, and economical, with a specific focus on the requirements of Industry 4.0. The complete potential of multimodal sensing can solely be actualized by employing groundbreaking developments in sensor technology, data analytics, and system integration. This will enable the exploration of new territories in industrial automation and intelligent production.

5.5 Summary

Multi-modal sensing plays a crucial role in the constantly evolving field of industrial automation, providing a key method to achieve higher levels of efficiency, safety, and adaptability. This chapter outlines the importance and challenges of multimodal sensing in industrial automation. The use of this emerging and promising technology in IIoT, advanced robotics, big data analytics, cloud computing, AI, and AR is fast becoming popular. This requires several types of sensors including RF sensors, vision sensors, localization and tracking sensors, infrared sensors, proximity sensors, IMU sensors, level sensors, temperature sensors, and pressure sensors. The chapter sheds light on the use of these sensors in industrial automation. Furthermore, it also delves into the implementation of multimodal sensors including the design of AI-assisted sensing and radar sensing networks. The chapter concludes by highlighting the challenges and opportunities associated with the use of multimodal sensing in industrial automation.

References

1 Hall, D.L. and Llinas, J. (1997). An introduction to multisensor data fusion. *Proceedings of the IEEE* 85 (1): 6–23.
2 Tang, Q., Liang, J., and Zhu, F. (2023). A comparative review on multi-modal sensors fusion based on deep learning. *Signal Processing* 109165.
3 Rahate, A., Mandaokar, S., Chandel, P. et al. (2023). Employing multimodal co-learning to evaluate the robustness of sensor fusion for industry 5.0 tasks. *Soft Computing* 27 (7): 4139–4155.
4 de Gea Fernández, J., Mronga, D., Günther, M. et al. (2017). Multimodal sensor-based whole-body control for human – robot collaboration in industrial settings. *Robotics and Autonomous Systems* 94: 102–119.
5 Athish, S.R., Rajesh, H., Vishvanath, N. et al. (2024). Multi-modal driver behavior analysis and speed estimation using fusion of computer vision and InCar sensor data. 9th *IEEE International Students' Conference on Electrical, Electronics and Computer Science (SCEECS)*, Bhopal, India (24–25 February 2024), pp. 1–5.

6 Papanastasiou, S., Kousi, N., Karagiannis, P. et al. (2019). Towards seamless human robot collaboration: integrating multimodal interaction. *The International Journal of Advanced Manufacturing Technology* 105: 3881–3897.

7 Yang, J., Liang, N., Pitts, B.J. et al. (2023). Multimodal sensing and computational intelligence for situation awareness classification in autonomous driving. *IEEE Transactions on Human-Machine Systems* 53 (2): 270–281.

8 Duan, S., Shi, Q., and Wu, J. (2022). Multimodal sensors and ML-based data fusion for advanced robots. *Advanced Intelligent Systems* 4 (12): 2200213.

9 Abate, A.F., Cimmino, L., Cuomo, I. et al. (2022). On the impact of multimodal and multisensor biometrics in smart factories. *IEEE Transactions on Industrial Informatics* 18 (12): 9092–9100.

10 Lutz, B., Kisskalt, D., Regulin, D. et al. (2021). Material identification for smart manufacturing systems: a review. *4th IEEE International Conference on Industrial Cyber-Physical Systems (ICPS)*, Virtual Conference (10–12 May 2021), pp. 353–360.

11 Kalsoom, T., Ramzan, N., Ahmed, S., and Ur Rehman, M. (2020). Advances in sensor technologies in the era of smart factory and industry 4.0. *Sensors* 20 (23): 6783.

12 Lu, C., Saifullah, A., Li, B. et al. (2015). Real-time wireless sensor-actuator networks for industrial cyber-physical systems. *Proceedings of the IEEE* 104 (5): 1013–1024.

13 Xu, H., Yu, W., Griffith, D., and Golmie, N. (2018). A survey on industrial Internet of Things: a cyber-physical systems perspective. *IEEE Access* 6: 78238–78259.

14 Berger, C., Hees, A., Braunreuther, S., and Reinhart, G. (2016). Characterization of cyber-physical sensor systems. *Procedia Cirp* 41: 638–643.

15 Brecher, C., Müller, A., Dassen, Y., and Storms, S. (2021). Automation technology as a key component of the Industry 4.0 production development path. *The International Journal of Advanced Manufacturing Technology* 117: 2287–2295.

16 Gupta, A.K. and Arora, S.K. (2011). *Industrial Automation and Robotics*. Laxmi Publications.

17 Kaur, M. (2024). Future of industrial automation with AI and cloud robotics. In: *Shaping the Future of Automation With Cloud-Enhanced Robotics* (ed. R.R. Gatti and C. Singh), 1–19. IGI Global.

18 Arvind, R.V., Raj, R.R., Raj, R.R., and Prakash, N.K. (2016). Industrial automation using wireless sensor networks. *Indian Journal of Science and Technology* 9: 1–8.

19 Bangemann, T., Karnouskos, S., Camp, R. et al. (2014). State of the art in industrial automation. *Industrial Cloud-Based Cyber-Physical Systems: The IMC-AESOP Approach* 23–47.

20 Xu, X., Lu, Y., Vogel-Heuser, B., and Wang, L. (2021). Industry 4.0 and Industry 5.0 – Inception, conception and perception. *Journal of Manufacturing Systems* 61: 530–535.

21 Skobelev, P.O. and Borovik, S.Y. (2017). On the way from Industry 4.0 to Industry 5.0: From digital manufacturing to digital society. *Industry 4.0* 2 (6): 307–311.

22 Grabowska, S., Saniuk, S., and Gajdzik, B. (2022). Industry 5.0: improving humanization and sustainability of Industry 4.0. *Scientometrics* 127 (6): 3117–3144.

23 Sisinni, E., Saifullah, A., Han, S. et al. (2018). Industrial internet of things: challenges, opportunities, and directions. *IEEE Transactions on Industrial Informatics* 14 (11): 4724–4734.

24 Gupta, D., de Albuquerque, V.H.C., Khanna, A., and Mehta, P.L. (2021). Smart sensors for Industrial Internet of Things. *Springer International Publishing* 10: 973–978.

25 Li, X., Li, D., Wan, J. et al. (2017). A review of industrial wireless networks in the context of industry 4.0. *Wireless Networks* 23 (1): 23–41.

26 Ambika, P. (2020). Machine learning and deep learning algorithms on the Industrial Internet of Things (IIoT). *Advances in Computers* 117 (1): 321–338.

27 Grau, A., Indri, M., Lo Bello, L., and Sauter, T. (2020). Robots in industry: the past, present, and future of a growing collaboration with humans. *IEEE Industrial Electronics Magazine* 15 (1): 50–61.

28 Li, W., Hu, Y., Zhou, Y., and Pham, D.T. (2023). Safe human – robot collaboration for industrial settings: a survey. *Journal of Intelligent Manufacturing* 1–27.

29 Brogårdh, T. (2007). Present and future robot control development – an industrial perspective. *Annual Reviews in Control* 31 (1): 69–79.

30 Kong, L., Zhang, D., He, Z. et al. (2016). Embracing big data with compressive sensing: a green approach in industrial wireless networks. *IEEE Communications Magazine* 54 (10): 53–59.

31 Reis, M.S. and Gins, G. (2017). Industrial process monitoring in the big data/industry 4.0 era: from detection, to diagnosis, to prognosis. *Processes* 5 (3): 35.

32 Ding, X., Tian, Y., and Yu, Y. (2015). A real-time big data gathering algorithm based on indoor wireless sensor networks for risk analysis of industrial operations. *IEEE Transactions on Industrial Informatics* 12 (3): 1232–1242.

33 Yan, J., Meng, Y., Lu, L., and Li, L. (2017). Industrial big data in an industry 4.0 environment: challenges, schemes, and applications for predictive maintenance. *IEEE Access* 5: 23484–23491.

34 Haghnegahdar, L., Joshi, S.S., and Dahotre, N.B. (2022). From IoT-based cloud manufacturing approach to intelligent additive manufacturing: industrial Internet of Things – an overview. *The International Journal of Advanced Manufacturing Technology* 1–18.

35 Sokolov, B., Ivanov, D., and Dolgui, A. (2020). *Scheduling in Industry 4.0 and Cloud Manufacturing*, vol. 289. Springer.

36 Pandey, N.K., Kumar, K., Saini, G., and Mishra, A.K. (2023). Security issues and challenges in cloud of things-based applications for industrial automation. *Annals of Operations Research* 1–20.

37 Jin, J., Yu, K., Kua, J. et al. (2023). Cloud-fog automation: vision, enabling technologies, and future research directions. *IEEE Transactions on Industrial Informatics* 20 (2): 1039–1054.

38 P. K. R. Maddikunta Pham, Q.V., Prabadevi, B. et al., "Industry 5.0: a survey on enabling technologies and potential applications," *Journal of Industrial Information Integration*, vol. 26, no. February, p. 100257, 2022, https://doi.org/10.1016/j.jii.2021.100257.

39 Ghosh, A.M. and Grolinger, K. (2020). Edge-cloud computing for internet of things data analytics: embedding intelligence in the edge with deep learning. *IEEE Transactions on Industrial Informatics* 17 (3): 2191–2200.

40 Annanperä, E., Jurmu, M., Kaivo-oja, J. et al. (2021). From Industry X To Industry 6.0. *Allied ICT Finland* 1–38. https://helda.helsinki.fi/bitstream/handle/10138/335672/Industry_X_White_Paper.pdf?sequence=1. (accessed on 21 October 2024).

41 Javaid, M., Haleem, A., Singh, R.P., and Suman, R. (2022). Artificial intelligence applications for industry 4.0: a literature-based study. *Journal of Industrial Integration and Management* 7 (11): 83–111.

42 Peres, R.S., Jia, X., Lee, J. et al. (2020). Industrial artificial intelligence in industry 4.0-systematic review, challenges and outlook. *IEEE Access* 8: 220121–220139.

43 Rao, T.V.N., Gaddam, A., Kurni, M., and Saritha, K. (2021). Reliance on artificial intelligence, machine learning and deep learning in the Era of industry 4.0. In: *Smart Healthcare System Design: Security and Privacy Aspects* (ed. S.K.H. Islam and D. Samanta), 281–299. Wiley.

44 Yao, X., Zhou, J., Zhang, J., and Boër, C.R. (2017). From intelligent manufacturing to smart manufacturing for industry 4.0 driven by next generation artificial intelligence and further on. *5th International Conference on Enterprise Systems (ES)*, Beijing, China (22–24 September 2017), pp. 311–318.

45 Azizi, A. (2019). *Applications of Artificial Intelligence Techniques in Industry 4.0.* Springer.

46 Dopico, M., Gómez, A., la Fuente, D. et al. (2016). A vision of industry 4.0 from an artificial intelligence point of view. *18th International Conference on Artificial intelligence (ICAI)*, Las Vegas, USA (25–28 July 2016).

47 Ribeiro, J., Lima, R., Eckhardt, T., and Paiva, S. (2021). Robotic process automation and artificial intelligence in Industry 4.0 – a literature review. *Procedia Computer Science* 181: 51–58.

48 Bécue, A., Praça, I., and Gama, J. (2021). Artificial intelligence, cyber-threats and Industry 4.0: challenges and opportunities. *Artificial Intelligence Review* 54 (5): 3849–3886.

49 B. Meindl, N. F. Ayala, J. Mendonça, and A. G. Frank, "The four smarts of Industry 4.0: evolution of ten years of research and future perspectives," *Technological Forecasting and Social Change*, vol. 168, no. March, p. 120784, 2021, https://doi.org/10.1016/j.techfore.2021.120784.

50 Yin, Y., Zheng, P., Li, C., and Wang, L. (2023). A state-of-the-art survey on augmented reality-assisted digital twin for futuristic human-centric industry transformation. *Robotics and Computer-Integrated Manufacturing* 81: 102515.

51 Voinea, G.-D., Gîrbacia, F., Duguleană, M. et al. (2023). Mapping the emergent trends in industrial augmented reality. *Electronics* 12 (7): 1719.

52 Tyagi, N., Chavan, S.S., Gangadharan, S.M.P. et al. (2024). Towards intelligent industrial systems: a comprehensive survey of sensor fusion techniques in IIoT. *Measurement: Sensors* 32: 100944.

53 Zand, P., Chatterjea, S., Das, K., and Havinga, P. (2012). Wireless industrial monitoring and control networks: the journey so far and the road ahead. *Journal of sensor and actuator networks* 1 (2): 123–152.

54 Meng, Z., Wu, Z., and Gray, J. (2018). RFID-based object-centric data management framework for smart manufacturing applications. *IEEE Internet of Things Journal* 6 (2): 2706–2716.

55 Martusewicz, J., Szewczyk, K., and Wierzbic, A. (2021). EFQM RADAR-based assessment of RFID system as part of Industry 4.0 implementation – a case study of a production plant. In: *Industry 4.0: A Global Perspective* (ed. J. Duda and A. Gąsior), 95–108. Routledge.

56 Xu, J., Li, Z., Zhang, K. et al. (2023). The principle, methods and recent progress in RFID positioning techniques: a review. *IEEE Journal of Radio Frequency Identification* 7: 50–63.

57 Pereira, E.S.S., Ordoñez, R.E.C., Beydoun, G., and Babar, A. (2024). Exploring the nexus of RFID and industry 4.0: bibliometric analysis to investigate the strategic themes and thematic evolution. *Journal of Industrial Engineering and Management* 17 (1): 1–34.

58 Meng, Z., Liu, Y., Gao, N. et al. (2021). Radio frequency identification and sensing: integration of wireless powering, sensing, and communication for IIoT innovations. *IEEE Communications Magazine* 59 (3): 38–44.

59 Wang, J., Chuang, J., Berweger, S. et al. (2023). Towards opportunistic radar sensing using millimeter-wave Wi-Fi. *IEEE Internet of Things Journal* 11 (1): 188–200.

60 Steinbaeck, J., Steger, C., Holweg, G., and Druml, N. (2017). Next generation radar sensors in automotive sensor fusion systems. *2017 Sensor Data Fusion: Trends, Solutions, Applications (SDF*, Bonn, Germany (10–12 October 2017), pp. 1–6.

61 Ramasubramanian, K. (2017). mmWave radar for automotive and industrial applications. *Texas Instruments*. https://www.ti.com/content/dam/videos/external-videos/en-us/2/3816841626001/5675916489001.mp4/subassets/Mmwave_webinar_Dec2017.pdf (accessed 22 October 2024).

62 Liu, Y., Shen, Y., Fan, L. et al. (2022). Parallel radars: from digital twins to digital intelligence for smart radar systems. *Sensors* 22 (24): 9930.

63 Saponara, S., Greco, M.S., and Gini, F. (2019). Radar-on-chip/in-package in autonomous driving vehicles and intelligent transport systems: opportunities and challenges. *IEEE Signal Processing Magazine* 36 (5): 71–84.

64 Baumann, D., Mager, F., Wetzker, U. et al. (2020). Wireless control for smart manufacturing: recent approaches and open challenges. *Proceedings of the IEEE* 109 (4): 441–467.

65 Roriz, R., Cabral, J., and Gomes, T. (2021). Automotive LiDAR technology: a survey. *IEEE Transactions on Intelligent Transportation Systems* 23 (7): 6282–6297.

66 Cheng, Q. (2021). Application of computer vision technology in industrial automation. *Journal of Physics: Conference Series* 2037: 12015.

67 Kumar, K., Kumar, P., Kshirsagar, V. et al. (2023). A real-time object counting and collecting device for industrial automation process using machine vision. *IEEE Sensors Journal* 23: 13052–13059.

68 Ruan, H., Dorneanu, B., Arellano-Garcia, H. et al. (2022). Deep learning-based fault prediction in wireless sensor network embedded cyber-physical systems for industrial processes. *IEEE Access* 10: 10867–10879.

69 Lesi, V., Jakovljevic, Z., and Pajic, M. (2021). Security analysis for distributed IoT-based industrial automation. *IEEE Transactions on Automation Science and Engineering* 19 (4): 3093–3108.

70 Varga, P., Plosz, S., Soos, G., and Hegedus, C. (2017). Security threats and issues in automation IoT. *2017 IEEE 13th International Workshop on Factory Communication Systems (WFCS)*, Trondheim, Norway (31 May–2 June 2017), pp. 1–6.

71 Rameez, M., Pettersson, M.I., and Dahl, M. (2022). Interference compression and mitigation for automotive FMCW radar systems. *IEEE Sensors Journal* 22 (20): 19739–19749.

6

Hybrid Neuromorphic-Federated Learning for Activity Recognition Using Multi-modal Wearable Sensors

Ahsan Raza Khan, Habib Ullah Manzoor, Fahad Ayaz, Muhammad Ali Imran, and Ahmed Zoha

James Watt School of Engineering, University of Glasgow, Glasgow, UK

Human activity recognition (HAR) is crucial in developing real-time systems for patient care and lifestyle management. However, traditional deep learning methods used for HAR with multi-modal data face several challenges, such as data heterogeneity, privacy issues, high communication costs, and the need for efficient on-device processing. To address these challenges, a new federated learning (FL) framework has been proposed, which combines the computational efficiencies of spiking neural networks (SNNs) with the sequential data-processing strengths of long short-term memory (LSTM) to create a hybrid model called spiking-LSTM (S-LSTM). The proposed S-LSTM model employs surrogate gradient learning alongside backpropagation through time to facilitate end-to-end learning for multi-modal time series data. It combines the event-driven operational advantage of SNNs with the sequence modeling proficiency of LSTMs. The effectiveness of the proposed S-LSTM scheme is evaluated using two publicly available datasets. The results show that the S-LSTM model outperformed conventional LSTM, CNN, and S-CNN models in terms of accuracy and energy efficiency. For instance, the S-LSTM model achieved an accuracy of 97.36% for indoor environments and 89.69% for outdoor settings. Furthermore, the results showed a significant 32.30% enhancement in energy efficiency compared to traditional LSTM models. Moreover, this study highlights the importance of personalization in HAR. It shows that fine-tuning the model using local data can increase accuracy by up to 9% for individual users. This finding advances the field toward more personalized, efficient, and privacy-conscious monitoring solutions in healthcare applications.

Multimodal Intelligent Sensing in Modern Applications, First Edition.
Edited by Masood Ur Rehman, Ahmed Zoha, Muhammad Ali Jamshed, and Naeem Ramzan.

Published 2025 by John Wiley & Sons, Inc.

6.1 Multi-modal Human Activity Recognition

Wearable sensing technology has enabled significant advancements in human activity recognition (HAR), which has implications for individual well-being across various domains such as healthcare, smart homes, public safety, and athletic performance [1]. The emergence of wearable devices like smartphones and smartwatches has revolutionized HAR, providing a wealth of contextual data that can be used in applications ranging from remote patient monitoring to lifestyle management. These advancements are particularly promising in elderly care, where continuous monitoring and timely interventions can help foster independence and minimize risks [2]. Additionally, wearable sensors help track outdoor activities, providing insights for personal health monitoring, environmental exposure assessment, and improving the performance and safety of athletes. They offer comprehensive lifestyle insights for users, contributing to a holistic understanding of human well-being and performance in diverse environments [3].

The advancement of data acquisition technologies has brought HAR to a crossroads, where the selection of methodologies like vision-based or sensor-based significantly impacts the system's effectiveness and practicality. Vision-based methods rely on high-resolution cameras and advanced computer vision techniques. However, they encounter concerns regarding privacy and environmental limitations like lighting conditions and camera resolution [4]. In contrast, sensor-based approaches offer solutions that can be wearable or non-wearable. Non-wearable sensors, especially those employing radio frequencies (RFs) such as channel state information (CSI) or received signal strength indicator (RSSI), are gaining popularity for monitoring indoor HAR owing to their unobtrusive and privacy-preserving attributes [5, 6]. Meanwhile, sensor-based techniques that fuse data from wearable devices like accelerometers and gyroscopes present a viable alternative. Wearable sensors directly capture movement data and provide unparalleled precision in activity detection, rendering them indispensable for dynamic, real-world environments where activities can range from simple to complex and vary in intensity [3, 7]. Navigating the choice among different data acquisition methods requires a nuanced consideration of their strengths and limitations, focusing on achieving an optimal balance between accuracy, privacy, and practicality for the intended use case.

Despite the potential of high-volume multi-modal sensor data, the traditional HAR systems predominantly rely on a centralized architecture, as shown in Figure 6.1, where data from various sensors, collected from numerous participants, is collected and processed on a central server or cloud infrastructure for analysis [8]. This centralized approach poses significant challenges, such as scalability issues, privacy risks, and increased computational demands. These challenges are becoming even more complex due to the increasing quantity and

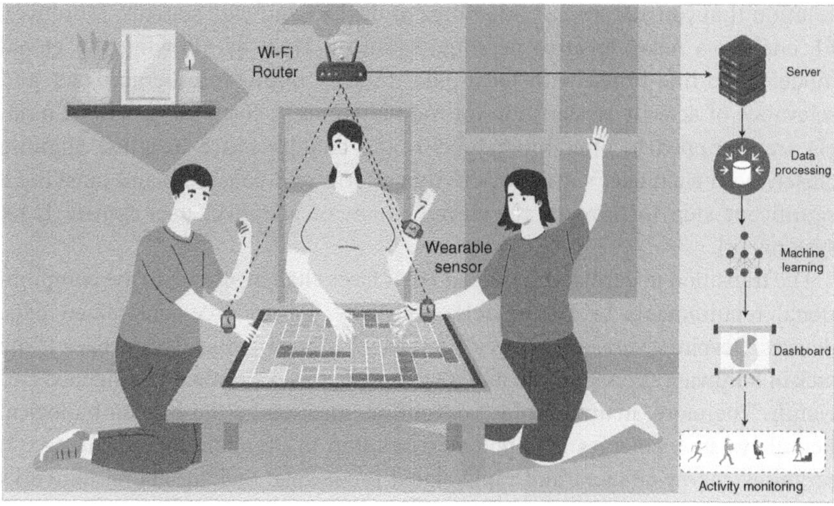

Figure 6.1 Conceptual framework of centralized indoor HAR using wearable sensors. *Source:* gstudioimagen1/Freepik.

variety of data sources. Additionally, the advent of deep learning (DL) technologies has significantly advanced the field of HAR, enabling the automated extraction of complex patterns from raw sensor data, thus eliminating the need for manual feature engineering. Techniques such as convolutional neural networks (CNNs) [9] and recurrent neural networks (RNNs) [10, 11] have demonstrated superior performance in HAR classification over traditional machine learning methods. However, the deployment of DL in HAR is hindered by various factors, including the lack of annotated data necessary for model training and manual data annotation processes that require extensive domain expertise [4]. Moreover, the centralized data processing model incurs significant communication and storage costs and can lead to considerable latency in real-time activity recognition tasks, further exacerbating privacy concerns in an era where data protection regulations are increasingly stringent [12, 13].

Therefore, a new distributed learning approach, federated learning (FL), has been introduced to address the critical challenges centralized learning faces. The core idea behind FL is to enable collaborative learning, where multiple participants can train a shared model without sharing data. This approach significantly reduces the communication overhead associated with transferring raw data [14]. Additionally, FL mitigates privacy and scalability issues and enhances HAR systems' computational efficiency by leveraging decentralized datasets. The FL framework inherently supports real-time processing by distributing computational tasks across various devices, offering a scalable solution that can adapt to a wide range of devices and data sources. Moreover, FL enables a novel level of personalization in HAR by allowing the global model to be fine-tuned with local data, thereby improving the accuracy and relevance of activity recognition for individual users. Such personalized models are equipped to offer more precise and context-aware insights, aligning closely with each user's unique activity patterns and behaviors and marking a significant step forward in the development of adaptive, user-centric HAR systems [8].

The transition toward decentralized HAR faces challenges due to the computational requirements of DL models, especially for resource-constrained edge devices. Despite several advantages offered by FL, one significant drawback is the lack of hardware support in the market to handle this distributed intelligence efficiently. Therefore, neuromorphic computing, inspired by the human biological neural systems, emerges as a potential solution to overcome this challenge. It promises energy-efficient and rapid signal processing. Spiking neural networks (SNNs), a subset of neuromorphic learning, are gaining attention because of their unique event-driven processing of binary inputs, known as "spikes" [15]. Unlike traditional DL models, SNNs operate on a temporal, event-driven paradigm, making them particularly suitable for on-device learning. The real-time and continuous nature of activity data in HAR accentuates the potential advantages of neuromorphic computing, highlighting the necessity for models that can adeptly capture the temporal dynamics of human activities. Although SNNs are computationally efficient, the traditional DL models, such as long short-term memory (LSTM) networks, are more effective in processing sequential data.

This chapter introduces a novel hybrid spiking-LSTM (S-LSTM) model offering a scalable, privacy-preserving and reliable solution for HAR. The proposed hybrid neuromorphic federated learning (HNFL) framework is specifically designed for time series HAR data. It combines the strengths of SNNs and LSTM and multi-modal data fusion to improve activity recognition accuracy while ensuring privacy and reducing computational demands [16, 17]. This study rigorously evaluates the S-LSTM model using two publicly available datasets covering various environmental settings and activity scenarios. The study compares the performance of the S-LSTM model with that of traditional LSTM, spiking CNN (S-CNN), and simple CNN models. The results highlight the strength and robustness of the proposed HNFL framework and demonstrate how multi-modal data fusion can significantly improve HAR accuracy within a FL paradigm. Furthermore, this research analyses how randomly selecting clients affects the performance of the HAR model and provides valuable insights into finding the best balance between computational and communication efficiency in relation to the accuracy of the HAR model. The findings offer a strategic framework for selecting clients, which can help develop more effective and efficient FL implementations in HAR systems.

6.2 Machine Learning Methods in Multi-modal Human Activity Recognition

In recent years, HAR has made significant progress, especially with the popularity of wearable sensors. Initially, HAR systems relied on traditional machine-learning techniques, but now they have shifted toward more advanced DL models. The use of DL enables the extraction of complex patterns directly from raw sensor data, eliminating the need for manual feature engineering and resulting in improved accuracy and reliability. However, modifying these models to ensure user privacy and maintain computational efficiency in a decentralized and privacy-focused digital landscape is essential. Therefore, this section provides an overview of the latest developments in (i) a centralized learning-based HAR system and (ii) a FL-based HAR.

6.2.1 Centralized Learning-based HAR Systems

There is a shift from traditional machine learning techniques to more sophisticated DL models in centralized HAR systems. This shift is driven by the increased availability and capabilities of wearable sensing technologies. For instance, the study in [18], introduced a novel heterogeneous convolution operation tailored for HAR, showcasing how the division of convolutional filters can enhance feature diversity and improve performance across multiple HAR datasets. Similarly, the authors of [19] leveraged LSTM networks with neural structured learning to achieve remarkable recall rates of 99% on publicly available datasets, far surpassing those of conventional CNN and RNN methods, having a recall rate of approximately 94%. Moreover, the study in [20] employed a hybrid approach utilizing a bi-convolutional recurrent neural network (Bi-CRNN) enhanced by an auto-fusion technique for multi-sensor data processing, achieving significant accuracy improvements [20]. Despite its computational demands, the Bi-CRNN achieved a classification accuracy of 94.7%. The quest for efficiency has also led to the development of fast and robust DL models like fast and robust deep convolutional neural network (FR-DCNN), which optimizes training speed without compromising the accuracy of activity recognition [21]. The FR-DCNN model achieved a prediction accuracy of 95.27%, which remained impressive even with compressed datasets at 94.18%. Furthermore, the merging-squeeze-excitation (MSE) technique presents another novel approach to HAR, focusing on feature recalibration during data fusion to refine the model's adaptability and precision [22]. When tested with DL models such as LeNet5, AlexNet, and VGG16 for feature extraction, the proposed MSE method consistently achieved high accuracy across three different datasets.

The use of data fusion methods has made significant progress in the field of fall detection to improve detection accuracy. The literature broadly categorizes the discussion on data fusion into three main levels: input, feature, and decision. The challenges inherent in heterogeneous data sources have often restricted the input-level fusion in HAR systems to single-modal data. [13]. However, various studies have been conducted on data fusion to improve the reliability of HAR using wearable sensors. For instance, the authors of [23] have explored innovative fusion methods, such as combining 2D contour projections to reconstruct 3D volumes or integrating human skeletal sequences from RGB images to improve detection accuracy. This method uses a threshold-based mechanism to detect falls, but it solely relies on visual data, which limits its potential compared to traditional 2D imaging methods. In contrast, the authors in [24] proposed categorizing fall incidents in the UP-Fall dataset by merging skeletal sequences and facial points from RGB imagery. This method, which is simple but effective, only focuses on input-level fusion and does not involve multimodal data fusion.

On the other hand, the feature level analysis has been divided into two categories – manually extracted features and those derived from NN models. The authors in [3] studied the fusion process that combined acceleration data from pose sensors with skeleton sequences from video data. This study used the gradient-boosting decision trees (GBDT) algorithm to classify fall movements and demonstrated its effectiveness over traditional support vector machine (SVM) and NN approaches. Other studies have also shown the fusion of manually selected features, which were subsequently classified using bidirectional long short-term memory (Bi-LSTM) networks or CNNs [25, 26]. Although DL and fusion techniques in centralized HAR systems improve accuracy and adaptability, very limited studies are available which consider the privacy and energy efficiency of the system. Therefore, neuromorphic computing offers energy-efficient processing of HAR data with improved performance and user privacy. Combining these techniques can lead to more efficient and effective health monitoring solutions.

6.2.2 Federated Learning-based HAR Systems

The HAR system has shifted toward a decentralized model training approach enabled by FL in response to privacy concerns and the need for data decentralization. FL allows for model training to be carried out directly on edge devices (i.e., participants) without transferring raw data to centralized servers. This approach safeguards user privacy and significantly reduces communication overhead [13, 4]. Recent explorations into FL-based HAR systems reveal a concerted effort to leverage multi-modal data fusion for enhancing detection accuracy, particularly in fall detection scenarios [13]. The study proposed a new technique that transforms time-series sensor data into images to detect anomalies. This approach also leverages visual data from cameras and input-level data fusion within FL frameworks to achieve a classification accuracy of 89.76%.

The challenge of nonindependently and identically distributed (non-IID) data across devices remains a pivotal concern in real-world FL deployments, potentially compromising the generalizability and performance of HAR systems. In response, novel frameworks such as ProtoHAR [27] and ClusterFL [28] have been developed, focusing on mitigating biases and enhancing model performance through strategic data handling and model training techniques. These frameworks ensure privacy and aim to optimize model accuracy and convergence speed, offering robust solutions to the non-IID dilemma in federated settings. Similarly, novel FL via augmented knowledge distillation (FedAKD) was designed for the collaborative training of heterogeneous DL models [29]. FedAKD exhibited superior communication efficiency compared to the Federated Averaging (FedAvg) algorithm, with a 200-fold increase. It also achieved 20% higher accuracy than other knowledge distillation-based FL methods.

The increasing demand for on-device processing capabilities, coupled with the computational intensity of conventional DL algorithms, necessitates the exploration of energy-efficient computing paradigms. Neuromorphic computing, mainly through the utilization of SNNs, emerges as a promising avenue, demonstrating the potential to augment-federated environments with improved energy efficiency and processing accuracy [16, 30, 31]. These advancements highlight the synergistic potential of SNNs and FL in addressing the computational challenges of HAR systems.

Although HAR has evolved dramatically, transitioning from centralized learning to more privacy-aware FL architecture. This evolution reflects a growing emphasis on improving communication efficiency and bolstering privacy in FL paradigms. Notably, while the exploration of energy efficiency in SNNs has been a focal point in contemporary research, the integration of DL and SNNs for HAR, especially in a manner that capitalizes on data fusion from multiple sensor modalities, remains under-explored. The increasing trend toward the miniaturization of HAR devices underscores the critical need for models that support on-device processing and optimize energy efficiency without compromising accuracy. To address this gap, we advocate for a novel hybrid model, the S-LSTM, that synergistically combines the energy-efficient processing attributes of SNNs with the adept sequential data handling capabilities of LSTMs. This model is designed to leverage multi-modal data fusion to achieve an optimal balance between energy efficiency and accuracy in HAR applications, thereby facilitating a more nuanced and comprehensive understanding of human activities through wearable sensors.

6.3 System Model

This section provides a detailed discussion of the fundamental principles of FL and SNNs, along with the hybrid S-LSTM model used for HAR in distributed settings.

6.3.1 Federated Learning Framework

FL emerges as a paradigm facilitating distributed machine learning, allowing for model training across an extensive network of decentralized nodes, as shown in Figure 6.2. Each participant or node holds only a subset of the dataset, ensuring privacy. The learning process is collaborative, with a central federated server (FS) coordinating the training process and aggregating the model parameters from all participants (N) to refine the global model. The training continues until the global model reaches a specified level of accuracy or a predetermined number of iterations.

Without losing the generality, for each participant $i = 1, \ldots, N$, a localized dataset is denoted as $|D^{(i)}| \equiv D$, with $D^{(i)}$ representing the subset of the dataset at the ith device, and the cumulative dataset expressed as $D = \sum_{i=1}^{N} |D^{(i)}|$. In this schema, given the model parameter w and a local loss function $p(w,x)$ applicable to any data sample x, the local empirical loss at the ith participant is reformulated as:

$$p^{(i)}(w) = \frac{1}{|D^{(i)}|} \sum_{x \in D^{(i)}} p(w,x) \tag{6.1}$$

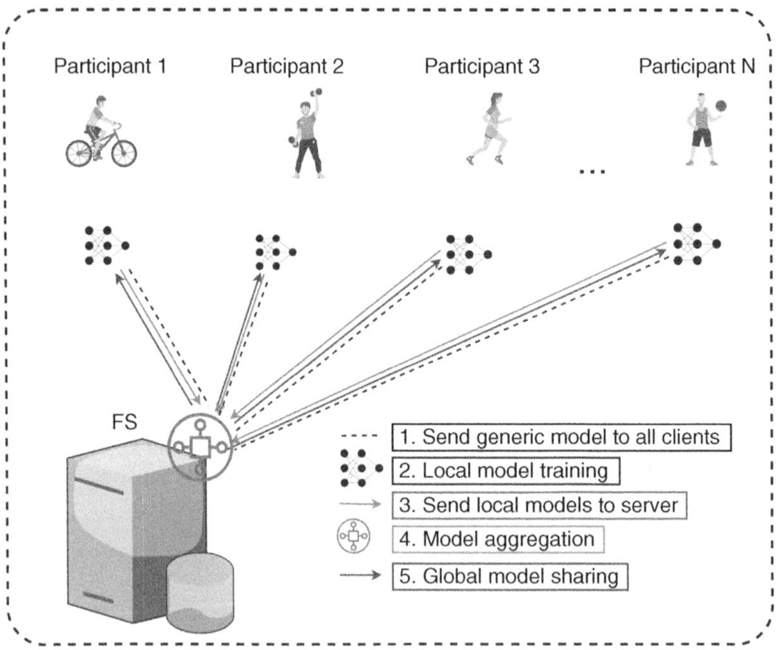

Figure 6.2 Conceptual FL framework for HAR using wearable sensing in the outdoors.

where $p^{(i)}(w)$ signifies the participant-specific loss function, predicated on the parameter vector w. The core job of FL is to optimize a global loss function P on the FS, which is mathematically defined as:

$$\min_{w} P(w) := \frac{1}{\sum_{i=1}^{N} |D^{(i)}|} \sum_{i=1}^{N} |D^{(i)}| p^{(i)}(w) \tag{6.2}$$

where $L(w)$ represents the global loss function based on the aggregate dataset D. FL is an iterative process that necessitates each participant to compute the local gradient at each time iteration $t = 1, \ldots, T$, following the equation:

$$w^{(i)}(t) = \tilde{w}^{(i)}(t-1) - \eta \nabla p\left(\tilde{w}^{(i)}(t-1), x^{(i)}(t)\right) \tag{6.3}$$

where $w^{(i)}(t)$ denoting the model parameters for the ith participant, η as the learning rate, $w^{(i)}(t-1)$ as the model parameters from the previous iteration, and $\nabla p(\tilde{w}^{(i)}(t-1), x^{(i)}$ as the gradient of the loss function p with respect to the model parameters for the data point $x^{(i)}(t)$. Upon receiving all local updates, the FS conducts model aggregation, expressed as:

$$w(t) = \frac{1}{\sum_{i=1}^{N} |D^{(i)}|} \sum_{i=1}^{N} |D^{(i)}| w^{(i)}(t) \tag{6.4}$$

6.3.2 Spiking Neural Network

SNNs are inspired by biological neural networks that use discrete events "spike" for information processing, as shown in Figure 6.3. Unlike ANNs, which process information in a continuous manner, SNNs utilize discrete events to encode and transmit information, embodying a more energy efficient [16]. The operation of SNNs relies on the idea of event-driven computation. This means that the neurons in the network stay inactive until incoming spikes trigger them. This mechanism makes the computational model highly efficient because only a few neurons are active at any given time, achieving sparsity [32]. Additionally, the binary nature of spike signals (0s and 1s) facilitates processing with low-precision arithmetic, further reducing the computational load. These characteristics collectively endow SNNs with the capacity for high-efficiency processing, particularly on resource-constrained edge devices, a feature that sets them apart from conventional DL models [16, 32].

In every spiking neuron, the membrane potential accumulates spike signals received from preceding neurons. This potential varies over time t, potentially increasing, decreasing, or remaining unchanged based on the incoming spikes. When the membrane potential surpasses a predefined threshold v_{th}, it triggers the

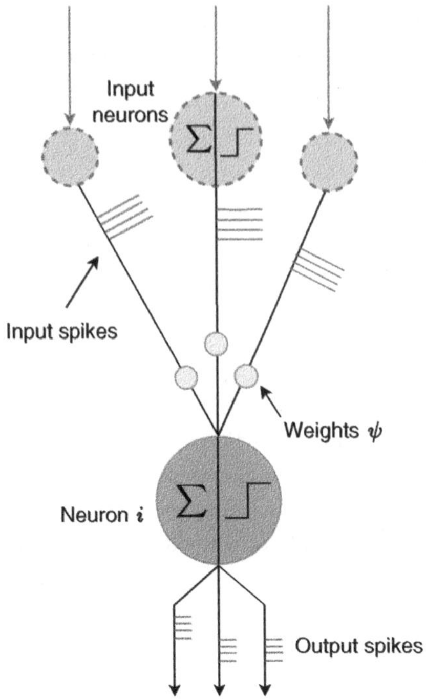

Figure 6.3 Spiking neurons propagation process.

neuron to emit a spike signal, which is then propagated to subsequent layers in the network. Following spike generation, the neuron undergoes a refractory period during which its membrane potential is temporarily invariant to further spikes, ensuring a period of inactivity post-firing. This dynamic spike generation and transmission process is modeled using the leaky integrate-and-fire (LIF) model. The LIF model is celebrated for its simplicity and effectiveness in capturing the essential characteristics of neuronal spiking behavior, making it a standard framework for simulating SNNs [33]. This model is closely analogous to an electrical circuit comprising a capacitor Q, a resistor Z, a power source V, and an input current J. The LIF model in layer l for a neuron with index number i can be mathematically represented as follows [34]:

$$\tau_q \frac{dV_i^{(l)}(t)}{dt} = -\left(V_i^{(l)}(t) - V_{res}\right) + Z J_i(t) \tag{6.5}$$

where $\tau_q = Q \cdot Z$ denotes the membrane potential time constant decay, $V_i^{(l)}(t)$ is the neuron's membrane potential at time t, Z is the resistance, and $J_i(t)$ represents the input current at time t, which denotes the pre-synaptic inputs. When $V_i^{(l)}(t)$ exceeds the given threshold v_{th}, the neuron generates a spike and resets its membrane voltage to V_{res}. Hence, for a specific layer l and neuron index i, the membrane potential $V_i^{(l)}$ at time step t can be expressed as [34, 35].

$$V_i^{(l)}(t) = \beta V_i^{(l)}(t-1) + \sum_j w_{ij} o_j^{(l-1)}(t), \tag{6.6}$$

where β $(0 < \beta < 1)$ is the leakage factor, j represents the number of neurons in the previous layer $l-1$, w_{ij} is the weight from neuron j in layer $l-1$ to neuron i in layer l, and $o_j^{(l-1)}(t)$ is the binary output of neuron j in layer $l-1$ at time t. The spike sequence emitted by neuron i in layer l at time t, when $V_i^{(l)}(t)$ is greater than the firing threshold v_{th}, can be defined as [34]

$$o_i^{(l)}(t) = \begin{cases} 1, & \text{if } V_i^{(l)}(t) \geq v_{th} \\ 0, & \text{otherwise} \end{cases} \tag{6.7}$$

The inherent challenge in training SNNs lies in the non-differentiability of the spike function, a hurdle for traditional gradient descent methods, like backpropagation, which rely on continuous and differentiable activation functions [36]. The main problem with the step function is that its gradient is either zero or undefined, which makes it impossible to update weights during training. To solve this issue, surrogate gradient methods are used. These methods employ a smooth, differentiable approximation of the step function during the backward pass, which makes it possible to compute the gradient. This allows the network to be trained using traditional techniques while still using the unique spiking behavior of SNNs during inference. The surrogate piecewise linear function, aligning with the previously established notation, is mathematically represented as [16, 34]:

$$\frac{\partial o_i^{(l)}(t)}{\partial V_i^{(l)}(t)} = \gamma \max\left(0, 1 - \frac{|V_i^{(l)}(t) - v_{th}|}{v_{th}}\right) \tag{6.8}$$

where γ is a scaling factor controlling the SNNs' update magnitude and v_{th} is the threshold voltage for spiking. The backpropagation method in SNNs mirrors that of ANNs, except for using a surrogate function to approximate the non-differentiable threshold function. Hence, the weight update rule on the local participant for layer l at a given time is mathematically represented as [35]

$$w_{ij}^{(l)}(t) = w_{ij}^{(l)}(t-1) - \eta \frac{\partial p(t)}{\partial w_{ij}^{(l)}(t-1)} \tag{6.9}$$

where $w_{ij}^{(l)}(t-1)$ represent the model parameters at $t-1$ for neuron j in layer l to neuron i. Numerous strategies have been developed to effectively utilize the capabilities of SNNs and overcome their inherent challenges, especially the non-differentiability of spike operations. Among these strategies, spike-timing-dependent plasticity (STDP) is particularly useful for unsupervised learning, as it leverages the temporal dynamics of spikes and is well-suited for handling unlabeled data [37]. On the other hand, supervised learning scenarios benefit from the backpropagation through time (BPTT) technique, which modifies traditional backpropagation to allow for error correction over sequential time steps [16], thus facilitating the analysis of spatiotemporal data. Additionally, surrogate gradient learning introduces a means to approximate the non-differentiable spike function with a continuous, differentiable surrogate, enabling the use of gradient-based optimization techniques [16]. Parallelly, reward-modulated STDP integrates principles from reinforcement learning, applying rewards to modulate STDP in a feedback-driven learning process [31]. Our investigation focuses on the supervised training of SNNs with diverse datasets. This has led to developing a hybrid model that combines BPTT with surrogate gradient methods, as given in [38]. This approach addresses two pivotal challenges: it overcomes the spike function's non-differentiability, a significant obstacle for conventional gradient-based optimization, and it equips the model to learn temporal patterns in HAR data proficiently. By extending BPTT to facilitate backward error propagation through time, the model optimizes a loss function p, thereby improving its predictive accuracy. The formulation for the gradient of the loss p with respect to synaptic weights w_{ij} and the neuron's membrane potential is given as:

$$\frac{\partial p(t)}{\partial w_{ij}} = \sum_{t} \frac{\partial p(t)}{\partial V_i^{(l)}(t)} \cdot o_j^{(l-1)}(t) \tag{6.10}$$

$$\frac{\partial p(t)}{\partial V_i^{(l)}(t)} = \frac{\partial p(t)}{\partial o_i^{(l)}(t)} \cdot \frac{\partial o_i^{(l)}(t)}{\partial V_i^{(l)}(t)} + \beta \frac{\partial L(t)}{\partial V_i^{(l)}(t+1)} \tag{6.11}$$

where $o_j^{(l-1)}(t)$ signifies the output from neuron j in the preceding layer $l-1$ at time t, influencing the input to neuron i in layer l. This formulation underscores our system's ability to navigate the complexities inherent in training SNNs, enabling detailed learning of temporal dependencies critical for HAR applications.

6.3.3 Proposed S-LSTM Model

Our proposed S-LSTM model seamlessly combines LSTM units with the spiking behavior of LIF neurons, as shown in Figure 6.4. Initially, the framework employs an LSTM layer composed of 100 neurons dedicated to analyzing the input data. This particular layer is designed to return sequences, capturing the essential temporal correlations present within the data. Subsequently, the spiking layer with LIF replaces the conventional activation functions. The spiking

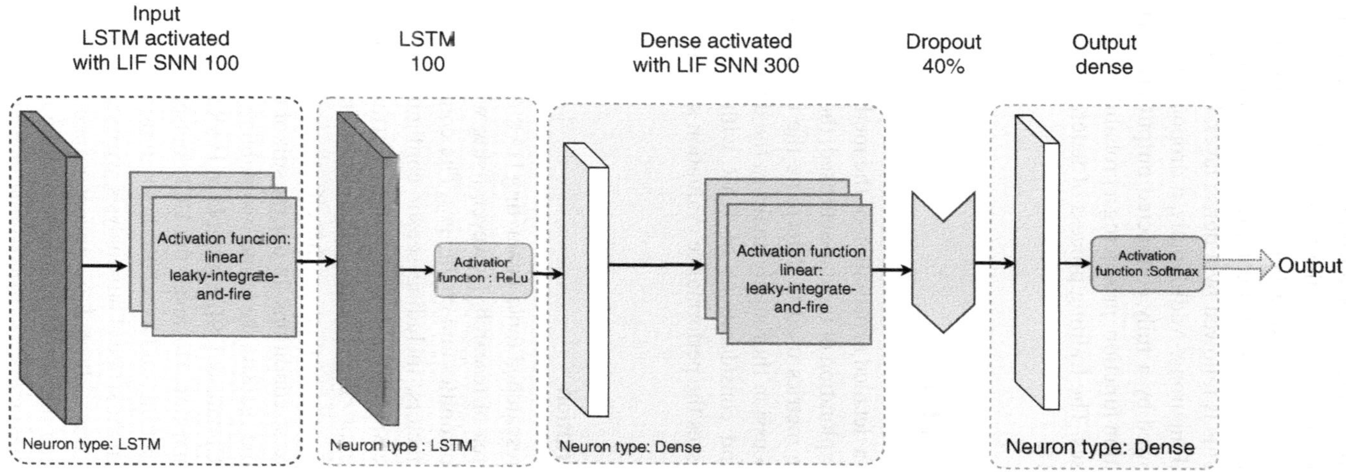

Figure 6.4 Proposed hybrid S-LSTM model where input LSTM layer activated by LIF.

layer has a trainable threshold that determines neuron firing and uses a surrogate gradient to approximate the gradient during backpropagation due to the non-differentiable nature of spiking behavior. Following the spiking neural layer, an additional LSTM layer, also comprising 100 neurons, further refines the data sequences. This is followed by a dense layer, integrating 300 neurons, which also adopts LIF neurons. Additionally, a dropout layer is added to mitigate overfitting, followed by a fully connected output layer. The output layer uses a SoftMax activation function, producing a probability distribution over the possible activity classes. The training process of federated S-LSTM for HAR is given in Algorithm 6.1.

6.4 Simulation Setup

This section provides a detailed explanation of the methodologies we employed, offering an in-depth exploration of the datasets used, the criteria used to evaluate performance, and the metrics used to determine the effectiveness of our proposed model. Our discussion thoroughly analyses the datasets, highlighting the unique challenges and considerations of HAR with wearable sensor data. Furthermore, it discusses the performance evaluation strategy and metrics used in the study.

6.4.1 Dataset Description

Despite being a well-researched topic, evaluating HAR using smartphone data is a recent and active area of research. However, using wearable sensing data for HAR offers both opportunities and challenges. The complexity of such datasets arises from various factors, including sensor configurations, sampling rates, accessibility, and diversity of the collected data. Furthermore, distinct activity patterns among different classes lead to significant class imbalances, making HAR an ideal domain for evaluating the effectiveness of neuromorphic FL approaches under diverse conditions. In our pursuit of datasets that offer reproducibility, diversity, and realism, we found two publicly accessible datasets. The first, known as the UCI dataset [39], is a staple in HAR research due to its widespread use in benchmarking. However, the UCI dataset is collected in a highly controlled laboratory setting, and its relatively small sample size presents certain limitations. To broaden the scope of our analysis, we incorporated the Real-World dataset [40], which was collected in unconstrained outdoor environments. Further details on each dataset, including its unique characteristics and challenges, are discussed below.

Algorithm 6.1 Federated S-LSTM Training with Surrogate Gradient and BPTT

Input: Initial model parameters w_0, clients $i = 1, 2, \ldots, N$
Output: Trained model parameters w
Procedure: Initialization
for each client i **in parallel do**
$D^{(i)} \leftarrow$ local dataset
$w^{(i)} \leftarrow w_0$ {Initialize local model}
end for
Procedure: FL training
for round $r = 1, 2, \ldots, R$ **do**
for each client $i = 1, 2, \ldots, N$ **in parallel do**
$\Delta w^{(i)} \leftarrow$ LocalTraining($w^{(i)}$, $D^{(i)}$)
end for
$w_{r+1} \leftarrow$ ServerUpdate$\left(\{\Delta w^{(i)}\}_{i=1}^{N}\right)$ {Aggregate updates}
broadcast w_{r+1} to clients
end for
Procedure: LocalTraining(w, D)
Initialize local parameters $w^{(i)}$, learning rate η
for each time step $t = 1, \ldots, T$ **do**
Compute local gradient using surrogate gradient and BPTT {Based on Equations (6.8)-(6.11)}
Update local model parameters $w^{(i)}(t)$
end for
return model update Δw
Procedure: ServerUpdate$\left(\{\Delta w^{(i)}\}_{i=1}^{N}\right)$
$w \leftarrow$ **aggregate**$\left(\{\Delta w^{(i)}\}\right)$ Based on Equation
return updated model parameters w

6.4.1.1 UCI Dataset

The UCI dataset was collected using Samsung Galaxy S II smartphones worn by a diverse group of 30 participants of various ages and genders. These individuals engaged in six everyday activities, including walking, walking upstairs, walking downstairs, sitting, standing, and lying down. The activities were performed under varied conditions, with the smartphones positioned on the left wrist and at a location of the participant's choosing to simulate real-world usage scenarios. The embedded sensors in the smartphones, specifically the accelerometers and gyroscopes, were instrumental in capturing data on triaxial linear acceleration and angular velocity, achieving a sampling rate of 50 Hz. This dataset was extensively processed to remove noise and enhance signal quality using advanced filtering techniques [39]. A total of 17 distinct features were meticulously extracted from each signal, covering a broad spectrum of time and frequency domain characteristics, such as signal magnitude, jerk, and the application of the fast Fourier transform (FFT).

The collected signals were segmented into discrete windows lasting 2.56 seconds each to facilitate a detailed analysis, with a 50% overlap between consecutive windows [39]. This segmentation process yielded 561 unique features for each window, drawn from various statistical and frequency-domain analyses. The comprehensive dataset encompasses over 10 299 instances, thoughtfully divided into training (70%) and testing (30%) subsets to support robust model evaluation. However, the dataset was intentionally partitioned into five distinct subsets to simulate localized datasets for individual participants, mirroring real-world FL scenarios where data distribution is inherently uneven. Each participant's data was further divided, allocating 80% for training and 20% for testing purposes, with the testing portions aggregated to create a global test set. This organized approach emphasizes the adaptability of the dataset in FL research and demonstrates the meticulous technique used for data preprocessing and merging sensor data. The preprocessing methods, such as noise reduction and segregating sensor signals into significant components, highlight the dataset's usefulness in capturing and analyzing intricate human activities through multi-sensor integration.

6.4.1.2 Real-World Dataset

While the UCI dataset is widely used in HAR research, it has limitations because it was collected in a controlled laboratory setting. Additionally, due to the small sample size, the true potential of FL could not be explored. Therefore, we opted for a more realistic dataset collected by Sztyler and Stuckenschmidt [40]. This dataset comprises data from 15 individuals (8 males and 7 females) embodying a broad spectrum of daily activities such as climbing stairs, jumping, lying, standing, sitting, running, and walking. Distinctively, this dataset captures accelerometer, GPS, gyroscope, light, magnetic field, and sound levels. We used the accelerometer and gyroscope data in which sensors were placed on seven strategic body locations: the chest, forearm, head, shin, thigh, upper arm, and waist – thereby providing a comprehensive view of bodily movements during various activities.

The data collection utilized standard smartphones and a smartwatch affixed to these body positions, recording at a frequency of 50 Hz. The devices used were synchronized using network time services to guarantee accuracy in time-stamping the collected data [40]. To mirror real-life usage scenarios as closely as possible, the devices were attached to the body using everyday items such as sports armbands and pockets. This method ensured comfort and realism, simulating how wearable devices are typically worn during daily activities. The data collection spanned various real-world locations, including urban and natural settings, to capture a diverse range of movement patterns and environmental influences on sensor data.

The innovative aspect of our study is multi-sensor fusion, incorporating accelerometer and gyroscope data of HAR in federated settings. The preprocessing and feature extraction stages are pivotal in transforming the raw sensor data into a format suitable for analysis. The data was segmented into windows, each spanning one second and overlapping by half to ensure continuity and capture transitions between activities effectively [40]. Various time and frequency-based features were extracted from these windows, including applying a discrete Fourier transform to translate time-based signals into the frequency domain. Our data processing technique included the unique feature of computing gravity-based characteristics to determine the position of the wearable device on the body. A low-pass filter was applied to segregate the acceleration data into gravitational and body movement components, allowing us to compute the orientation of the device. This orientation data helped us understand the context of activity recognition. To avoid over-fitting and improve the generalizability of the model, we categorized these orientations into predefined groups [40].

The Real-World dataset was suitable for the HAR study because it was captured in a natural environment and had a realistic class imbalance. For example, the jumping activity only made up 2% of the data, while standing accounted for 14% of the total data. Moreover, the high-class imbalance and the availability of separated user data made it a perfect fit for a comprehensive study on FL approaches for HAR.

6.4.2 Performance Metrics

HAR is a multi class classification where multiple metrics are used to assess the effectiveness of the DL model. The commonly used metric accuracy, the proportion of correctly identified activities among all classifications, is a fundamental metric. However, the utility of accuracy is often limited in datasets characterized by a significant imbalance among classes. This imbalance can lead to the accuracy paradox, where models might exhibit high accuracy by predominantly predicting the majority class, thus neglecting the nuanced detection of less frequent activities. We employ additional metrics, precision, recall, and the F1-score, to address these limitations and provide a more comprehensive evaluation.

- **Precision:** This metric evaluates the accuracy of positive predictions, given by the formula:

$$\text{Precision} = \frac{\text{TP}}{\text{TP} + \text{FP}} \tag{6.12}$$

where TP represents true positives and FP denotes false positives.

- **Recall (sensitivity):** This measure assesses the model's ability to identify all relevant instances, defined as:

$$\text{Recall} = \frac{\text{TP}}{\text{TP} + \text{FN}} \tag{6.13}$$

with FN indicating false negatives.

- **F1-score:** Offering a balance between precision and recall, this metric is particularly valuable in the presence of class imbalance. It is calculated as:

$$\text{F1} - \text{score} = 2 \times \frac{\text{Precision} \times \text{Recall}}{\text{Precision} + \text{Recall}}. \tag{6.14}$$

For a comprehensive evaluation, all experimental procedures were conducted in a simulated environment. This allowed us to evaluate its effectiveness using two strategies: global performance evaluation and personalized model assessment. The global assessment evaluates the model's capability across the entire dataset by leveraging a global test set to infer its generalization potential. On the other hand, personalized evaluations are conducted at an individual participant level, using local data to refine the global model and tailor predictions more closely to individual patterns. This bifurcated approach enables a comparison between the model's universal applicability and its customized effectiveness. In addition, we have expanded our evaluation criteria to include energy efficiency, which is a crucial factor in FL situations where computational and communication resources are limited. The energy efficiency metric is based on the computational requirements of local training and the amount of data transmitted during each communication round, which is represented by the following formula [41, 42]:

$$E_{est} = R\left[\left(\alpha * t_{com}\right) + N\left(\beta * P_{trn}\right)\right] \tag{6.15}$$

where α is the computation constant having dimensions of energy per second and β is the communication constant with dimensions of energy per kilobyte. R represents the number of communication rounds; N is the number of participants; t_{com} is the computation time, which is dependent on the device type; and P_{trn} is the data payload size per communication round.

6.5 Results and Discussion

In this chapter, the proposed hybrid neuromorphic S-LSTM model is rigorously evaluated using two distant publicly available datasets (UCI and Real-World), focusing on applying HAR indoors and outdoors. This study aims to assess the performance of the proposed S-LSTM in terms of accuracy, energy efficiency, and adaptability within the context of wearable sensor data fusion. The evaluation begins with the UCI dataset, which provides a controlled indoor environment to benchmark the capabilities of the proposed S-LSTM model against traditional DL architectures such as LSTM and CNN. The UCI dataset, with its structured activities and controlled settings, offers an ideal scenario for scrutinizing the nuanced differences between the models. It allows us to explore the effectiveness of the S-LSTM model in capturing and learning from the temporal sequences inherent in human activities. Following the UCI dataset, we extend our evaluation to the Real-World dataset, which presents a more challenging and varied set of outdoor activities, thus testing the robustness and generalizability of our model in scenarios closer to everyday human behaviors.

6.5.1 UCI Results

As previously discussed, the UCI dataset covers various indoor human activities, making it a suitable option to evaluate our model's effectiveness in a controlled setting. This dataset is divided among five participants acting as the edge nodes capable of model training without data sharing. To analyze the results, we used an 80–20 split for training and testing the model on each participant, creating separate local training and testing sets. Later, we aggregated the test set of each participant to create a global test set, which allowed us to assess the robustness of the proposed architecture.

It is important to note that this evaluation is done in simulations, where each model training spans 500 communication rounds in a federated setting. Each model is trained for three epochs locally and shares the model parameters with FS, where FedAVG is adopted for model aggregation. In our first comparison, the results in Figure 6.5 represent the accuracy of the learning curve obtained using a global test set in the training process. This learning curve exhibits a typical pattern of a rapid increase in accuracy with more steady-state behavior as training approaches its performance limits. It is noteworthy that the hybrid S-LSTM model performed better than both LSTM and CNN models, reaching a maximum accuracy of 97.36%. The LSTM model came in a close second with a peak accuracy of 96.30%, while the CNN and S-CNN models had slightly lower results, peaking at 95.14% and 93.25%, respectively.

Confusion matrices provide a comprehensive predictive accuracy metric across activities to analyze model performance effectively on the UCI HAR dataset. The results in Figure 6.6 provide an insight into the predictive accuracy of the CNN, S-CNN, LSTM, and S-LSTM models for individual classes. Each figure represents a normalized confusion matrix, where the diagonal elements indicate the percentage of correct predicted labels for each class. It is worth noting that class 6 (lying) had a true positive rate of 100% across all models, as it is an easily distinguishable

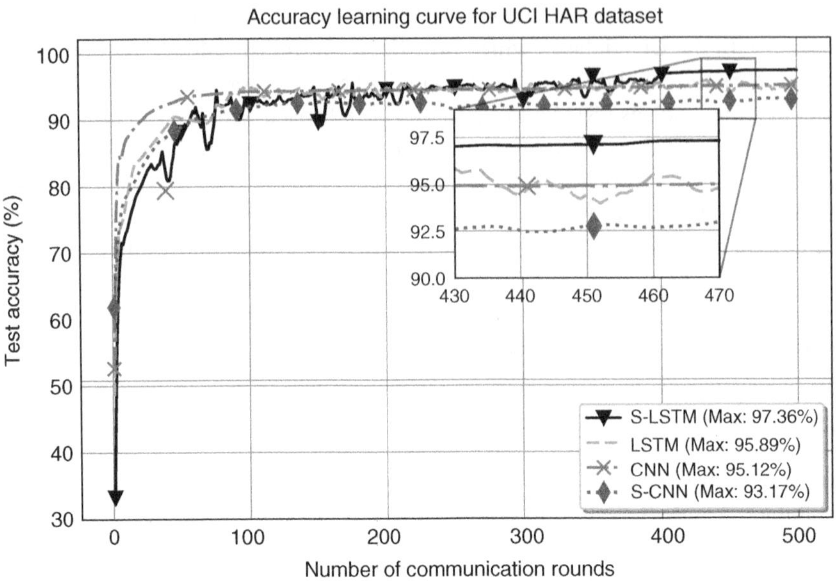

Figure 6.5 Learning curve representing the accuracy for UCI dataset obtained using global test set.

activity. The results in Figure 6.6d show that the S-LSTM model performed well and distinguished between classes 1, 2, and 3, with true positive rates exceeding 0.99%. Additionally, the model showed impressive proficiency in distinguishing other classes like 4 (sitting) and 5 (standing), with a misclassification rate of only 0.06%. This demonstrates its enhanced ability to identify the subtle temporal patterns that differentiate these activities. The CNN and S-CNN models displayed slightly higher misclassification rates of 0.12% and 0.11%, respectively, for classes 4 and 5, while the LSTM showed a rate of 0.11%. These models performed well but were not as proficient as the S-LSTM model in distinguishing between the relevant classes.

The results presented in Table 6.1 provide a detailed comparison of the performance metrics, including precision, recall, and F1-score. The models were evaluated based on their classification capabilities for six common daily activities. The proposed S-LSTM model outperforms others, with high F1-scores of 0.99% for activities like walking and walking upstairs and a perfect F1-score of 1.00 for other activities like walking downstairs and lying. The uniform success in identifying the lying activity across all models can be attributed to the unique motion patterns associated with this activity. However, the analysis has identified a challenge in distinguishing between sitting and standing activities. The proposed

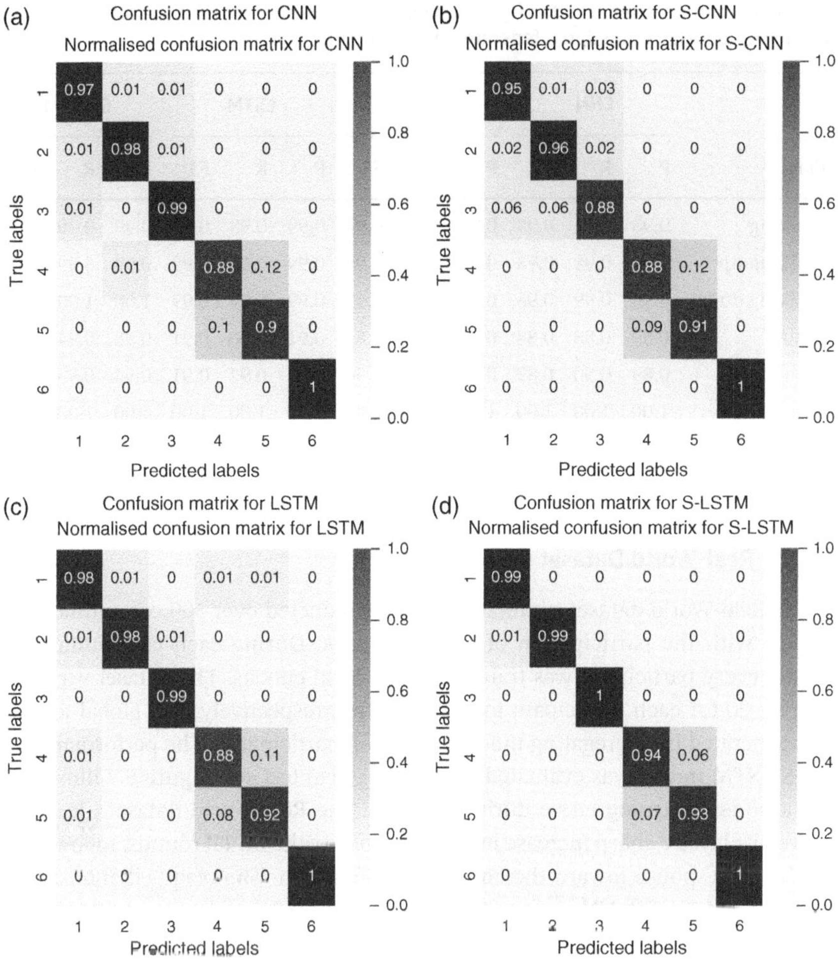

Figure 6.6 The confusion matrix for four DL models compared in this study for UCI HAR dataset, where the index number represents the activity. The labels corresponding to the activities are (1) walking, (2) walking upstairs, (3) walking downstairs, (4) sitting, (5) standing, and (6) lying.

S-LSTM model has a small advantage with F1-scores of 0.93% for sitting and 0.94% for standing. Moreover, the CNN and S-CNN models have lower efficacy, with F1-scores ranging between 0.88 and 0.90. This pattern suggests that the motion characteristics of sitting and standing are similar, posing challenges for models, especially CNN and S-CNN, in accurately distinguishing between these two activities.

Table 6.1 Comparative results of global models for CNN, S-CNN, LSTM, and S-LSTM for the UCI dataset trained in a federated environment.

	CNN			S-CNN			LSTM			S-LSTM		
Class	**P**	**R**	**F1**	**P**	**R**	**F1**	**P**	**R**	**F1**	**P**	**R**	**F1**
Walking	0.98	0.97	0.98	0.93	0.95	0.94	0.99	0.98	0.99	0.99	0.99	0.99
Walking up	0.98	0.98	0.98	0.93	0.96	0.95	0.99	0.99	0.99	0.99	0.99	0.99
Walking down	0.98	0.99	0.98	0.93	0.88	0.91	0.99	0.99	0.99	1.00	1.00	1.00
Sitting	0.89	0.88	0.88	0.90	0.88	0.89	0.91	0.90	0.91	0.93	0.94	0.93
Standing	0.89	0.90	0.89	0.89	0.91	0.90	0.91	0.92	0.91	0.94	0.93	0.94
Lying	1.00	1.00	1.00	1.00	1.00	1.00	1.00	1.00	1.00	1.00	1.00	1.00

Here **P, R,** and **F1** represents precision, recall, and F1-score.

6.5.2 Real-World Dataset Results

In the Real-World dataset, simulations were conducted over 500 communication rounds with the participation of 15 edge nodes. During each communication round, every participant was trained for five local epochs. The dataset was split into 80–20 for each participant to train and test, respectively. The global test set was generated by aggregating the test sets of all participants. The performance of the S-LSTM model was evaluated using this global test set. Figure 6.7 illustrates the accuracy learning curve during training. The Real-World dataset's learning curve displayed a sharp increase in accuracy during the initial rounds, followed by a steadier response toward the end of the simulation. The results demonstrated that S-LSTM and S-CNN perform better than their traditional counterparts, LSTM and CNN, with the highest accuracy of 89.69% and 86.90%, respectively. In contrast, the LSTM and CNN models achieved accuracies of 85.85% and 84.97%, respectively. This comparison of learning curves highlighted the superior performance of hybrid spiking models and their potential to capture complex human activities in less controlled outdoor environments.

To further consolidate the discussion, the confusion matrix of the four models considered in this study is illustrated in Figure 6.8. More specifically, the results in Figure 6.8d provide the classification ability of S-LSTM for eight outdoor activities in the Real-World HAR dataset. The results show that the S-LSTM model is proficient in classifying more prominent activities, with a score of 0.96% for jumping, 0.86% for running, and 0.91% for walking. However, the confusion matrix also reveals that the S-LSTM model encountered challenges in classifying activities with subtle motion patterns, such as climbing down 0.88%, climbing up 0.87%,

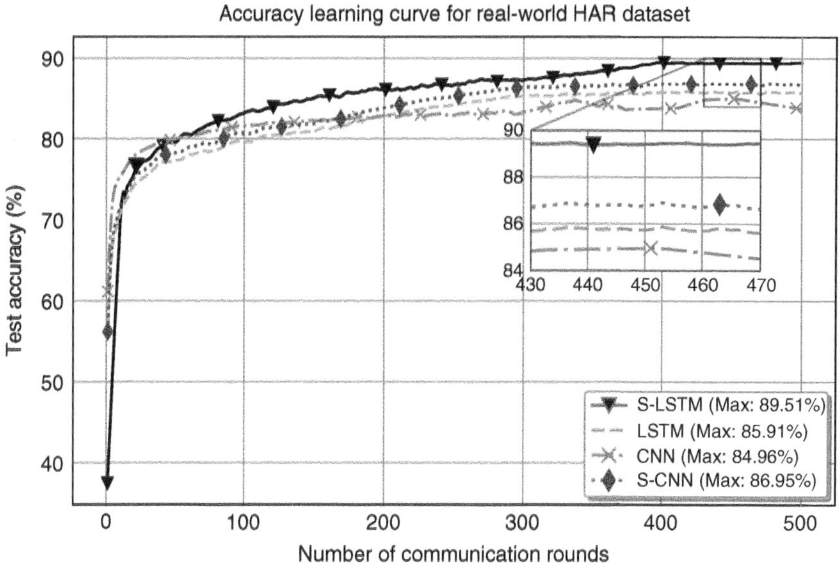

Figure 6.7 Learning curve of accuracy obtained using a global test set for Real-World dataset spanning 500 communication rounds.

and standing 0.80%. Misclassifications mainly occur between activities that share similar motion characteristics. For instance, climbing down was occasionally misclassified as walking, climbing up was confused with running, and sitting was sometimes classified as standing.

The results in Table 6.2 extensively analyze various models trained using the Real-World HAR dataset in a FL context. The proposed S-LSTM has consistently demonstrated superior performance among the models, surpassing the other models in precision, recall, and F1-score. For instance, when classifying the walking activity, the S-LSTM model achieved a precision of 0.94%, a recall of 0.93%, and an F1-score of 0.94%, which sets a benchmark that surpasses other models in this study. The S-LSTM model performs better in accurately identifying and classifying complex temporal activity patterns. This is evident from its remarkable F1-scores of 0.94% for climbing down and 0.93% for climbing up. These scores demonstrate the model's proficiency in distinguishing and categorizing intricate motion sequences. It is worth noting that all models examined encountered challenges in differentiating between the "sitting" and "standing" activities. The F1-scores in these categories remained relatively modest, ranging from 0.73% to 0.85%. However, the S-LSTM model achieved marginally higher scores in these challenging classifications, confirming its

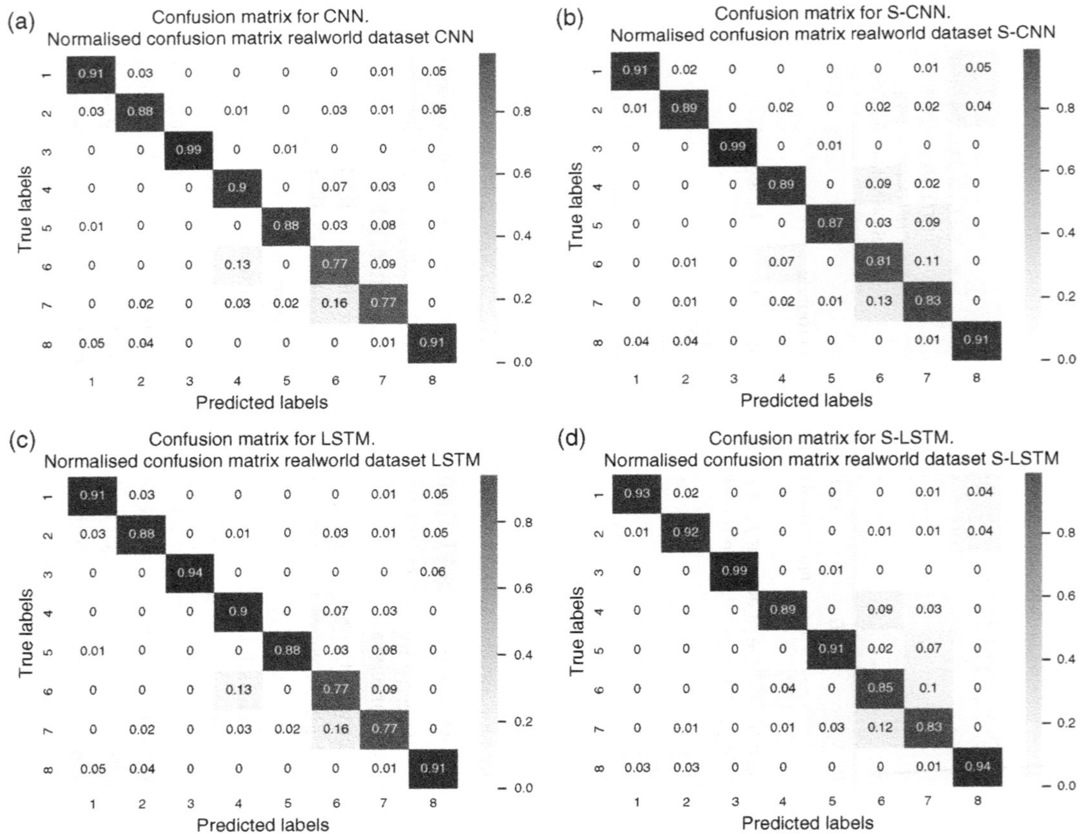

Figure 6.8 The confusion matrix CNN, S-CNN, LSTM, and S-LSTM models for Real-World data set. The index represents the activity, where the label corresponding to the activities are: (1) climbing down, (2) climbing up, (3) jumping, (4) lying, (5) running, (6) sitting, (7) standing, and (8) walking.

Table 6.2 Comparison of different DL techniques for Real-World dataset.

	CNN			S-CNN			LSTM			S-LSTM		
Class	**P**	**R**	**F1**	**P**	**R**	**F1**	**P**	**R**	**F1**	**P**	**R**	**F1**
Climb down	0.90	0.91	0.90	0.92	0.91	0.91	0.90	0.91	0.90	0.94	0.93	0.94
Climb up	0.90	0.88	0.89	0.92	0.89	0.90	0.90	0.88	0.89	0.93	0.92	0.93
Jumping	1.00	1.00	1.00	1.00	1.00	1.00	0.99	0.94	0.96	1.00	1.00	1.00
Lying	0.84	0.90	0.87	0.89	0.89	0.89	0.84	0.90	0.87	0.95	0.89	0.92
Running	0.98	0.88	0.93	0.98	0.87	0.93	0.98	0.88	0.93	0.97	0.91	0.94
Sitting	0.73	0.77	0.75	0.74	0.81	0.77	0.73	0.77	0.75	0.78	0.85	0.82
Standing	0.77	0.77	0.77	0.75	0.83	0.79	0.77	0.77	0.77	0.79	0.83	0.81
Walking	0.91	0.91	0.91	0.92	0.91	0.91	0.91	0.91	0.91	0.93	0.94	0.93

Here **P**, **R**, and **F1** represents precision, recall, and F1-score, respectively.

robustness and the effectiveness of its learning architecture in complex scenario discrimination.

6.5.3 Energy Efficiency Comparison

To further enhance our comparative study, we implemented a strategy where only 50% of edge nodes would participate randomly during the training process. The idea behind randomly selecting clients is to reduce communication overheads and assess the trade-off between accuracy and communication cost. For comparison, the results shown in Figure 6.9 illustrate the accuracy of the learning curve obtained during the training process using a global test set of 500 rounds. This approach was used to determine how reducing client participation affects model performance and communication cost benefits. The results indicate that the proposed S-LSTM model achieved the highest accuracy of 88.11%, while the counterpart LSTM reached a maximum of 84.43%. The CNN and S-CNN models achieved accuracies of 83.98% and 85.20%, respectively. More specifically, the random client participation significantly reduced communication costs in this case, with only a slight variation in performance.

To further analyze the energy efficiency of the proposed hybrid model, rigorous evaluations were conducted by quantifying energy estimates as defined by Equation (6.15). The energy efficiency is calculated for a single communication round with all participants and a randomly selected 50% of participants. For the simplicity of our analysis, the computation constant α and communication constant β were set to 0.003 and 0.0001, respectively, as adopted from past

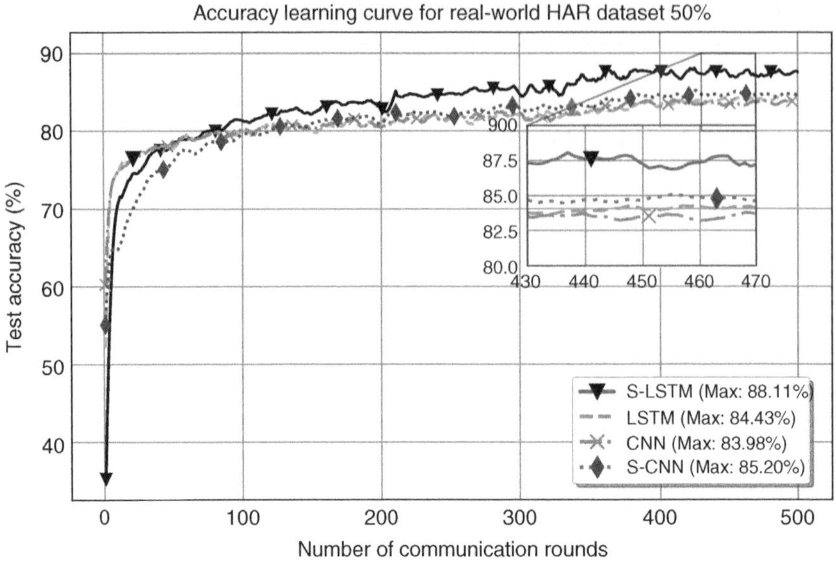

Figure 6.9 Learning curve for Real-World dataset, with 50% random client participant trained for 500 communication rounds.

Table 6.3 Comparison of energy efficiency using single Eq. (6.15).

Model	Model parameter P_{trn} (KB)	Computation time t_{com} (s)	Energy estimate E_{est} (W)
CNN	25321	258	38.24
		143	15.73
S-CNN	19418	252	29.38
		136	15.67
LSTM	5231	220	8.07
		137	4.32
S-LSTM	3931	208	6.10
		121	3.27

In this case, the 50% random participant selection was made, and the energy estimate was calculated for single communication round.

literature [42, 41]. The results in Table 6.3 present the energy estimates E_{est}, which are dependent on both computation time and model parameters P_{trn}, under the specified client selection criteria. Notably, the S-LSTM model had a minimum computation time of 208 seconds with all participants, further reduced to 121 seconds with 50% client participation. Regarding energy consumption, the S-LSTM model demonstrated superior efficiency, with the lowest energy estimate of 6.10 W. This represents a 24.41% decrease compared to the LSTM model, which had an energy estimate of 8.07 W. The results confirm that selecting 50% of participants at random significantly improved the model's energy efficiency. However, it is important to consider the accuracy-efficiency trade-off when using DL models, especially for energy-constrained applications. Finding the optimal balance between accuracy and efficiency is critical for deploying these models.

6.5.4 Personalized Model Comparison

The previous results were obtained and discussed using the global model and test set. However, using local data, participants can customize the model to their needs. Therefore, the global model is fine-tuned using local data to create a personalized model. The personalized performance is then compared to the global model. The results in Figure 6.10 compare the global and personalized models on test accuracy across 15 participants. The x-axis represents the client number, while the y-axis plots the test accuracy. For each client, two bars represent the global and personalized model accuracy. The results show that personalization significantly improved the local test accuracy compared to the global model for all models under consideration. Specifically, the proposed S-LSTM personalized model achieved the highest average accuracy of 97.12% across clients, substantially higher than its 89.69% global accuracy. The other models showed similar trends of increased accuracy with personalization, with averages of 95.39% (LSTM), 95.96% (S-CNN), and 95.10% (CNN).

6.6 Summary

In this chapter, we proposed an HNFL framework that synergizes the computational efficiency of SNNs with the dynamic temporal learning capabilities of LSTM networks for HAR using multi-model data from wearable sensors. This integrated S-LSTM model capitalizes on LSTM layers to adeptly capture temporal dependencies within time-series sensor data while incorporating spiking layers to facilitate event-driven processing, thereby enhancing energy efficiency in federated settings. The model training leverages surrogate gradient learning and BPTT, facilitating supervised end-to-end learning. Our approach has been tested on two publicly available HAR datasets – UCI and Real-World. The UCI dataset is for indoor environments, while the Real-World dataset is for outdoor settings. Our evaluations demonstrated the superior performance of

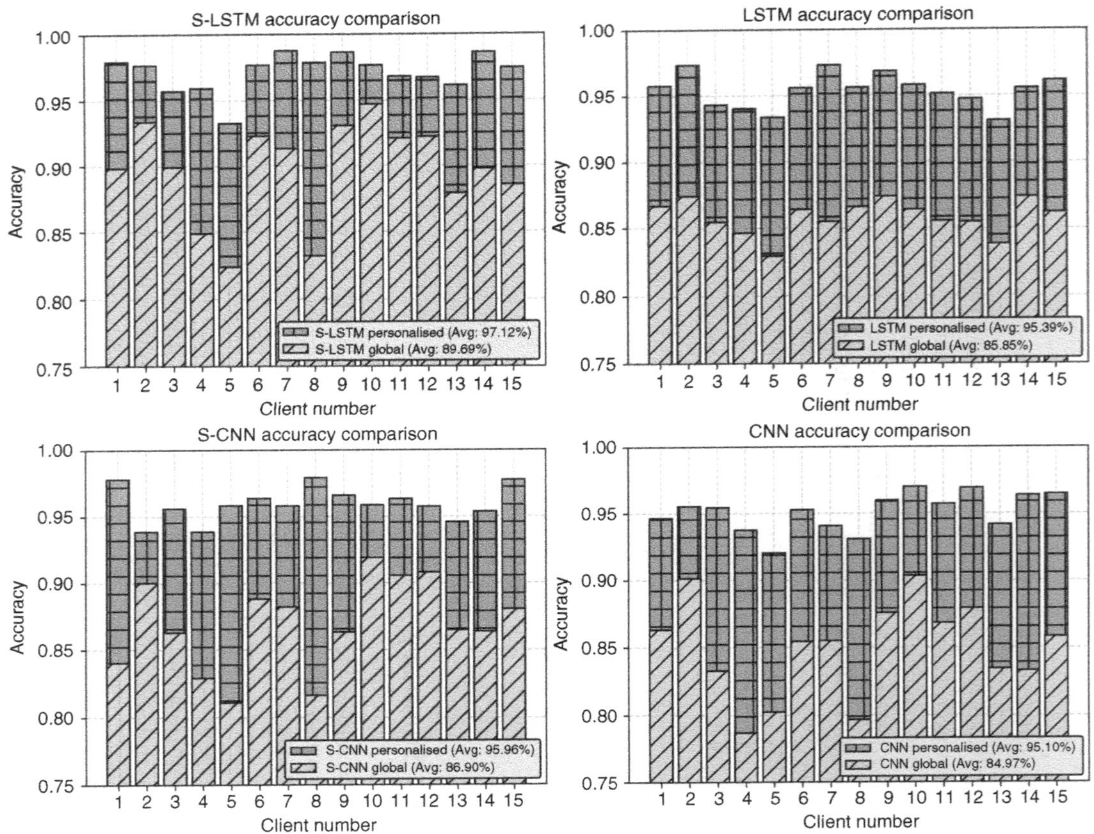

Figure 6.10 Accuracy comparison graph for global and personalized models for participants. The personalized accuracy was obtained after fine-tuning using the local dataset.

the proposed S-LSTM model in an FL paradigm compared to traditional models like LSTM, CNN, and S-CNN.

The simulation results show concrete evidence that the S-LSTM model outperforms the LSTM model in accuracy. In controlled indoor environments represented by the UCI dataset, the S-LSTM model outperformed the LSTM model by 1.06%. However, its performance in the complex and diverse outdoor settings of the Real-World dataset was even more impressive, with a significant 3.84% increase in accuracy over the LSTM model. This improvement highlights the S-LSTM model's robustness and ability to handle the unpredictable nature of real-world human activities. Moreover, our findings highlighted a significant 32.30% improvement in energy efficiency compared to the LSTM model. Randomly selecting participants for model training improves energy efficiency but affects accuracy. However, personalizing the global model by fine-tuning it with local data can significantly increase performance. On average, this approach improved accuracy by 9% across participants.

References

1 Diraco, G., Rescio, G., Siciliano, P., and Leone, A. (2023). Review on human action recognition in smart living: sensing technology, multimodality, real-time processing, interoperability, and resource-constrained processing. *Sensors* 23 (11): 5281.

2 Kalabakov, S., Jovanovski, B., Denkovski, D. et al. (2023). Federated learning for activity recognition: a system level perspective. *IEEE Access* 11: 64442–64457.

3 García-Requejo, A., Pérez-Rubio, M.C., Villadangos, J.M., and Hernández, Á. (2023). Activity monitoring and location sensory system for people with mild cognitive impairments. *IEEE Sensors Journal* 23 (5): 5448–5458.

4 Yu, H., Chen, Z., Zhang, X. et al. (2021). FedHAR: Semi-supervised online learning for personalized federated human activity recognition. *IEEE Transactions on Mobile Computing* 22: 3318–3332.

5 Bokhari, S.M., Sohaib, S., Khan, A.R. et al. (2021). DGRU based human activity recognition using channel state information. *Measurement* 167: 108245.

6 Cheng, X., Cai, J., Xu, J., and Gong, D. (2022). High-performance strain sensors based on Au/graphene composite films with hierarchical cracks for wide linear-range motion monitoring. *ACS Applied Materials & Interfaces* 14 (34): 39230–39239.

7 Islam, M.M., Nooruddin, S., Karray, F., and Muhammad, G. (2022). Human activity recognition using tools of convolutional neural networks: a state of the art review, data sets, challenges, and future prospects. *Computers in Biology and Medicine* 106060.

8 Yao, S., Hu, S., Zhao, Y. et al. (2017). Deepsense: a unified deep learning framework for time-series mobile sensing data processing. *Proceedings of the 26th International Conference on World Wide Web*, Perth Australia (3–7 April 2017), pp. 351–360.

9 Wang, J., Chen, Y., Hao, S. et al. (2019). Deep learning for sensor-based activity recognition: a survey. *Pattern Recognition Letters* 119: 3–11.

10 Guan, Y. and Plötz, T. (2017). Ensembles of deep LSTM learners for activity recognition using wearables. *Proceedings of the ACM on Interactive, Mobile, Wearable and Ubiquitous Technologies* 1 (2): 1–28.

11 Singh, M.S., Pondenkandath, V., Zhou, B. et al. (2017). Transforming sensor data to the image domain for deep learning – an application to footstep detection. *2017 International Joint Conference on Neural Networks (IJCNN)*, Anchorage, AK, USA (14–19 May 2017), pp. 2665–2672. IEEE.

12 Hafeez, S., Khan, A.R., Al-Quraan, M. et al. (2023). Blockchain-assisted UAV communication systems: a comprehensive survey. *IEEE Open Journal of Vehicular Technology* 4: 558–580.

13 Qi, P., Chiaro, D., and Piccialli, F. (2023). FL-FD: federated learning-based fall detection with multimodal data fusion. *Information Fusion* 101890.

14 McMahan, B., Moore, E., Ramage, D. et al. Communication-efficient learning of deep networks from decentralized data. Aarti Singh, Jerry Zhu Artificial Intelligence and Statistics PMLR, Fort Lauderdale, FL, USA (20–22 April 2017); vol. 54 2017. p. 1273–1282.

15 Wang, Y., Duan, S., and Chen, F. (2023). Efficient asynchronous federated neuromorphic learning of spiking neural networks. *Neuro-computing* 557: 126686.

16 Venkatesha, Y., Kim, Y., Tassiulas, L., and Panda, P. (2021). Federated learning with spiking neural networks. *IEEE Transactions on Signal Processing* 69: 6183–6194.

17 Khatun, M.A., Yousuf, M.A., Ahmed, S. et al. (2022). Deep CNN-LSTM with self-attention model for human activity recognition using wearable sensor. *IEEE Journal of Translational Engineering in Health and Medicine* 10: 1–16.

18 Han, C., Zhang, L., Tang, Y. et al. (2022). Human activity recognition using wearable sensors by heterogeneous convolutional neural networks. *Expert Systems with Applications* 198: 116764.

19 Uddin, M.Z. and Soylu, A. (2021). Human activity recognition using wearable sensors, discriminant analysis, and long short-term memory-based neural structured learning. *Scientific Reports* 11 (1): 16455.

20 Jain, V., Gupta, G., Gupta, M. et al. (2023). Ambient intelligence-based multimodal human action recognition for autonomous systems. *ISA Transactions* 132: 94–108.

21 Qi, W., Su, H., Yang, C. et al. (2019). A fast and robust deep convolutional neural networks for complex human activity recognition using smartphone. *Sensors* 19 (17): 3731.

22 Laitrakun, S. (2023). Merging-squeeze-excitation feature fusion for human activity recognition using wearable sensors. *Applied Sciences* 13 (4): 2475.

23 Auvinet, E., Multon, F., Saint-Arnaud, A. et al. (2010). Fall detection with multiple cameras: an occlusion-resistant method based on 3-d silhouette vertical distribution. *IEEE Transactions on Information Technology in Biomedicine* 15 (2): 290–300.

24 Xie, L., Yang, Y., Zeyu, F., Naqvi, S.M. (2021). Skeleton-based fall events classification with data fusion. *2021 IEEE International Conference on Multisensor Fusion and Integration for Intelligent Systems (MFI)*, Karlsruhe Institute of Technology in Germany (23–25 September 2021). IEEE, pp. 1–6.

25 Amsaprabhaa, M. (2023). Multimodal spatiotemporal skeletal kinematic gait feature fusion for vision-based fall detection. *Expert Systems with Applications* 212: 118681.

26 Yang, Y., Yang, H., Liu, Z. et al. (2022). Fall detection system based on infrared array sensor and multi-dimensional feature fusion. *Measurement* 192: 110870.

27 Cheng, D., Zhang, L., Bu, C. et al. (2023). ProtoHAR: prototype guided personalized federated learning for human activity recognition. *IEEE Journal of Biomedical and Health Informatics* 27: 3900–3911.

28 Ouyang, X., Xie, Z., Zhou, J. et al. (2022). ClusterFL: a clustering-based federated learning system for human activity recognition. *ACM Transactions on Sensor Networks* 19 (1): 1–32.

29 Gad, G. and Fadlullah, Z. (2022). Federated learning via augmented knowledge distillation for heterogenous deep human activity recognition systems. *Sensors* 23 (1): 6.

30 Aouedi, O., Piamrat, K., and Südholt, M. (2023). HFedSNN: efficient hierarchical federated learning using spiking neural networks. *21st ACM International Symposium on Mobility Management and Wireless Access (MobiWac 2023)*.

31 Xie, K., Zhang, Z., Li, B. et al. (2022). Efficient federated learning with spike neural networks for traffic sign recognition. *IEEE Transactions on Vehicular Technology* 71 (9): 9980–9992.

32 Roy, K., Jaiswal, A., and Panda, P. (2019). Towards spike-based machine intelligence with neuromorphic computing. *Nature* 575 (7784): 607–617.

33 Gerstner, W., Kistler, W.M., Naud, R., and Paninski, L. (2014). *Neuronal Dynamics: From Single Neurons to Networks and Models of Cognition*. Cambridge University Press.

34 Kim, Y. and Panda, P. (2021). Optimizing deeper spiking neural networks for dynamic vision sensing. *Neural Networks* 144: 686–698.

35 Neftci, E.O., Mostafa, H., and Zenke, F. (2019). Surrogate gradient learning in spiking neural networks: Bringing the power of gradient-based optimization to spiking neural networks. *IEEE Signal Processing Magazine* 36 (6): 51–63.

36 Wu, Y., Deng, L., Li, G. et al. (2018). Spatio-temporal backpropagation for training high-performance spiking neural networks. *Frontiers in Neuroscience* 12: 331.

37 Lee, C., Srinivasan, G., Panda, P., and Roy, K. (2018). Deep spiking convolutional neural network trained with unsupervised spike-timing-dependent plasticity. *IEEE Transactions on Cognitive and Developmental Systems* 11 (3): 384–394.

38 Tumpa, S.A., Singh, S., Khan, M.F.F. et al. (2023). Federated learning with spiking neural networks in heterogeneous systems. *2023 IEEE Computer Society Annual Symposium on VLSI (ISVLSI),* Foz do Iguacu, Brazil (20–23 June 2023). IEEE, 1–6.

39 Anguita, D., Ghio, A., Oneto, L. et al. (2013). A public domain dataset for human activity recognition using smartphones. In: *21st European Symposium on Artificial Neural Networks*, Bruges, Belgium, April 24-25-26. ESANN 2013 – Proceedings.

40 Sztyler, T. and Stuckenschmidt, H. (2016). On-body localization of wearable devices: an investigation of position-aware activity recognition. *2016 IEEE International Conference on Pervasive Computing and Communications (PerCom),* Sydney, Australia (14–18 March 2016). IEEE, 1–9.

41 Manzoor, H.U., Khan, A.R., Flynn, D. et al. (2023). FedBranched: leveraging federated learning for anomaly-aware load forecasting in energy networks. *Sensors* 23 (7): 3570.

42 Mian, A.N., Shah, S.W.H., Manzoor, S. et al. (2022). A value-added IoT service for cellular networks using federated learning. *Computer Networks* 213: 109094.

7

Multi-modal Beam Prediction for Enhanced Beam Management in Drone Communication Networks

Iftikhar Ahmad[1], Ahsan Raza Khan[1], Rao Naveed Bin Rais[2], Muhammad Ali Imran[1], Sajjad Hussain[1], and Ahmed Zoha[1]

[1] *James Watt School of Engineering, University of Glasgow, Glasgow, UK*
[2] *Artificial Intelligence Research Center (AIRC), Ajman University, Ajman, UAE*

The rapid development of drone communication technologies requires innovative solutions for improved beam management in millimeter wave (mmWave) 6G networks. Beamforming can enhance signal integrity and improve the signal-to-noise ratio (SNR). However, traditional beam management methods, which rely on exhaustive searches across a codebook, are inefficient for applications that require high mobility due to extensive training requirements and high latency. To address this challenge, this study introduces a state-of-the-art technique for mmWave beam prediction that considers real-world visual and communication contexts. Our approach leverages computer vision (CV) and ensemble learning through stacking to combine multimodal vision sensing with positional information, enabling accurate drone location and orientation predictions. Specifically, the framework utilizes the "you only look once" version 5 (YOLO-v5) CV model to identify drones in RGB images through bounding boxes. The resulting vision and positional data are used to train two distinct neural network sets, whose outputs are combined in a meta-learner to predict K-beams from a predefined codebook. Our approach achieves outstanding performance with a top-1 accuracy rate of approximately 90%, surpassing the accuracy rates of vision-only and position-only models, which are 86% and 60%, respectively. Furthermore, our approach achieves near-perfect top-3 and top-5 accuracy rates of approximately 100%, significantly enhancing received signal strength.

Multimodal Intelligent Sensing in Modern Applications, First Edition.
Edited by Masood Ur Rehman, Ahmed Zoha, Muhammad Ali Jamshed, and Naeem Ramzan.

Published 2025 by John Wiley & Sons, Inc.

7.1 Drone Communication

Future wireless communication technologies, such as 5G Advanced and 6G, aim to accommodate highly mobile devices, including drones and autonomous vehicles. Drones, also known as unmanned aerial vehicles (UAVs), are poised to play a pivotal role in the evolution of future technologies. They offer promising solutions for extending the coverage of mmWave wireless networks, enabling applications with low latency, and improving security surveillance systems [1]. As highlighted in [2], UAVs have attracted considerable interest across various areas due to their flexible applications. In the military, UAVs are deployed for tasks like object detection, monitoring positions, gathering information, and surveillance. On the civilian front, their applications range from information dissemination and disaster response to transporting goods. Recently, the role of UAVs in wireless communications has become increasingly significant. Their mobility, adaptability, and easy deployment render them an effective means for enhancing connectivity, especially in remote or disaster-stricken areas. By being equipped with communication technologies, UAVs can function as airborne base stations or relay points, thereby broadening the reach and capacity of wireless networks. The efficacy of UAV-supported wireless communications has seen significant improvements through the establishment of line-of-sight (LoS) connections in air-to-ground (AtG) transmissions. These advancements have led to decreased propagation loss and enhanced link quality of service (QoS). Moreover, mmWave communication technologies have demonstrated considerable effectiveness in emergencies and environments with complex topographies or high levels of topological change [3]. Given the LoS facilitation by AtG links, the exploration of mmWave communications from aerial platforms has emerged as a growing field of study. UAVs offer the advantages of immediate deployment, adaptable network configuration, and a high probability of maintaining LoS communication channels. UAVs are frequently utilized as aerial base stations or relays to enhance network capacity and offer more flexible coverage [4]. To support these advanced operations, drones are likely to be outfitted with mmWave communication devices for high-speed data transmission. The use of mmWave communication systems in drones, however, is constrained by the requirement for large antenna arrays and narrow directive beams to maintain a sufficient signal-to-noise ratio (SNR). These conditions lead to a high overhead for beam training, making it difficult to support highly mobile drones. Consequently, it is necessary to explore alternative solutions to overcome these obstacles and improve mmWave communication for highly mobile drones [5].

7.2 Beam Management

In recent years, significant efforts have been dedicated to addressing the challenges associated with beam training overhead in mmWave systems. Initially, research concentrated on two primary strategies: (i) adaptive beam training using a codebook-based approach, and (ii) channel estimation employing compressive sensing to capitalize on the channel's sparsity. The article in [6] introduced the concept of using exhaustive or adaptive methods for adaptive beam training to pinpoint optimal beams for both transmitters and receivers. Conversely, the authors of [7] presented a method for mmWave channel estimation that treats it as a sparse reconstruction challenge, taking advantage of mmWave channels' inherent sparsity. Further innovation is seen in [8], which proposes a novel three-dimensional (3D) beam training technique for UAV-assisted mmWave communications, utilizing the inverse discrete-space Fourier transform to create a training beam with a flat-topped characteristic. This approach also explores a hybrid beamforming (BF) system, employing a greedy geometric (GG) method to select the optimal beam. Although these methodologies have contributed to reducing beam training overhead, their effectiveness is generally limited to a single order of magnitude. Such a reduction is not adequate for managing scenarios involving multiple users with high mobility. Due to the limitations of conventional systems in effectively handling multiuser environments with high mobility, machine learning-based approaches are making progress and gaining momentum. These innovative systems use historical data and various types of sensing information, such as user location [9], camera or visual imagery [10, 11], LiDAR outputs [12], and radar signals [13], to improve performance. In particular, the combination of drones equipped with cameras and millimeter wave (mmWave) technology, as discussed in a study [5], offers significant improvements in wireless communications. The study introduces a deep learning strategy that uses computer vision to predict wireless beam directions accurately. This method enables drones to maintain a consistent connection, even in motion. In [14], researchers introduced a cutting-edge method called "latency-aware vision-aided federated wireless networks (VFWN)" for predicting beam blockages using both vision and wireless sensing data. The VFWN framework utilizes distributed learning across edge nodes for data processing and model training, with federated averaging algorithms to synthesize a global model. This approach boasted a remarkable 99% accuracy rate while reducing communication costs and latency by 81.31% and 6.77%, respectively. Additionally, the authors of [15] outlined an innovative strategy to address connectivity challenges in drone-assisted networks through a sophisticated drone clustering algorithm. This method specifies a protocol for selecting cluster heads

(CHs) and executing group handovers (HOs) that cover network selection, execution, and initiation phases. However, it is noted that most current beam prediction models are designed for scenarios involving user equipment (UE) such as humans, vehicles, or robots, which typically move in two dimensions and are relatively straightforward to predict. The authors of [16] suggest that incorporating multiple data modalities for beam prediction facilitates a comprehensive environmental understanding, leading to more accurate forecasts. By integrating inputs from various sensors, the precision of predictions is significantly enhanced. Fusion algorithms further refine performance by dynamically adjusting the contribution of each data modality, thus optimizing prediction accuracy.

The need for fast and reliable communication systems has led to the exploration of mmWave frequency bands for drone communication. Despite their potential, these high-frequency bands face significant path loss and limited penetration abilities, making it difficult to maintain a stable link between drones and ground stations. Beamforming has emerged as a promising technique to strengthen this communication link. However, determining the optimal beam direction in dynamic drone environments poses a significant challenge due to the directional and highly mobile characteristics of mmWave signals. In this chapter, we introduce a new method for predicting beams that use multi-modal data fusion to reduce the overhead associated with mmWave beam training. A new beam prediction model has been developed, leveraging computer vision (CV) and ensemble learning techniques, specifically stacking, to combine data from multiple modalities, such as vision and position sensing. This approach is designed for training models that enhance mmWave drone communications in real-world scenarios. The YOLO-v5 algorithm has been fine-tuned with real-time annotations to identify and extract valuable information from vision-sensing data. This information includes identifying the class of objects (e.g., whether a drone or a distractor) and the bounding box coordinates of the detected objects. The outputs from the vision sensing and position sensing neural network models are combined to train a meta-learner. This step aims to optimize beam prediction accuracy, demonstrating the proposed framework's effectiveness through sensing-aided UAV beam prediction analysis. The evaluation utilizes the DeepSense 6G dataset [17], a publicly available resource containing extensive multi-dimensional sensor and wireless communication data, to test the framework's efficacy.

7.3 System Model

This study investigates a communication system in which a mmWave base station serves a highly mobile UAV at varying heights in a real-world wireless environment. This section outlines the model used in the wireless communication system, including the scenario, problem formulation, and dataset description. This research delves into a wireless communication setup as depicted in Figure 7.1. In this environment, a mmWave base station establishes communication with a highly mobile UAV that is equipped with a GPS receiver for real-time location tracking. The UAV is outfitted with a single-antenna transmitter. Conversely, the base station is equipped with a uniform linear array (ULA) consisting of M elements and an RGB camera. The communication system utilizes orthogonal frequency division multiplexing (OFDM) featuring K subcarriers and a cyclic prefix of length D. Additionally, the base station employs a predetermined codebook $F = \{f_n\}_{n=1}^{N}$, with $f_n \in \mathbb{C}^{MX1}$, vector and N indicating the total count of beamforming vectors. The OFDM technology enables efficient data transmission from the UAV by partitioning the signal across various subcarriers, while the cyclic prefix serves to mitigate inter-symbol interference arising from multipath propagation. In scenarios where data is transmitted from the base station to the UAV (downlink), the wireless channel at a given time t and k_{th} subcarrier is represented by $h_k[t] \in \mathbb{C}^{MX1}$. Consequently, the signal received by the UAV is expressed as follows:

$$y_k[t] = h_k^T[t] f_n[t] x + z_k[t] \tag{7.1}$$

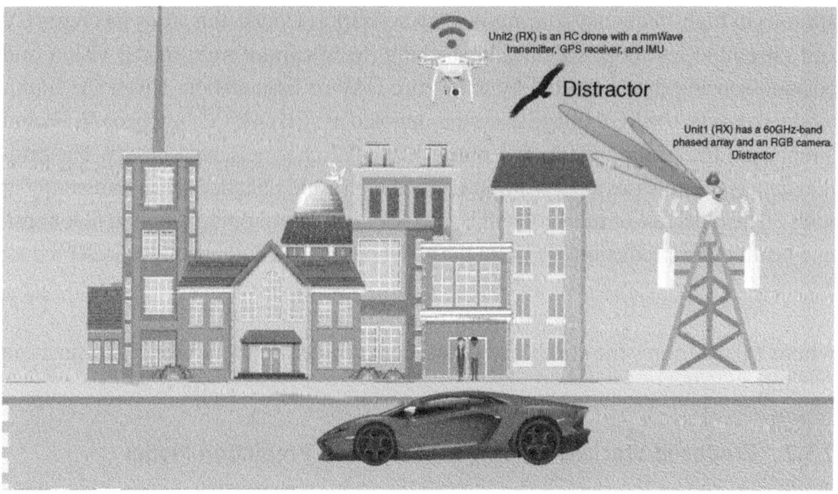

Figure 7.1 A real wireless communication scenario, the mmWave base station serves a UAV (radio-controlled drone) that is outfitted with both GPS and an inertial measurement unit (IMU). To ensure seamless connectivity, the base station utilizes additional sensing information, such as vision data and GPS coordinates, to identify the optimal beam for communication with the UAV.

where $f \in \mathcal{F}$ is the beamforming vector; $z_k[t]$ is the noise level. The selected beamforming vector $f^*[t] \in \mathcal{F}$ at time t is to minimize average SNR denoted by the following Eq. (7.1):

$$f^*[t] = \arg \max_{f_n[t] \in \mathcal{F}} \frac{1}{K} \sum_{k=1}^{K} \text{SNR} \left| h_k^T[t] f_n[t] \right|^2 \tag{7.2}$$

Millimeter-wave (mmWave) technology employs high-frequency electromagnetic waves to facilitate high-speed communication across short distances. The mmWave base station utilizes beamforming techniques to direct the communication signal towards the UAV, thereby ensuring a stable and efficient communication link. To optimize beamforming, the base station accounts for various factors, including the UAV's current position, its movement patterns, and environmental conditions.

7.3.1 Problem Formulation for Beam Prediction

To select the optimal beam $f^*[t]$ from the predefined codebook $\mathcal{F}$, the transmitted symbol x in Eq. (7.1) must adhere to the constraint $\mathbb{E}\left[|x|^2 \right] = P$, where P represents the average power per symbol, and maximize P as per Eq. (7.2). In conventional mmWave systems, determining the optimal beam involves either exhaustively searching through a given codebook or utilizing explicit channel state information. While exhaustive search incurs higher training overhead, acquiring channel information is arduous in high-frequency communication scenarios. Hence, this study leverages CV and ensemble learning, specifically stacking, to integrate multi-modal vision and position-sensing data collected by either the UAV or base station. Given the multi-modal datasets, the RGB image dataset is denoted as $X[t] \in \mathbb{R}^{HXWX3}$, where H, W, and 3 represent the height, width, and number of color channels, respectively. The position data g[t] comprises a 2D position vector, height, and distance of the transmitter at time t. The objective of this problem is to determine the mapping function f_Θ for optimal beam index prediction $f\hat{[t]}$ from the codebook $\mathcal{F}$, mathematically represented as:

$$f_\Theta : \mathcal{D}[\sqcup] \rightarrow f\hat{[t]} \tag{7.3}$$

where $\mathcal{D}[\sqcup]$ denotes the combined dataset encompassing both vision and position sensing.

7.3.2 Proposed Stacked Vision-Assisted Beam Prediction Model

This section introduces an ensemble stacking classifier designed to exploit two distinct data modalities: RGB images and positional data. The objective is to create a model that can accurately execute the mapping described in Eq. (7.3). The process involves stacking the outputs of YOLO-v5 with those of a subsequent neural network. Specifically, the outputs from YOLO-v5, which include bounding boxes and class probabilities for detected objects within an image, serve as the inputs for another neural network. This subsequent neural network undertakes additional processing and generates predictions based on these inputs.

The rationale behind stacking YOLO-v5 with a neural network lies in enhancing the overall performance of the system. For instance, the secondary neural network could leverage the detailed information provided by YOLO-v5 to either make comprehensive final predictions or refine object detection capabilities by adjusting the bounding box dimensions. This synergistic approach aims to harness the strengths of both components to achieve optimal prediction accuracy and system efficiency. To summarize, stacking YOLO-v5 with a neural network is an effective technique for developing advanced systems capable of object detection and other operations. This approach takes advantage of the combined capabilities of YOLO-v5 and neural networks to outperform the accuracy and performance possible with either technology alone. The combination of YOLO-v5's precise object detection and neural networks' adaptive processing power enables the development of extremely efficient and accurate prediction models. Figure 7.2 depicts the architecture and operational dynamics of this proposed stacking approach, demonstrating how different technologies interact to improve system capabilities.

7.4 Simulation and Analysis

To evaluate the proposed beam prediction methodology within a millimeter-wave (mmWave) drone communication framework, the DeepSense 6G dataset [17] is employed. The following provides a comprehensive overview of the simulation setup utilized in this study.

Algorithm 7.1 Stacked Neural Network for Optimal Beam Prediction

---

Require: Training data $(\mathbf{x}_{img}^{(i)}, \mathbf{x}_{pos}^{(i)}, \mathbf{y}^{(i)})\ i = 1^N$, where $\mathbf{x}_{img}^{(i)}$ and $\mathbf{x}_{pos}^{(i)}$ are the image and position vector inputs respectively for the ith sample and $\mathbf{y}^{(i)}$ is its corresponding true label vector.

Ensure: Meta learner $f(\cdot)$

Train image-based model to obtain predictions $\mathbf{y}_{img}^{(i)}$ for each sample $\mathbf{x}_{img}^{(i)}$
Train position-based model to obtain predictions $\mathbf{y}_{pos}^{(i)}$ for each sample $\mathbf{x}_{pos}^{(i)}$
for $i = 1$ to N **do**
Concatenate predictions to form input vector $\mathbf{x}^{(i)} = [\mathbf{y}_{img}^{(i)}, \mathbf{y}_{pos}^{(i)}]$
Obtain prediction vector $\hat{\mathbf{y}}^{(i)}$ using meta learner $f(\cdot)$
end for
Train meta learner using cross-entropy loss:
$\mathcal{L} = -\frac{1}{N}\sum_{i=1}^{N}\sum_{j=1}^{64} y_j^{(i)} \log(\hat{y}_j^{(i)})$

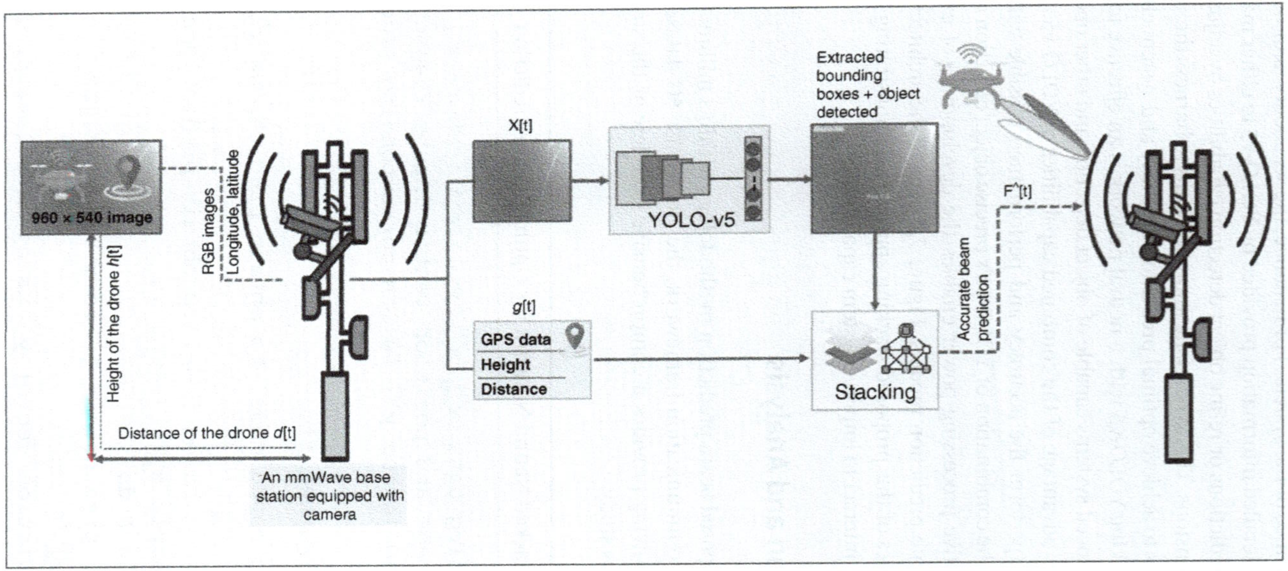

Figure 7.2 A schematic illustration of the stacking model architecture for optimizing the drone's wireless communication. The model accepts drone images as input, utilizes YOLO-v5 for object detection to identify the drone within the image, and subsequently generates bounding boxes around the detected drone. These bounding boxes, along with the wireless and position data, are then inputted into a stacking model, which predicts the optimal beam for wireless communication.

7.4.1 Description of the Dataset

The DeepSense 6G dataset, which is publicly accessible, encompasses a diverse range of multimodal data types, including vision sensing (images), GPS data, LiDAR, and radar sensing. Acquired from an actual wireless communication testbed, this dataset is particularly applicable for investigating the practicality and efficacy of the proposed beam prediction solution. This section aims to briefly review the scenario covered by the DeepSense 6G dataset and delve into the composition of the final dataset employed for the development and validation of the sensing-assisted beam prediction methodology. For exploring high-frequency wireless communications with UAVs, Scenario 23 from the DeepSense 6G dataset has been selected. This scenario features a stationary base station (Unit1 – RX) equipped with a 60-GHz band-phased array and an RGB camera of standard resolution. The phased array, designed to capture signals, is configured with an oversampled set of 64 predefined beams ($Q = 64$) and 16 elements ($M = 16$), enhancing its capability to receive a wide range of signals accurately.

To broaden the field of view (FoV) of the base station, both the RGB camera and the mmWave phased array are strategically placed on a table, approximately 1.5 m above ground level, oriented upwards. This setup ensures comprehensive coverage and signal reception from various angles. Unit2 (RX), an RC drone, is equipped with GPS, a mmWave transmitter, and inertial measurement units (IMU). The drone's transmitter, featuring a quasi-omnidirectional antenna, consistently operates at the 60 GHz frequency, ensuring continuous communication with the base station.

The scenario is specifically designed to augment the dataset's diversity, allowing the drone to operate at varying heights, distances, and speeds relative to the base station. This variability introduces a wide range of conditions and challenges typical of real-world UAV operations in high-frequency wireless communication settings, making it an invaluable resource for testing and validating the proposed beam prediction solution.

7.4.2 Configuration for Simulation

The simulation of the sensing-aided beam prediction model leverages the diverse data collected across different sensing modalities, including position, height, distance, and visual information. For analytical purposes, the data related to position, height, and distance are amalgamated into a single modality, while visual data is treated as an independent modality. The dataset is partitioned into training and validation subsets, adhering to a 70/30 split to facilitate the model's training and subsequent evaluation.

7.4.2.1 YOLO-v5 Training Process

The YOLO-v5 framework is employed to discern the UAV and pinpoint its location (via bounding boxes) within the input imagery, which also features distractor elements. These input images are standardized to dimensions of 960×540 pixels. For the training phase, YOLO-v5 is configured with a batch size of 8 and undergoes 100 epochs of training. The dataset earmarked for this process comprises 600 images for the training segment and 29 images dedicated to validation purposes. Post-training, the YOLO-v5 model is tasked with identifying the UAV and its bounding boxes across the entirety of the DeepSense image dataset, ensuring a comprehensive application of the trained model for drone detection and localization.

Training neural network: To train the neural network, inputs include bounding box coordinates, height, GPS (longitude, latitude), and distance from wireless sensors. The neural network consists of two dense layers, each with 512 neurons, and uses a rectified linear unit activation function. The output layer has 64 classes and uses softmax activation with sparse categorical cross-entropy as the loss function. The neural network is trained with a 90% train-to-10% test split across 100 epochs with a batch size of 32. The learning rate is set to 0.01, with LR decay at epochs 20, 40, and 80 with a reduction factor of 0.1. We fine-tune the hyper-parameters of the networks based on the values in Table 7.1 of the study. For the vision-aided approach, we use a stacking model that combines the YOLO-v5 and Neural Network models to predict the optimal beam indices using the same training and validation sets as the prior modality-specific models. To evaluate the proposed models, we employ a holdout test set that was not used during training or validation. We evaluate the effectiveness of the proposed models by comparing them to state-of-the-art approaches. For this examination, we employ conventional metrics like as precision, recall, accuracy, and the F1-score. Overall, the simulation scenario aims to evaluate the efficacy of the proposed models in increasing beam prediction accuracy across various sensing modalities.

Table 7.1 Hyper-parameters for design and training.

Parameters	YOLO-v5 Training	Neural Network
Input	960×540 Images	Bounding box
Batch size	8	32
Epochs	100	100
Learning rate	0.01	0.01

7.4.3 Results and Analysis

The evaluation of the proposed model's performance reveals noteworthy outcomes in terms of precision, recall, F1-score, and overall accuracy. Specifically: the model exhibits a precision rate of 0.8888, signifying that its positive predictions are accurate 88.88% of the time. This high precision underscores the model's effectiveness in correctly identifying true positive instances among all positive predictions. With a recall rate of 0.8855, the model demonstrates its capability to correctly identify 88.55% of all actual positive cases. This measure is crucial for scenarios where missing a positive instance could have significant consequences, indicating the model's robustness in detecting relevant signals. The model's F1-score stands at 0.8853, as detailed in Table 7.2. The F1-score, being the harmonic mean of precision and recall, offers a balanced view of the model's performance, especially in situations where an equilibrium between precision and recall is desired. The overall accuracy of the model is reported at 0.8910, meaning it correctly classifies 89.10% of all instances. This metric highlights the model's general effectiveness across a variety of conditions and instances.

These results, visualized in Figure 7.3, collectively demonstrate that the proposed beam prediction model is both reliable and efficient. Its high precision and recall rates suggest that it can serve effectively in various practical applications, such as monitoring, surveillance, and search-and-rescue operations, by reliably locating and identifying drones in challenging outdoor environments. The F1-score and accuracy further reinforce the model's robustness and its potential as a valuable tool in the context of mmWave drone communication systems.

The performance of the proposed model in predicting instances of the target variable stands out as exceptionally effective, as evidenced by its accuracy and recall scores of 0.8888 and 0.8855, respectively. The model's F1-score of 0.8853 and the accuracy score of 0.8910 further highlight its excellent predictive capabilities. When compared to alternative approaches, such as position-based and vision-based models, the proposed stacking model demonstrates excellent performance, with recall scores of 0.5603 and 0.8549, and accuracy scores of 0.6034 and 0.8632, respectively.

Table 7.2 Model evaluation metrics.

Model	Precision	Recall	F1-Score	Top-1 Accuracy
Vision	0.8567	0.8549	0.8587	0.8632
Position	0.5803	0.5603	0.5788	0.6034
Proposed	0.8888	0.8855	0.8853	0.8910

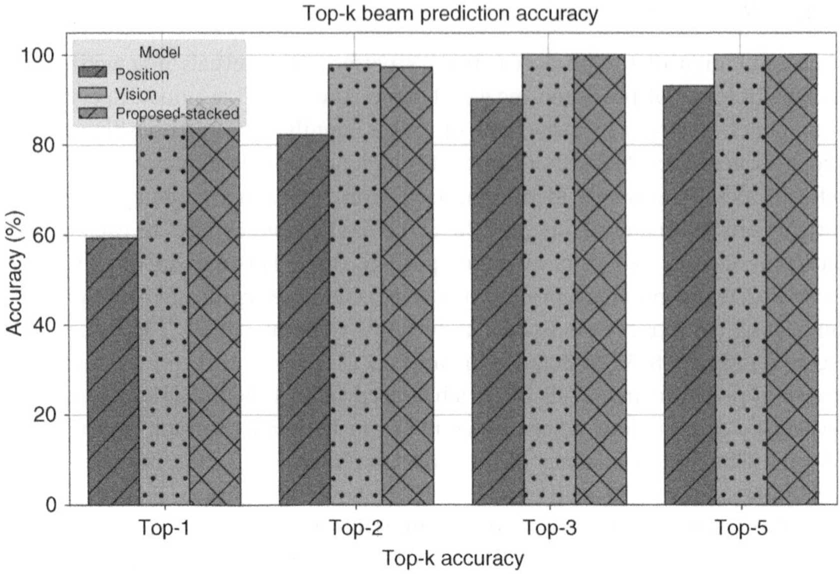

Figure 7.3 The plot displays the accuracy scores for position-based prediction, vision-based prediction, and the proposed stacked model for predicting the top K-beams.

These results underscore the proposed model's enhanced predictive power and its ability to accurately classify instances of the target variable with a high degree of reliability. Particularly, the model achieves higher precision, recall, F1-score, and accuracy compared to both vision-based and position-based models. This dominance indicates that the proposed model is more adept at identifying instances of the target variable, making it a more reliable choice for accurate predictions across a range of applications.

In conclusion, the assessment metrics given for the proposed model demonstrate its high effectiveness in predicting target variable occurrences, significantly outperforming vision-based and position-based models. This shows the model's potential as a highly reliable approach to accurate beam prediction in mmWave drone communication systems, with important implications for monitoring, surveillance, and search-and-rescue missions.

The graph illustrated in Figure 7.4, involving four distinct methodologies – actual beam power, vision-based, position-based, and the proposed stacking method – reveals notable enhancements in beam gains when applying the stacking approach over other single-modality strategies such as vision and position alone. This analysis underscores the efficacy of the proposed stacking method, although it also points out minor discrepancies between the estimated beam gains and those initially designed for various UAV positions.

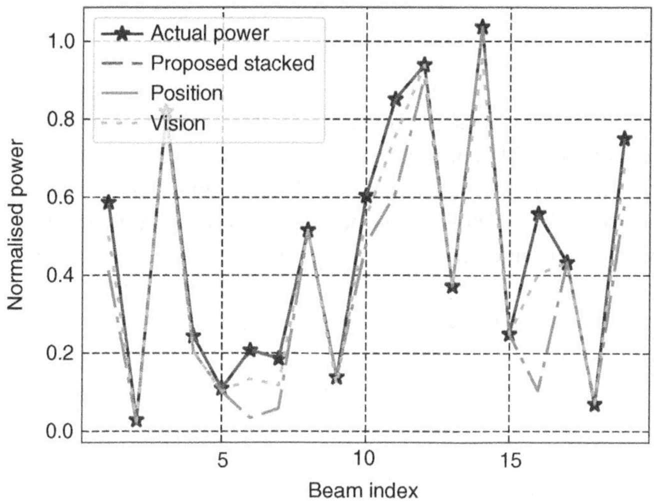

Figure 7.4 The comparison of the top-1 normalized power across different approaches – vision, position, and the proposed method – against the actual power derived from a codebook illustrates the effectiveness of each approach in accurately predicting beam alignment in mmWave communication systems. This comparison is essential for understanding how well each method can approximate the ideal beamforming vectors needed to maximize communication efficiency and reliability.

The graph provided in Figure 7.4 shows the power levels of both actual and predicted beams, indicating that a higher top-1 accuracy leads to predictions closely mirroring the actual beam power graph. Conversely, lower top-1 accuracy introduces a marked divergence between actual and predicted power, leading to potential performance declines, which are less than ideal.

These insights are particularly compelling, emphasizing the advantage of integrating position and vision-based methods to refine beam prediction in drones operating within mmWave communication frameworks. By adopting such sophisticated techniques, it is possible to enhance the precision and dependability of the communication process, thereby facilitating more effective data transmission and reception. This advancement holds significant promise for high-speed, low-latency communication needs found in applications such as autonomous vehicles, remote healthcare systems, and automated manufacturing processes.

Leveraging the combined strengths of mmWave technology and drones through these innovative methods opens a plethora of new possibilities, setting the stage for groundbreaking developments in the future. This approach not only optimizes the utilization of mmWave technology but also paves the way for exploring new horizons in various high-tech applications, ensuring that the potential of such advanced communication systems is fully realized.

7.5 Summary

This chapter discusses a potential solution to the challenge of establishing reliable communications in drones. A novel approach is introduced to address the challenge of establishing reliable communications in UAVs through a sensing-based prediction model that integrates visual and communication aspects to enhance beam prediction. The proposed method employs a stacking technique to predict K-beams with remarkable accuracy, achieving 90% accuracy for the top-1 beams and nearly 100% accuracy for the top-3 and top-5 beams. This research contributes to the body of knowledge on drone communication by presenting a solution that enhances the efficiency and dependability of communication in drones. The results demonstrate how the proposed method accurately predicts K-beams and enhances the performance of mmWave drone communication networks overall.

This work can be further enhanced by exploring alternative machine-learning techniques to improve the accuracy of prediction models. Additionally, the suggested technology could be integrated with different approaches, such as employing multiple antennas and adaptive modulation, to further enhance the overall efficiency of mmWave drone communication systems.

References

1 Charan, G., Hredzak, A., Stoddard, C. et al. (2022). Towards real-world 6G drone communication: position and camera aided beam prediction. In: *GLOBECOM 2022–2022 IEEE Global Communications Conference*, Rio de Janeiro, Brazil, 2951–2956. IEEE. doi: 10.1109/GLOBECOM48099.2022.10000718.

2 Abubakar, A.I., Ahmad, I., Omeke, K.G. et al. (2023). A survey on energy optimization techniques in UAV-based cellular networks: from conventional to machine learning approaches. *Drones* 7 (3): 214.

3 Hua, Z., Lu, Y., Pan, G. et al. (2023). Computer vision aided mmWave UAV communication systems. *IEEE Internet of Things Journal* 10: 12548–12561.

4 Ahmad, I., Kaur, J., Abbas, H.T. et al. (2022). UAV-assisted 5G networks for optimised coverage under dynamic traffic load. *2022 IEEE International Symposium on Antennas and Propagation and USNC-URSI Radio Science Meeting (AP-S/URSI)*, Denver, CO, USA, 1692–1693. IEEE. doi: 10.1109/AP-S/USNC-URSI47032.2022.9886848.

5 Charan, G, Hredzak, A., and Alkhateeb, A. (2023). Millimeter wave drones with cameras: Computer vision aided wireless beam prediction. *2023 IEEE International Conference on Communications Workshops (ICC Workshops)*, Rome, Italy, 1896–1901. IEEE. doi: 10.1109/ICCWorkshops57953.2023.10283784

6 Hur, S., Kim, T., Love, D.J. et al. (2011). Multilevel millimeter wave beamforming for wireless backhaul. *2011 IEEE Globecom Workshops (Gc Wkshps)*, Houston, TX, USA, 253–257. IEEE. doi: 10.1109/GLOCOMW.2011.6162448.

7 Alkhateeb, A., El Ayach, O., Leus, G., and Heath, R.W. (2014). Channel estimation and hybrid precoding for millimeter wave cellular systems. *IEEE Journal of Selected Topics in Signal Processing* 8 (5): 831–846.

8 Zhong, W., Gu, Y., Zhu, Q. et al. (2020). A novel 3D beam training strategy for mmWave UAV communications. *2020 14th European Conference on Antennas and Propagation (EuCAP)*, Copenhagen, Denmark, 1–5. IEEE. doi: 10.23919/EuCAP48036.2020.9135449

9 Morais, J., Behboodi, A., Pezeshki, H., and Alkhateeb, A. (2022). Position aided beam prediction in the real world: how useful GPS locations actually are? *arXiv preprint arXiv:220509054.*

10 Charan G, Osman T, Hredzak A et al. (2022). Vision-position multi-modal beam prediction using real millimeter wave datasets. *2022 IEEE Wireless Communications and Networking Conference (WCNC)*, Austin, TX, USA, 2727–2731. IEEE. doi: 10.1109/WCNC51071.2022.9771835

11 Charan, G., Demirhan, U., Morais, J. et al. (2022). Multi-modal beam prediction challenge 2022: towards generalization. *arXiv preprint arXiv:220907519.*

12 Jiang, S., Charan, G., and Alkhateeb, A. (2022). Lidar aided future beam prediction in real-world millimeter wave v2i communications. *IEEE Wireless Communications Letters* 12: 212–216.

13 Demirhan, U. and Alkhateeb, A. (2022). Radar aided 6G beam prediction: deep learning algorithms and real-world demonstration. *2022 IEEE Wireless Communications and Networking Conference (WCNC)*, Austin, TX, USA, 2655–2660. IEEE. doi: 10.1109/WCNC51071.2022.9771564

14 Khan, A.R., Ahmed, I., Mohjazi, L. et al. (2023). Latency-aware blockage prediction in vision-aided federated wireless networks. *Frontiers in Communications and Networks* 4: 1130844.

15 Skondras, E., Kosmopoulos, I., Michailidis, E.T. et al. (2022). A group handover scheme for supporting drone services in IoT-based 5G network architectures. *Drones* 6 (12): 425.

16 Roy, D., Salehi, B., Banou, S. et al. (2022). Going beyond RF: how AI-enabled multimodal beamforming will shape the nextg standard. *arXiv preprint arXiv:220316706.*

17 Alkhateeb, A., Charan, G., Osman, T. et al. (2022). DeepSense 6G: A large-scale real-world multi-modal sensing and communication dataset. *arXiv preprint arXiv:221109769.*

8

Multi-modal-Sensing System for Detection and Tracking of Mind Wandering

Sara Khosravi, Haobo Li, Ahsan Raza Khan, Ahmed Zoha, and Rami Ghannam

James Watt School of Engineering, University of Glasgow, Glasgow, UK

Mind wandering, a common occurrence where attention shifts from external stimuli to internal thoughts, significantly impacts students' learning outcomes. It hampers comprehension, lowers academic performance, obstructs critical thinking, and fosters disengagement in classroom activities. This research endeavors to introduce a novel approach for identifying and monitoring mind-wandering episodes in university students. Employing wearable sensors such as galvanic skin response (GSR), photoplethysmography (PPG), eye trackers, and advanced machine learning techniques, the study presents a proof of concept for this multimodal methodology. The investigation delves into the correlation between prolonged fixation duration and mind-wandering instances and explores how the instructor's presence influences fixation patterns, consequently affecting the frequency and occurrence of mind-wandering episodes. Additionally, the study assesses the feasibility of utilizing eye trackers alongside GSR and PPG sensors within a wearable multisensory data collection framework. The effectiveness of the wearable multisensory device is evaluated by 10 participants, university students aged between 21 and 30, revealing that with sensor fusion, the support vector machine (SVM) and gated recurrent unit (GRU) models achieve maximum accuracies of 86.53% and 89.86%, respectively. Furthermore, the research observes that participants tend to fixate on instructors more frequently just before experiencing mind-wandering episodes.

Multimodal Intelligent Sensing in Modern Applications, First Edition.
Edited by Masood Ur Rehman, Ahmed Zoha, Muhammad Ali Jamshed, and Naeem Ramzan.

Published 2025 by John Wiley & Sons, Inc.

8.1 Mind Wandering

In today's modern world, the abundance of distractions poses a significant challenge to maintaining focus and concentration. Despite educators' efforts to create stimulating and interactive learning environments, a considerable number of students continue to grapple with disengagement and lack of attentiveness [1]. This prevalent occurrence, characterized by the mind veering off course from the present task and becoming immersed in unrelated thoughts, is commonly referred to as mind wandering. This phenomenon not only disrupts the learning process but also undermines academic performance and overall cognitive engagement. As such, understanding and effectively addressing mind-wandering behaviors have become crucial endeavors in enhancing educational experiences and maximizing learning outcomes.

Mind wandering, a phenomenon closely intertwined with learning, has been extensively studied in the context of lectures [2, 3]. Researchers have employed two primary methodologies for data collection in mind-wandering studies: the "self-report" and "probe-caught" techniques [4, 5]. In the "probe-caught" approach, participants are intermittently interrupted and asked to reflect on their current mental state, while the "self-report" method relies on participants voluntarily reporting instances of mind wandering. This investigation adopts the "probe-caught" methodology, integrating a multisensory framework encompassing biological, physiological, and gaze-tracking sensors to facilitate data collection during experimental sessions.

In recent years, there has been a growing interest among researchers in leveraging innovative technologies to track and comprehend mind wandering in educational contexts, with the aim of enhancing learning outcomes. Efforts have included the utilization of mobile devices like smartphones and tablets to collect self-reported data from students, as well as the deployment of sensors capable of detecting biological indicators such as temperature and pressure. However, these methodologies can be intrusive for participants and necessitate extensive data analysis. In contrast, employing a multisensory approach offers a more reliable solution by integrating multiple data sources that can be validated through experimental validation. Therefore, a multisensory solution utilizing wearable sensors such as galvanic skin response (GSR), photoplethysmography (PPG), and eye trackers combined with machine learning presents a more efficient means of monitoring mind wandering in students [6]. These wearable sensors are comfortable to wear and enable real-time, rapid data analysis, as depicted in Figure 8.1. Moreover, they can be seamlessly integrated with data analysis algorithms, facilitating swift and real-time data analysis processes.

Students' responses to educational materials, including lecture slides have been explored by analyzing their visual attention using eye-tracking technology [7–9].

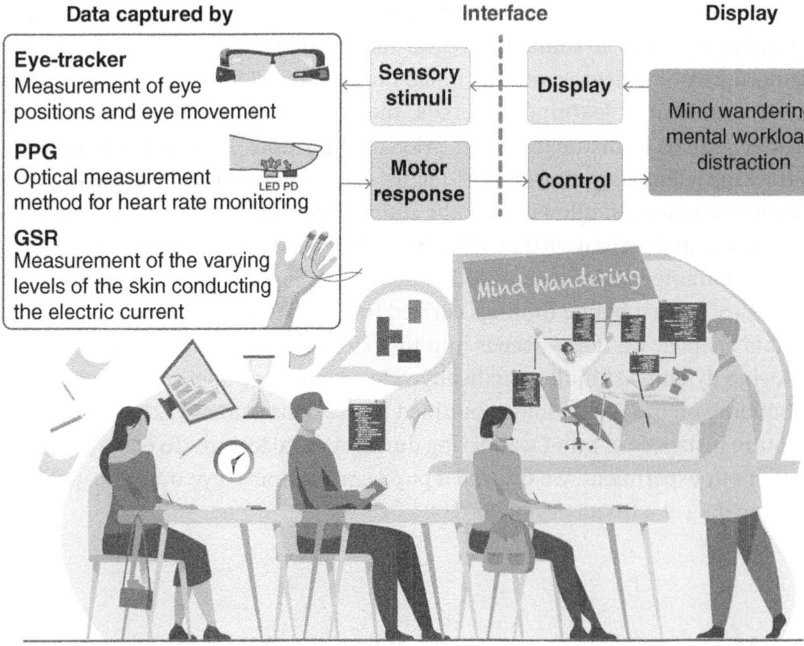

Figure 8.1 The conceptual framework outlines a comprehensive multisensory methodology designed to detect instances of mind wandering within the educational setting. This innovative approach integrates various sensory modalities, including PPG, GSR, and eye-tracking technology, to provide a holistic understanding of cognitive states during learning activities. At its core, the methodology employs a sophisticated block diagram that orchestrates the collection and analysis of physiological signals obtained from students. The envisioned block diagram illustrates the seamless integration of PPG, GSR, and eye-tracking devices, each contributing unique insights into the individual's cognitive processes. The PPG sensor captures real-time changes in blood volume, offering valuable data on physiological arousal levels. Concurrently, the GSR sensor measures skin conductance, providing insights into emotional responses and cognitive engagement. Furthermore, the eye-tracking technology monitors gaze patterns and fixation durations, elucidating attentional shifts and cognitive focus. By synergistically combining data from these sensory modalities, the methodology aims to create a robust framework for identifying and characterizing mind-wandering episodes in students. This comprehensive approach not only enhances our understanding of cognitive dynamics during learning but also opens avenues for developing personalized interventions to optimize educational outcomes.

However, it is important to recognize that visual attention represents just one facet of attentiveness. Even when an individual seems to be concentrating on a task and their gaze is fixated on the intended subject, their thoughts may be elsewhere. The relationship between eye movements and cognitive processes is unquestionably significant.

In measuring cognitive load, attention, focus, and mind wandering, analyzing gaze behavior offers a valuable avenue for monitoring visual attention [10]. Advanced wearable devices now facilitate the tracking of gaze movements across specific segments of learning materials, thereby enabling the identification of potential sources of distraction. Our prior investigations have utilized wearable eye-tracking technology to observe students' visual attention while engaging with educational materials, allowing for the differentiation of behaviors within text-based and graphic-based visuals [11]. Nonetheless, since visual attention alone does not guarantee complete task focus, the integration of an additional indicator is necessary to corroborate findings derived from wearable eye-tracking devices.

In this chapter, we delve deeper into the phenomenon of mind wandering by employing dynamic stimuli, specifically video content, as opposed to static stimuli. Building on the methodology outlined by previous studies [12], our goal was to capture instances of mind wandering during a video lecture. To ensure the precision of our experiment, we utilized a pupil core wearable eye tracker capable of recording data at a frequency of 200 Hz.

8.2 Multi-modal Wearable Systems for Mind-Wandering Detection and Monitoring

Aimed at enhancing learning and teaching experiences, utilizing a multisensory system comprising various sensors for collecting and processing diverse physiological and neurological data, wearable devices hold promise for improving the accuracy of attention and mind-wandering monitoring. In recent studies, the multisensory approach has been employed to assess mental workload, identify emotional states, and analyze the learning and information processing processes.

Various physiological biomarkers have been utilized for the mind-wandering detection such as heart rate and skin conductance [13]. For instance, Campisi et al. [14] investigated the mental workload of students while web browsing using augmented reality, Mutlu-Bayraktar et al. [15] examined the effects of split attention in multimedia learning environments using eye-tracking and EEG sensors, Saffaryazdi et al. [16] analyzed people's emotional state during stimuli exposure using GSR and PPG sensors, and Srivastava et al. [17] investigated the process of information learning and visual attention using eye trackers. Among these, pressure sensors [18], EEG, GSR, PPG, and eye trackers have been identified as the most accurate options [19]. However, each technology presents its own advantages and limitations, as summarized in Table 8.1. The choice of technology depends on the specific context and environment. The selection of the technology to be used depends on the specific context and environment.

Table 8.1 Pros and cons of various sensors for measuring mind wandering.

Sensors	Pros	Cons
Respiration/ pressure	Unique to its specific purpose	Lack of movement, discomfort
EEG	Provides data on cognitive state	Sensitive to environment
Heart rate sensors	Ease of access	Need for validation
GSR	Measuring the emotional state	Sensitive to movement
Eye tracker	Can be used with moving target	Only collect visual data
PPG	Provides a more indirect measurement	Inaccuracy during daily activities

In this chapter, we aim to demonstrate how a fusion of sensors can be utilized to assess mind wandering. In practical applications, it is impractical to measure all signals simultaneously. Therefore, in our study, we chose to focus on analyzing GSR, PPG, and eye-tracking sensors. GSR serves as a widely utilized physiological measure of emotional arousal, offering an indirect indication of emotional arousal through the measurement of sweat gland activity. Increased sweat gland activity is often correlated with heightened emotional arousal. However, it is crucial to acknowledge that GSR is not a direct measure of emotions but rather captures physiological arousal that may or may not be linked to emotional states. Moreover, GSR readings can be influenced by various factors such as skin hydration levels, temperature, and skin pressure [19]. These signals are well suited for integration into a simple wearable device, allowing for data collection from students without causing discomfort or privacy concerns.

In our earlier research, we conjectured that the duration of visual attention or fixation duration would be prolonged during episodes of mind wandering. Additionally, we hypothesized that the presence of an instructor in video content could amplify student mind wandering. This study aims to validate both hypotheses using a combination of multimodal sensors. The primary objective is to leverage eye tracking, PPG, and GSR sensors and process their data through machine learning techniques such as support vector machine (SVM) and gated recurrent unit (GRU). These methods offer practical applicability in classroom settings compared to EEG, which poses challenges in terms of accessibility and convenience for students. The utilization of wearable devices incorporating eye-tracking, PPG, and GSR signals ensures data collection without causing discomfort or privacy concerns to students.

8.2.1 Wearable Eye Trackers for Gaze Measurements

Most desktop eye trackers utilize a webcam to measure participant eye data, offering affordability and ease of use, leading to their widespread adoption by researchers. However, desktop eye trackers, operating at a distance beyond 50 cm, may be susceptible to interference from background lighting such as sunlight. In contrast, wearable eye trackers, positioned closer to the eye, can precisely monitor pupil movement. They also have the capability to track and record gaze points occurring outside the monitor's frame, enabling the identification of sources of distraction.

In this experiment, data were recorded using Pupil Core eye-tracking glasses at a resolution of 400×400 pixels and a sampling rate of 200 Hz [7]. Pupil Core eye trackers are capable of measuring gaze behavior in both 2D and 3D formats. The 3D gaze collection utilizes pye3d for 3D pupil detection [8]. Additionally, a confidence level is provided to establish a threshold for accurate pupil detection, ranging from 0 (indicating no pupil detection) to 1 (highest probability of pupil detection). To compute fixations, the Pupil Core eye tracker employs a dispersion-based algorithm [9] that can be implemented in both online and offline modes. For this study, an online method was utilized for both pupil and gaze detection, with fixation calculation based on a dispersion level of 1.5° of visual angle and a minimum duration of 100 ms [10]. Data with a confidence value below ~0.6 were excluded from the analysis.

Our recent research aimed to investigate differences in gaze behavior between text and graphic representations in lecture slides, recognizing the pivotal role of information presentation format in student outcomes. Drawing from a study outlined in [12], we conducted an experiment featuring an 18-minute video lecture on International Comparisons in Education. Prior to viewing the video, participants responded to 5 questions, followed by 18 additional questions afterward. To delineate between the slide content and the teacher's window, we defined areas of interest (AOIs). In gaze-tracking technology, AOIs denote defined regions or specific parts of a visual stimulus that capture a person's gaze. These AOIs enable the analysis and identification of which parts of an image are being attended to, as well as the duration of attention to each area.

8.2.2 Wearable GSR and PPG Sensors for Physiological Measurements

We constructed emotion classification models employing compact and lightweight PPG and GSR Shimmer3 sensors. Our development included creating a comprehensive application that incorporated a database for data storage and utilized images from the Geneva Affective Picture Database (GAPED) to evoke emotions in participants. The post-processing phase entailed extracting statistical parameters and power spectral density (PSD) as features, followed by employing SVM and k-nearest neighbors (KNNs) algorithms as classifiers.

In previous research, other physiological multi-sensors, such as EMG and GSR signals, have been employed for real-time emotion detection in gaming scenarios [21]. Our study focused on categorizing emotions using arousal and valence scales, with arousal divided into normal, high, and very high categories, and valence categorized as positive or negative. Additionally, another study described in [22] utilized EEG, GSR, EMG, BVP, EOG, and skin temperature signals to develop effective databases for emotion recognition. These databases contained speech, visual, or audio–visual data, with emotional labels including neutral, anxiety, amusement, sadness, joy, disgust, anger, surprise, and fear. Furthermore, in a related investigation, EEG, ECG, GSR, and facial activity were measured to explore the relationship between emotional attributes and personality traits [23]. Emotions were elicited using various stimuli, such as audio and video, and expressed through different modalities, including facial expressions, speech, and physiological responses, necessitating the creation of a multimodal database for implicit personality and affect recognition.

8.3 Design of Multi-modal Wearable System

The proposed multi-modal system is based on a number of sensors, multiple subjects, data fusion, and acquisition and analysis.

Two hypotheses will be tested utilizing this system to explore the influence of mind wandering:

Hypothesis 1: We anticipate a positive correlation between participants' fixation duration and their engagement in mind wandering during video-based learning sessions.

Hypothesis 2: We predict that the presence of an instructor in video-based teaching and learning environments will result in a higher incidence of mind wandering during lectures.

8.3.1 Selection of Sensor

Electrodermal activity, indicating the skin's resistance to electrical flow, is measurable through a GSR sensor. This process involves positioning two electrodes on the skin and administering a small electric charge, typically around 1.0 V, while recording the ensuing current between the electrodes. The recorded current is then converted into micro-ohms or microamperes, offering insight into the skin's resistance to electrical flow. Emotional stimuli can modulate this resistance, rendering it a valuable metric for discerning changes in emotional arousal.

The sympathetic nervous system, governing the body's "fight-or-flight" response, regulates sweat gland function. GSR sensors are adept at detecting even subtle fluctuations in electrical conductivity, enabling the identification of emotional stimuli without overt sweating. However, it is essential to recognize that GSR data alone cannot pinpoint the cause of emotional arousal. The interpretation of results necessitates consideration of other physiological and contextual factors. Consequently, meticulous control of external variables in experimental design is imperative for accurately interpreting GSR outcomes.

Elevated sweat gland activity correlates with increased perspiration and decreased skin resistance. In GSR measurements, conductance serves as the standard unit, calculated as the inverse of resistance: $Conductance = 1/Resistance$. This simplifies signal interpretation, as heightened sweat gland activity corresponds to higher skin conductance levels.

For emotional research purposes, the most prevalent method to measure a GSR signal involves employing a constant voltage system, also known as the exosomatic method. In this approach, the GSR sensor applies a steady voltage, typically around 0.5 V, to the two electrodes in contact with the skin. A minor resistance is inserted in series with the voltage supplier and the electrodes to gauge skin conductance and its fluctuations using Ohm's law ($Voltage = Current \times Resistance = Current/Conductance$). By maintaining a constant voltage, any variations in the current flow through the electrodes indicate changes in the skin's electrical properties, particularly sweat gland activity.

The link between sympathetic nervous activity, assessed through skin conductance (SC) and skin temperature (ST), and attentional states has prompted the utilization of physiology to monitor mind wandering [24]. Previous studies have demonstrated that heightened levels of mind wandering correlate with overall reduced SC levels [25]. Despite these findings, there have been no endeavors to devise automated mind-wandering detectors based on SC or ST signals, nor has there been any exploration of the relationship between ST and mind wandering. To address this gap, we compiled a substantial dataset wherein students were periodically prompted to report instances of mind wandering while engaging in computerized learning from instructional texts. Subsequently, machine learning models were trained on these signals to forecast mind-wandering occurrences.

8.3.2 Selection of Participant Groups and Testing Environment

Ten participants voluntarily participated in our experiments to evaluate the wearable multisensory device. These participants were postgraduate engineering students at the university, comprising both males and females aged between 21 and 30 years. The experiments were conducted with approval from the Glasgow University ethics committee, and participants provided their consent by completing a consent form.

8.3.3 Data Collection Process

The stimulus consisted of an 18-minute lecture on international education [12]. Prior to the lecture, participants completed a pre-lecture questionnaire comprising five questions related to their familiarity with the lecture concept. Subsequently, participants viewed the lecture presented as a video. Following the lecture, participants completed a post-lecture questionnaire comprising 18 questions pertaining to the content of the lecture, along with 2 questions soliciting participants' overall experience ratings.

Figure 8.2 depicts the schematic diagram of the experimental setup. Participants were instructed to sit comfortably in front of a computer screen. Eye-tracking glasses, along with a GSR sensor and a PPG sensor, were positioned on the participant's

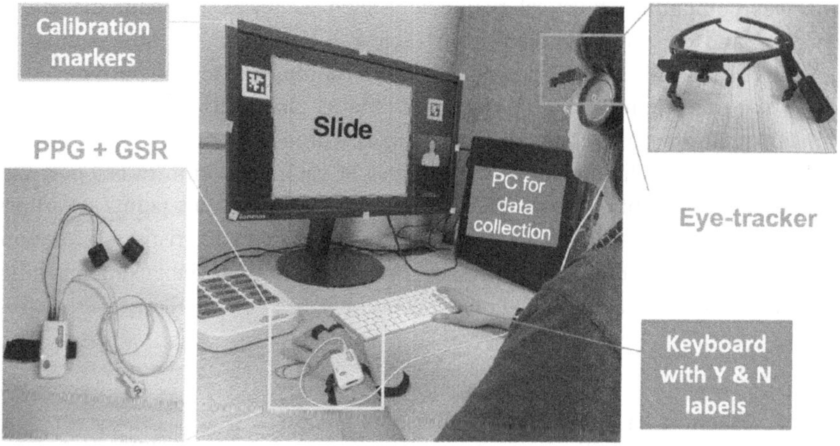

Figure 8.2 The experimental configuration encompasses a wearable device that integrates a comprehensive suite of multisensory components, including eye-tracking, GSR, and PPG sensors. The setup, as depicted in the accompanying figure, illustrates the calibration process alongside the key components: the Pupil Core headset for eye-tracking, Shimmer3 sensors for GSR and PPG measurements, and an individual participant engaged in the testing procedure. During the calibration phase, meticulous adjustments are made to ensure precise alignment and synchronization of the sensors with the participant's physiological responses. The Pupil Core headset facilitates real-time tracking of eye movements, capturing valuable insights into visual attention patterns. Meanwhile, the strategically positioned Shimmer3 sensors monitor changes in skin conductivity (GSR) and blood volume (PPG), indicating arousal levels and cardiovascular activity, respectively. As the participant undertakes the test, the wearable device seamlessly captures and processes a wealth of multimodal data, offering a holistic perspective on cognitive engagement and emotional arousal. This integrated approach enables a comprehensive analysis of the interplay between visual attention, physiological arousal, and cognitive states, laying the groundwork for advanced insights into human behavior and cognitive processes.

head and hand, respectively. Calibration and validation procedures were performed for the eye tracker. Two GSR sensors and one PPG sensor were affixed to the participant's fingers on the nondominant hand.

Participants received instructions regarding the test procedure, including guidelines on answering questions and identifying instances of mind wandering. Figure 8.3 displays the raw data collected from each individual's sensors (eye tracker, GSR, and PPG) both with and without occurrences of mind wandering. The acquisition frequency for the eye tracker is 200 Hz, while for the GSR and PPG sensors, it is 50 Hz.

The characteristic information extracted from the data presented in Figure 8.3 includes signal attributes like amplitude, shape, frequency, as well as the exact positions and values of signal peaks. Through a feature-extraction process, the SVM classifier derives this information from the original sensor data. This process involves computing statistical parameters such as mean, standard deviation, median, and other metrics detailed in Table 8.2. Subsequently, these computed features are furnished to the classifier for both training and testing purposes.

On the other hand, the GRU network obtains characteristic information by directly analyzing the temporal dependencies within the original signal. This is achieved through the update and reset gates of each GRU unit. In essence, the GRU network identifies the temporal relationships between data points at various time instances and utilizes them to encapsulate the characteristic information inherent in the raw data. The processing and fusion of these data using machine learning methods are elaborated on in Section 8.4.

8.3.4 Machine Learning and Multisensory Fusion

Before classification, numerical features are extracted from the raw sensor data measurements. The features utilized in this study are outlined in Table 8.2. The feature pool encompasses standard statistical parameters like mean, median, minimum, and standard deviation, along with higher-order statistical parameters such as skewness and kurtosis. Additionally, correlation-based features and signal energy features are incorporated into the feature set.

Two distinct machine learning methods, namely, a conventional SVM and GRU, are employed as classification models. The SVM, recognized for its robustness in indoor human activity recognition, gesture recognition, and biometric identification [26, 27], aims to construct a hyperplane to segregate feature points of different classes based on their distribution. Support vectors, located close to the decision boundary, influence the position and orientation of the hyperplane, while maximizing the margin between the positive and negative hyperplanes.

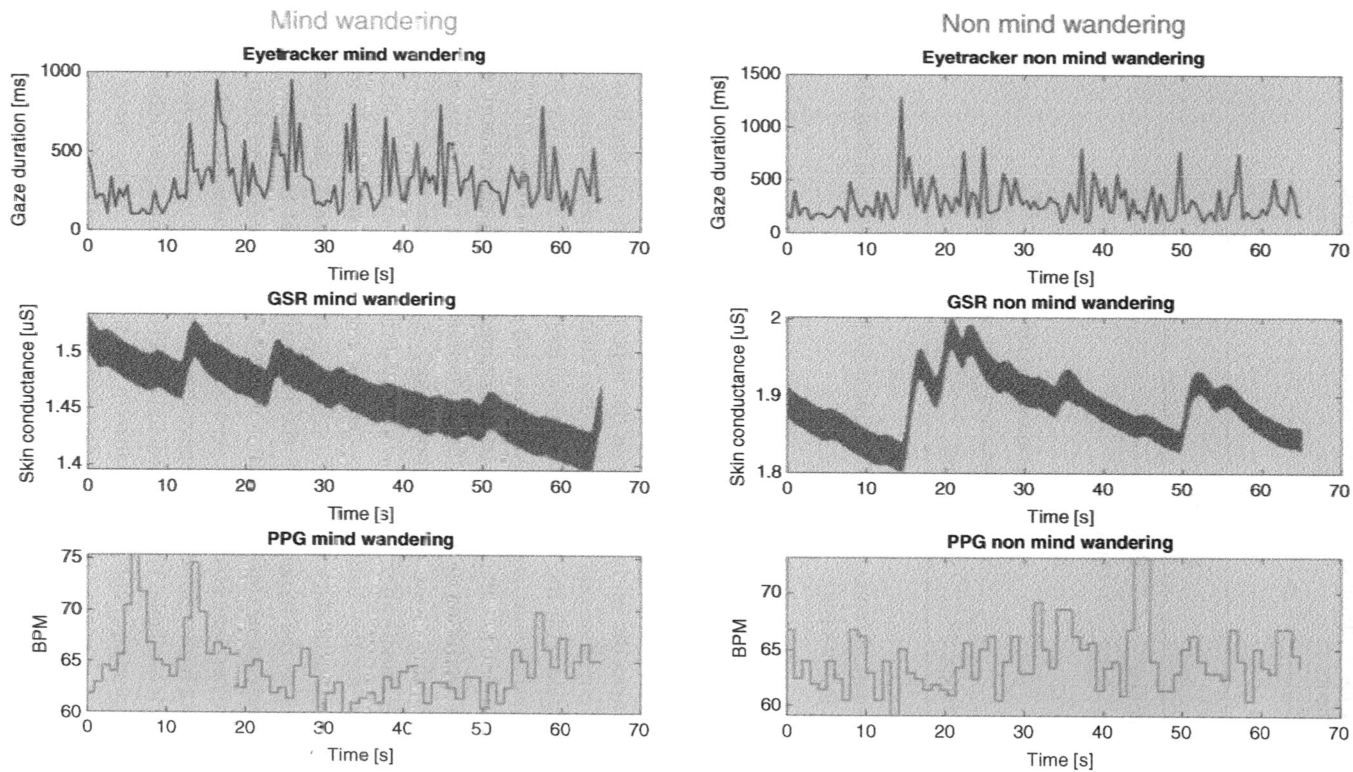

Figure 8.3 The raw data were collected from each sensor of the Pupil Core eye tracker, Shimmer3 GSR, and PPG sensors, respectively. Graphs on the left show the collected data with mind wandering, whereas the graphs on the right are for data with non-mind wandering.

Table 8.2 The list of numerical features extracted from the raw sensor data measurements.

Features	
Mean	**Skewness**
Standard deviation	Kurtosis
Median	Minimum
Mad	Range
25th quantile	Mean of the autocorrelation
75th quantile	Standard deviation of the autocorrelation
Inter quartile range	Signal energy

The mathematical representation of a linear SVM hyperplane and its objective function is as follows [28]:

$$h: x'W + b = 0 \tag{8.1}$$

$$\min_{W,b,\xi_i} \frac{1}{2} \| W \|^2 + C\sum_{i=1}^{n} \xi_i \left(C > 0, \xi_i \geq 0\right) \tag{8.2}$$

where W denotes the normal vector to the hyperplane and b is the bias value. C refers to the regularization parameter, also known as penalty factor, which is highly correlated with the tolerance of misclassification. The penalty factor is always greater than zero, and the larger factor will create a hard margin, and vice versa (soft margin); its value needs to be determined carefully since a hard margin may result in overfitting of the classifier. In our case, the penalty factor is set as one in the training of the classification model. ξ_i represents the slack variable related to the classification error, the SVM algorithm automatically allocates a slack variable for the feature points between the hyperplane and its margin, whereas the value of slack variable ($0 \leq \xi_i \leq 1$) is proportional to the distance of feature points to the hyperplane. In the circumstance that the feature points beyond the hyperplane (misclassification), the slack variable is larger than one.

If a linear hyperplane is not able to separate the feature points, the features can be mapped to a higher-dimensional space through a kernel function, where a linear boundary is available. The conventional kernel function includes higher-order polynomial (quadratic, cubic) and Gaussian function, whereas the choice of the kernel function depends on the data distribution and the optimal hyperplane to separate them. For this work, I chose the quadratic kernel function. SVM algorithm is suitable to implement on a multi-class problem by utilizing multiple binary classifiers via "one vs. one" approach, for instance, if there are N classes to distinguish, $N(N-1)/2$ times binary SVM will be computed to construct hyperplanes between each individual class. GRU is the improved version of the regular recurrent neural network (RNN) [29–31]. A standard GRU consists of two different gates, namely, the update gate and reset gate; compared to other variants of RNN, such as long short-term memory (LSTM), GRU has comparable performance with a simpler architecture, faster processing speed and less memory cost. The block diagram of a simple GRU is illustrated in Figure 8.4, and the mathematical expression of GRU is given below [29, 30]:

$$r(t)=\sigma\left(W_{rh}h_{t-1}+W_{rx}x_t+b_r\right) \tag{8.3}$$

$$u(t)=\sigma\left(W_{uh}h_{t-1}+W_{ux}x_t+b_u\right) \tag{8.4}$$

$$c(t)=\tanh\left(W_{ch}\left(r(t)\odot h(t-1)\right)+W_{cx}x_t+b_c\right) \tag{8.5}$$

$$h(t)=z(t)\odot c(t)+\left(1-z(t)\right)\odot h(t-1) \tag{8.6}$$

where $r(t)$ and $u(t)$ represent the output of the reset and update gate, respectively. σ denotes the sigmoid activation function. W refers to the weight index of gated units, whereas b is the bias value. ⊙⊙ refers to the Hadamard product of two vectors. $c(t)$ represents the output of the tanh operator, which receives a linear combination of the current input x(t) and the result of the Hadamard product between $r(t)$ and $h(t-1)$. $h(t)$ represents the current output of the GRU. The update gate $u(t)$ controls the ratio of current input information and historical information. When $u(t)$ is close to 1, most of the historical information is forgotten, and a larger volume of input information is taken from the current moment, and vice versa. In this paper, as GRU has the ability to extract useful time-dependent information from raw data, the input of GRU is a time-series signal rather than features.

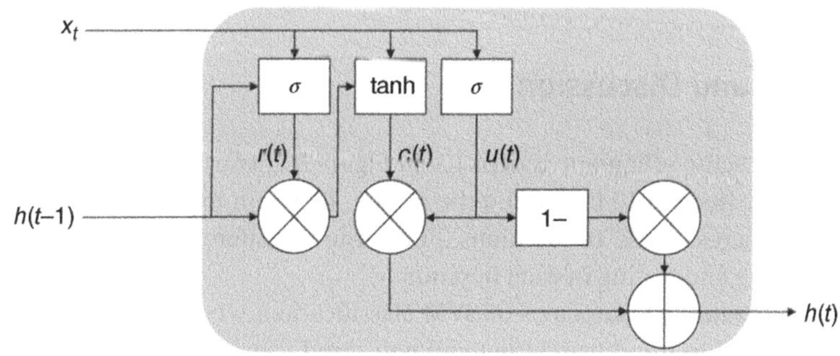

Figure 8.4 The block diagram of a simple GRU to process the recorded raw data with faster processing speed and less memory cost.

The confidence level of the classifier is a probability matrix with its size equal to nXm, n is the number of samples and m is the number of classes. It is used to measure the certainty of classifier decision-making, whereas, for each sample, the class yielding the highest confidence level will be chosen as the output class. The value of the confidence level is converted from the unnormalized classifier output through a softmax function, as shown below:

$$P_c = \frac{e_c}{\sum_{k=1}^{K} e_k} \tag{8.7}$$

where class c is the class of interest, P_c the confidence level of class c, e_c, and e_k denote the unnormalized classifier output of class c and class $k(k \leq K)$, respectively, and k is the number of classes. In this work, we have developed a decision-level fusion method to combine the confidence level of all sensing approaches. The fusion process [27, 32] is depicted in Eq. (8.8):

$$P_{Fusion}(s,c) = \sum_{n=1}^{N} P_n(s,c) \tag{8.8}$$

where P_n denotes the confidence level matrix for classifier n, P_{Fusion} is the confidence level matrix after fusion, s refers to the index of samples, and c indicates the class number. The confidence level matrix of different sensors shares the same dimension. The fusion of data from multiple sensors is the straightforward accumulation of their respective output confidence level matrices. The new prediction label is the class with the highest fusion confidence level. In our paper, n equates to 3, representing g classification results from eye tracker, GSR, and PPG, respectively. It is important to notice that the fusion of data is unrelated to the acquisition frequency of sensors because data fusion occurs at the decision level, following classification.

8.4 Results and Discussion

The experimental setup schematic is depicted in Figure 8.2. The training and testing procedure was iterated 10 times, and the results, shown in Figures 8.5, represent the average across these 10 iterations. Furthermore, different datasets were chosen for training and testing in each iteration.

Figure 8.5 illustrates the outcomes of SVM classification, where "class 1" and "class 2" correspond to mind wandering and non-mind wandering, respectively. The row elements of the confusion matrix represent the output class, while the column elements indicate the target class. The sum of the target class should

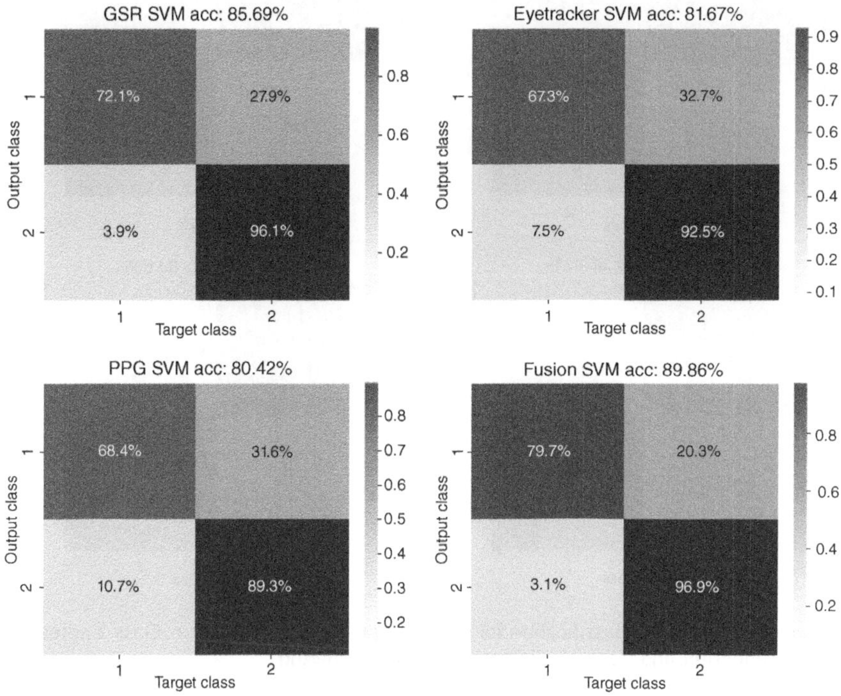

Figure 8.5 The SVM classification for GSR, eye tracker, PPG, and fusion. Class 1 notes mind wandering and class 2 indicates non-mind wandering.

equate to 100%. The right diagonal elements denote the correctly classified rate, whereas the left diagonal elements denote the misclassification rate. The processed results for individual sensors indicate accuracies of 80.97%, 76.81%, and 76.39% for GSR, eye tracker, and PPG, respectively. The SVM results demonstrate that the accuracy of sensor fusion is 86.53%.

In Figure 8.6, the outcomes of GRU classification are depicted, showcasing accuracies of 85.69%, 81.67%, and 80.42% for GSR, eye tracker, and PPG sensors, respectively, when processed individually. Employing GRU, the processing outcomes indicate that the accuracy of sensor fusion achieved is 89.86%. Fusion using GRU yields a subsequent enhancement of approximately 3.3% compared to fusion through the SVM algorithm.

Figure 8.7 displays the boxplot representing the 10 iterations of "training and testing." The blue circle signifies the mean value of the 10 different classification results, while the red line within the "box" denotes the median value. The upper and lower boundaries depict the maximum and minimum accuracy values of the 10 iterations of "training and testing" separately. The edges of the blue "box" represent the 25th and 75th percentiles of the classification results.

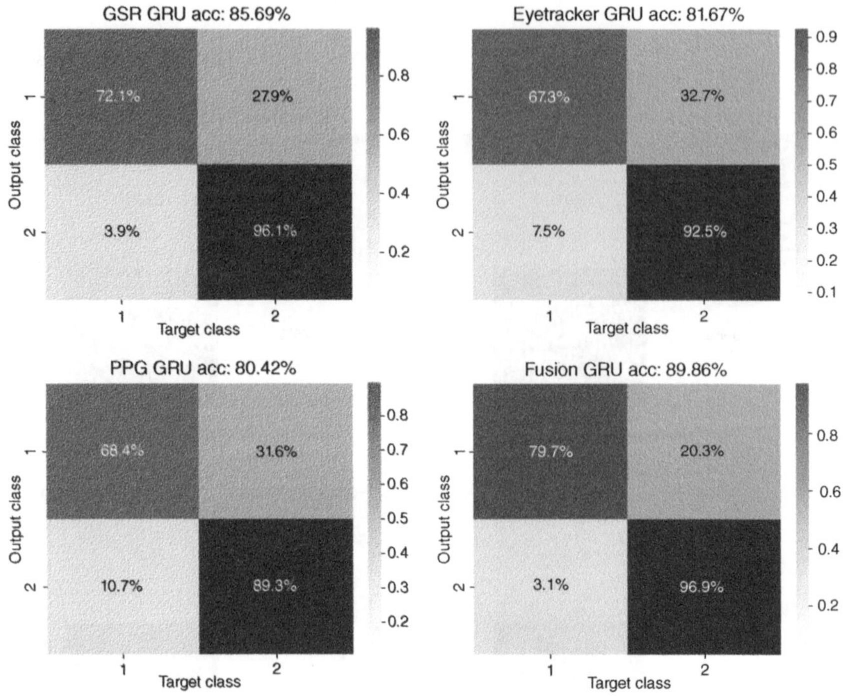

Figure 8.6 The GRU classification for GSR, eye tracker, PPG, and fusion. Class 1 notes mind wandering, and class 2 indicates non-mind wandering.

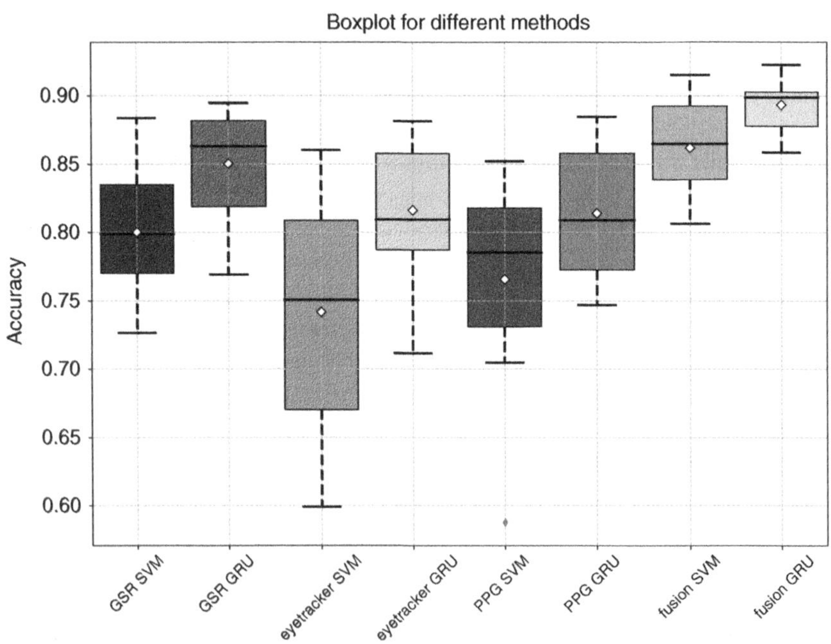

Figure 8.7 The boxplot of 10 iterations of training and testing using each individual sensor and their fusion using SVM and GRU.

Observations reveal that fusion with both SVM and GRU not only enhances the mean classification accuracy but also increases the variance of classification accuracy. In essence, the fusion process enhances the stability of the classification system. Moreover, compared to the SVM classifier, GRU-based RNNs yield a higher gain in the mean and variance of the 10 iterations of "training and testing." The fluctuations in performance exhibited by the wearable sensors in Figure 8.7 stem from the classifier's utilization of varying datasets for training and testing across the 10 iterations. Consequently, the classifier achieves high classification accuracy when exposed to favorable data segments and faces challenges when presented with less favorable data segments.

8.5 Summary

This chapter presents a novel approach where a proof-of-concept wearable multisensory system was developed and integrated with machine learning techniques to improve the detection of students' learning and concentration levels. The system incorporates multiple sensing modalities to capture various physiological signals associated with cognitive processes. A decision-level fusion method was devised to aggregate the confidence levels from all sensing techniques, enhancing the overall detection accuracy.

To evaluate the effectiveness of the wearable multisensory device, a pilot study was conducted with 10 participants, comprising university students of both genders aged between 21 and 30. Two distinct machine learning methods, SVM and GRU, were employed for the classification models. The results demonstrated that the fusion of sensor data using SVM and GRU achieved notable accuracies of 86.53% and 89.86%, respectively.

A larger participant pool and integration of an advanced multisensory device can further enhance this research. By involving more participants and refining the technology, the system's performance can be further optimized, offering valuable insights into enhancing students' learning experiences and cognitive engagement.

References

1 Khosravi, S., Li, H., Khan, A.R. et al. (2024). Exploring the elusive mind: a multimodal wearable sensor solution for measuring mind wandering in university students. *Advanced Sensor Research* 3 (1): 2300067.

2 Hutt, S., Wong, A., Papoutsaki, A. et al. (2023). Webcam-based eye tracking to detect mind wandering and comprehension errors. *Behavior Research Methods* https://dx.doi.org/10.3758/s13428-022-02040-x.

3 Chu, M.T., Marks, E., Smith, C.L., and Chadwick, P. (2023). Self-caught methodologies for measuring mind wandering with meta-awareness: a systematic review. *Consciousness and Cognition* 108: 103463.
4 Kane, M.J., Smeekens, B.A., Von Bastian, C.C. et al. (2017). A combined experimental and individual-differences investigation into mind wandering during a video lecture. *Journal of Experimental Psychology: General* 146 (11): 1649.
5 Risko, E.F., Anderson, N., Sarwal, A. et al. (2012). Everyday attention: variation in mind wandering and memory in a lecture. *Applied Cognitive Psychology* 26 (2): 234–242.
6 Khosravi, S., Bailey, S.G., Parvizi, H., and Ghannam, R. (2022). Wearable sensors for learning enhancement in higher education. *Sensors* 22 (19): 7633.
7 Khosravi, S., Khan, A.R., Zoha, A., and Ghannam, R. (2022). Self-directed learning using eye tracking: a comparison between wearable HeadWorn and webcam-based technologies. *IEEE Global Engineering Education Conference (EDUCON)*, Tunis, Tunisia (28–31 March 2022), 640–643. IEEE.
8 Khosravi, S., Khan, A.R., Zoha, A., Ghannam, R. (2022). Employing a wearable eye-tracker to observe mind-wandering in dynamic stimuli. *2022 29th IEEE International Conference on Electronics, Circuits and Systems (ICECS)*, Glasgow, United Kingdom (24–26 October 2022), 1–4. IEEE.
9 Brishtel, I., Khan, A.A., Schmidt, T. et al. (2020). Mind wandering in a multimodal Reading setting: behavior analysis & automatic detection using eye-Tracking and an EDA sensor. *Sensors* 20 (9): 2546.
10 Weinstein, Y. (2018). Mind-wandering, how do I measure thee with probes? Let me count the ways. *Behavior Research Methods* 50 (2): 642–661. https://link.springer.com/content/pdf/10.3758%2Fs13428-017-0891-9.pdf.
11 Khan, A.R., Khosravi, S., Hussain, S. et al. (2022). EXECUTE: exploring eye tracking to support E-learning. *2022 IEEE Global Engineering Education Conference (EDUCON)*, Tunis, Tunisia (28–31 March 2022), 670–676. IEEE.
12 Zhang, H., Miller, K.F., Sun, X., and Cortina, K.S. (2020). Wandering eyes: eye movements during mind wandering in video lectures. *Applied Cognitive Psychology* 34 (2): 449–464.
13 Smallwood, J. and Schooler, J.W. (2015). The science of mind wandering: empirically navigating the stream of consciousness. *Annual Review of Psychology* 66 (1): 487–518.
14 Campisi, C.A., Li, E.H., Jimenez, D.E., and Milanaik, R.L. (2020). *Augmented Reality in Medical Education and Training: From Physicians to Patients*, 111–138. Springer International Publishing.
15 Mutlu-Bayraktar, D., Ozel, P., Altindis, F., and Yilmaz, B. (2022). Split-attention effects in multimedia learning environments: eye-tracking and EEG analysis. *Multimedia Tools and Applications* 81 (6): 8259–8282.

16 Saffaryazdi, N., Goonesekera, Y., Saffaryazdi, N. et al. Emotion recognition in conversations using brain and physiological signals. *27th International Conference on Intelligent User Interfaces*, Helsinki Finland (22–25 March 2022). ACM.

17 Srivastava, N., Nawaz, S., Newn, J. et al. (2021). Are you with me? Measurement of Learners' Video-Watching Attention with Eye Tracking. *LAK21: 11th International Learning Analytics and Knowledge Conference*, Irvine, CA, USA (2–16 April 2021). ACM.

18 Zheng, Y., Wang, D., Zhang, Y., and Xu, W. (2019). Detecting mind wandering: an objective method via simultaneous control of respiration and fingertip pressure. *Frontiers in Psychology* 10. https://doi.org/10.3389/fpsyg.2019.00216.

19 McNeal, K.S., Zhong, M., Soltis, N.A. et al. (2020). Biosensors show promise as a measure of student engagement in a large introductory biology course. *CBE Life Sciences Education* 19 (4): ar50.

21 Nakasone, A., Prendinger, H., and Ishizuka, M. (2005). Emotion recognition from electromyography and skin conductance. *Proceedings of the 5th International Workshop on Biosignal Interpretation*. Citeseer, 219–222.

22 Soleymani, M., Lichtenauer, J., Pun, T., and Pantic, M. (2012). A multimodal database for affect recognition and implicit tagging. *IEEE Transactions on Affective Computing* 3 (1): 42–55. https://doi.org/10.1109/t-affc.2011.25.

23 Subramanian, R., Wache, J., Abadi, M.K. et al. (2018). ASCERTAIN: emotion and personality recognition using commercial sensors. *IEEE Transactions on Affective Computing* 9 (2): 147–160.

24 Wagner, J., Kim, J., and André, E. (2005). From physiological signals to emotions: implementing and comparing selected methods for feature extraction and classification. *2005 IEEE International Conference on Multimedia and Expo*, Amsterdam, Netherlands (6 July 2005), 940–943. IEEE.

25 Setz, C., Arnrich, B., Schumm, J. et al. (2009). Discriminating stress from cognitive load using a wearable EDA device. *IEEE Transactions on Information Technology in Biomedicine* 14 (2): 410–417.

26 Bufler, T.D. and Narayanan, R.M. (2016). Radar classification of indoor targets using support vector machines. *IET Radar, Sonar and Navigation* 10 (8): 1468–1476.

27 Li, H., Shrestha, A., Heidari, H. et al. (2019). Bi-LSTM network for multimodal continuous human activity recognition and fall detection. *IEEE Sensors Journal* 20 (3): 1191–1201.

28 Cortes, C. and Vapnik, V. (1995). Support-vector networks. *Machine Learning* 20: 273–297.

29 Wang, M., Zhang, Y.D., and Cui, G. (2019). Human motion recognition exploiting radar with stacked recurrent neural network. *Digital Signal Processing* 87: 125–131.

30 Wang, M., Cui, G., Yang, X., and Kong, L. (2018). Human body and limb motion recognition via stacked gated recurrent units network. *IET Radar, Sonar and Navigation* 12 (9): 1046–1051.

31 Cho, K., van Merrienboer, B., Gülçehre, C. et al. (2014). Learning phrase representations using RNN encoder-decoder for statistical machine translation. *arXiv preprint arXiv:14061078*.

32 Chen, C., Jafari, R., and Kehtarnavaz, N. (2015). A real-time human action recognition system using depth and inertial sensor fusion. *IEEE Sensors Journal* 16 (3): 773–781.

9

Adaptive Secure Multi-modal Telehealth Patient-Monitoring System

Muhammad Hanif[1], *Ehsan Ullah Munir*[1], *Muhammad Maaz Rehan*[1], *Saima Gulzar Ahmad*[1], *Tassawar Iqbal*[1], *Nasira Kirn*[2], *Kashif Ayyub*[1], *and Naeem Ramzan*[2]

[1] *Department of Computer Science, COMSATS University Islamabad, Wah Campus, Wah Cantt, Pakistan*
[2] *School of Computing, Engineering and Physical Sciences, University of the West of Scotland, Paisley, UK*

The advancement of technology has revolutionized the healthcare industry, enabling quick access to healthcare facilities, secure record management, and remote healthcare provisions. Industry 4.0 has propelled healthcare into a new era, leveraging advanced sensors, multi-modalities, and Internet of Things (IoT)-based solutions to transform traditional patient management systems (PMSs). Telehealth has emerged as a solution to existing challenges, incorporating modern technologies such as blockchain, machine learning (ML), and artificial intelligence (AI). Blockchain ensures secure and tamper-proof storage of healthcare data, while ML and AI algorithms aid in patient diagnosis. However, the adoption of these cutting-edge technologies has expanded the threat landscape, leading to cyberattacks, data breaches, and manipulation. Traditional cybersecurity frameworks have proven inadequate in addressing these modern threats, highlighting the need for innovative approaches. Machine learning (ML), though powerful, can be exploited for offensive purposes, necessitating a comprehensive security strategy. Recognizing the network as the fundamental building block of modern healthcare systems, zero-trust security (ZTS) principles have been introduced. These principles emphasize "never trust, always verify," least privilege access control, and "always assume breach" to enhance network security. While the implementation of ZTS requires strategic planning, it offers a promising approach to

Multimodal Intelligent Sensing in Modern Applications, First Edition.
Edited by Masood Ur Rehman, Ahmed Zoha, Muhammad Ali Jamshed, and Naeem Ramzan.

Published 2025 by John Wiley & Sons, Inc.

safeguarding hyperconnected healthcare networks in the face of evolving threats. Blockchain being distributed and sophisticated security protocol-enabled architecture is an important pillar for utilization as security controls for immutable storage, universal identity, and access management and anonymization.

This chapter comprehensively examines the security needs of multi-modal telehealth systems, explores existing solutions, underscores successful methodologies, and provides a detailed analysis of key security protocols. It discusses the viability of modern approaches to achieving optimal security measures considering the potential of blockchain-powered ZTS for telehealth-enabled patient monitoring systems (PMS), outlining processes and a roadmap for the implementation.

9.1 Healthcare Systems

Since the existence of humankind on the earth, health has been a concerning area of life. With the evolution of humans in different eras, healthcare facilities have been diversified with respect to the technology of the regime. Healthcare centers adopted database-led patient management system (PMS) for a long period of time since the evolution of the modern computing era [1]. These systems are now mature and working very efficiently according to the programmed set of functions. However, these systems are not aligned with the current boom of healthcare due to telehealth strategies. Due to advanced technologies, emergence and expanding facilities to the untapped areas especially undeveloped countries and locations are being the target of the health sector. The healthcare industry's growing dependence on digital technologies for enhancing patient care and operational efficiency introduces inherent cybersecurity risks [2]. Healthcare data, known for its reliability and permanence, often contains multiple immutable patient identifiers, amplifying the challenge of safeguarding sensitive information (Figure 9.1).

Healthcare systems are perceived as inherently less secure, making them prime targets for cyberattacks. Consequently, cybersecurity incidents have surged over the past decade, positioning healthcare as one of the most frequently targeted sectors [3]. According to a 2022 telehealth survey [4] conducted by the American Medical Association, 60% of participating physicians agreed or strongly agreed that telehealth enabled them to deliver high-quality care. Additionally, over 80% of respondents stated that telehealth improved patients' access to healthcare services. The traditional PMS are tuned as per the requirements of the legacy system requirements, however, modern systems powered by ML and Internet of Things (IoT)-based telehealth products require additional layer of security. Telehealth is dependent on the vital signs being transmitted from remote location to the centralized server. These vital signs are being generated through handheld sensors or IoT-based sensors which are centrally controlled [5]. The malfunction behavior at the edge would

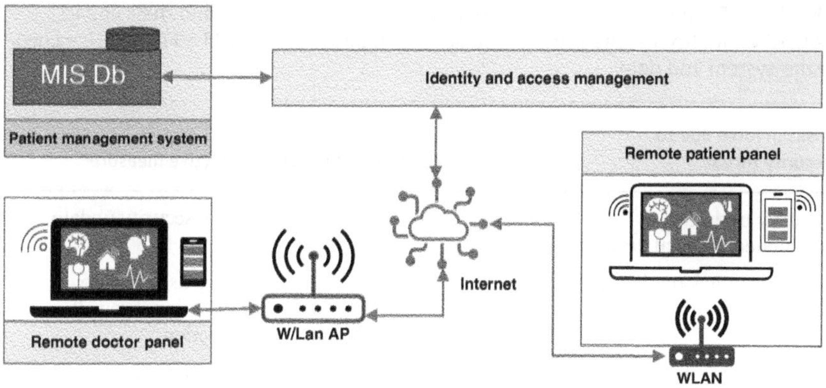

Figure 9.1 Typical client–server architecture of the telehealth system.

cause damage to the healthcare operation and diagnostic process. Therefore, security and trustworthiness of the edge are the prime requirements of telehealth.

9.1.1 Traditional Healthcare Systems

Traditional healthcare systems often face challenges such as limited access to healthcare facilities, long wait times for appointments, and difficulty in monitoring patients remotely. Traditional healthcare systems typically follow a centralized model, where patients visit healthcare facilities to receive diagnosis, treatment, and follow-up care. These systems are characterized by in-person consultations, paper-based medical records, and a hierarchical structure of healthcare delivery. While traditional models offer personalized care and direct physician–patient interaction, they often face challenges related to accessibility, affordability, and scalability, particularly in rural or underserved areas.

9.1.2 Multi-modal Telehealth Systems

In contrast, multi-modal telehealth systems leverage advanced technologies to deliver healthcare services remotely, overcoming geographical barriers and enhancing accessibility for patients. These systems encompass a range of telemedicine modalities, including virtual consultations, remote monitoring, and digital health platforms [6]. By integrating technologies such as telecommunication networks, cloud computing, and artificial intelligence (AI), telehealth systems enable real-time communication between patients and physician, personalized treatment plans, and continuous monitoring of patient health parameters. However,

Table 9.1 Telehealth systems risk due to security vulnerabilities and their impact, C = Confidentiality, I = Integrity, A = Availability, Impact (H = High, M = Medium, L = Low), on the system and data.

Risks involve due to security threat	C	I	A	Data	Impact	Corrective measure
Security update wise unpatched medical devices	✓	✓	✓	✓	H	Install security update
Malware attack on diagnostic device data	✓		✓		M	Antivirus/malware installation, firewall protection
Unauthorized device reprogramming impacting function	✓	✓	✓	✓	H	Firmware protection, system on chip devices adoption
Denial of service attacks rendering devices unavailable	✓		✓	✓	M	Firewall protection, rate limiting
Exfiltration of patient data from the network		✓	✓		M	Encryption at rest, watermarking, hashing
Unauthorized access to network, enabling device compromise	✓	✓	✓	✓	H	Single sign-on, access controls, least privilege access
Uncontrolled distribution of privileged passwords	✓	✓	✓	✓	H	Multifactor authentication, location, and authorization
Off-the-shelf software vulnerabilities	✓	✓	✓	✓	H	Software updates, replacement of software
Improper disposal of data				✓	L	Controlled access to data, provenance
Unauthorized firmware update on devices	✓		✓		M	Hardware access control, restrict admin access

telehealth-based PMS are at higher risk of modern cyber-attacks (Table 9.1) due to the increased need for hyper-connectivity and communication on public networks, which is essential to ensure the system's availability [7]. These security concerns are the major factor in non-adopting telehealth systems widely. The security of these systems is as important as the life of the human because the data stored through these systems are used for analysis and diagnosis of the patient and these records also comprise personal data as well. Traditional healthcare systems are being protected through firewalls, transport layer security (TLS), authentication, authorization, perimeter-based security, and identity management systems [8]. In the current security scenarios of cyberspace, these measures are not enough to stand against advanced cyber threats. Modernizing security controls is the only option on the table to deter cyber criminals. These advanced techniques are discussed in the following section.

9.2 Security in Healthcare Systems

The sensitive nature of healthcare data makes it an attractive target for cybercriminals seeking to exploit vulnerabilities in healthcare systems [9]. The Colonial Pipeline [10] case illustrates how user errors, such as the reuse of passwords across different systems, can lead to significant security breaches, even in networks with robust security measures like virtual private networks (VPNs). This incident underscores the ongoing importance of promoting cybersecurity awareness and implementing stringent security protocols to mitigate the risk of unauthorized access and data breaches. Unauthorized access to patient information not only compromises individual privacy but also undermines the integrity of healthcare services. Identity theft, medical fraud, and breaches of patient confidentiality can have far-reaching consequences, eroding trust in healthcare institutions and jeopardizing patient well-being. Thus, robust security measures are indispensable to mitigate these risks and safeguard the sanctity of patient data.

Security in telehealth systems can be divided into three areas: first is security of the data stored for decision-making and analytics; second, network security is the second pillar of the security of the system; and the third area includes the measures taken for security of edge from being compromised. All of these three areas cannot be covered through a single platform or solution. The most efficient framework proposed so far in the research and industry is the ZTS [11] (Figure 9.2).

By implementing robust security measures and leveraging advanced technologies such as blockchain, ML, and ZTS, healthcare organizations can mitigate security risks, protect patient privacy, and build trust in telehealth services. Consequently, enterprises are compelled to adopt novel security frameworks such as secure access service edge (SASE) [12] and ZTS, prioritizing data protection and resilience in alignment with the demands of the digital era. However, addressing security challenges requires a proactive and multifaceted approach, involving collaboration between healthcare providers, technology vendors, regulators, and other stakeholders.

The main objective of this chapter is to assess the current prevailing telehealth-enabled PMSs with respect to the security scenarios and finally recommend the best available solution for PMS.

9.2.1 Prevailing Techniques for Secure Telehealth

Telehealth has emerged as a rapidly expanding and invaluable resource in healthcare delivery, particularly amidst and post the COVID-19 pandemic [13]. Nevertheless, it also presents significant challenges and risks pertaining to privacy and security, given the transmission and storage of sensitive personal data across diverse networks and devices. Addressing these challenges and mitigating risks necessitates the implementation of robust security measures by telehealth

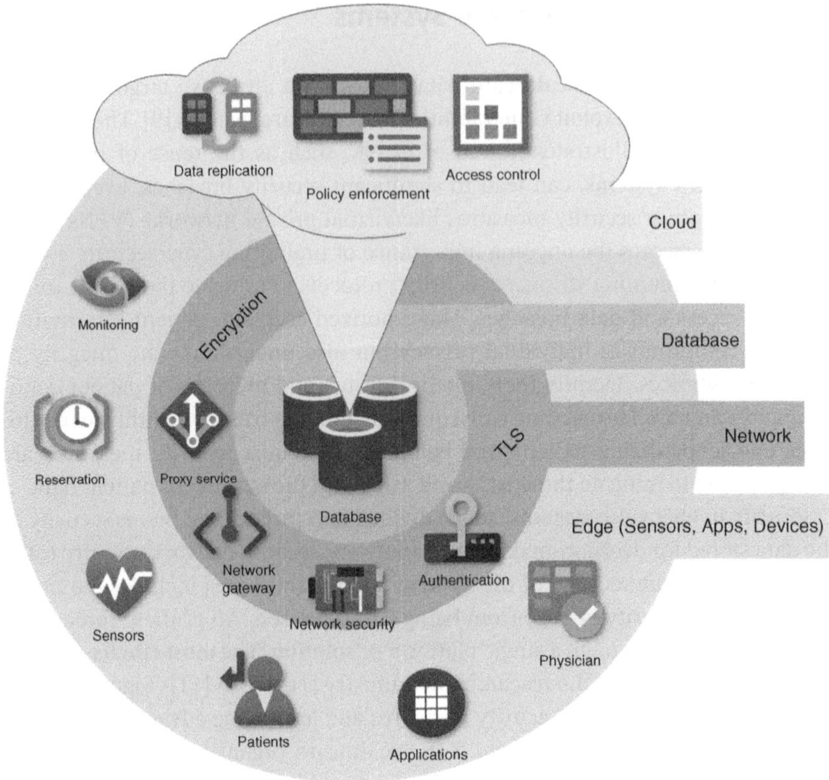

Figure 9.2 Prevailing traditional client–server architecture-based patient management system.

providers and users alike. Over the past two years, an alarming 80% of hospitals have fallen victim to security breaches, stemming from ransomware attacks [14]. Today, 277 days is the average time to identify and contain a cyber breach and each breach leads to an unacceptable average loss of US $4.35 million to organizations [15]. To safeguard against potential threats, it is imperative to adhere to established secure telehealth practices, including encryption, authentication, authorization, audit, and ongoing education [16]. These measures serve to fortify the confidentiality, integrity, and availability of telehealth systems and data. Moreover, these measures contribute to the overall enhancement of telehealth services by fostering quality, safety, and operational efficiency.

Advanced techniques such as continuous monitoring and threat detection (CM&TD), user behavior analysis (UBA), endpoint security and device management (EPS&DM), data loss prevention (DLP), and incident response and disaster

Table 9.2 Overview of the existing approaches being used to deter the cyberattacks in the health care regime.

Security issues	Existing solution	References
Attacks on end users, medical equipment, and access networks	Encryption, transport layer security, least privilege access, access control protocols implementation	[20–23]
Privacy breaches and unauthorized access to patient data	Encryption of data at rest and in transit, watermarking, hashing for data integrity, and detect tampering detection	[13, 24]
Vulnerabilities in communication channels and data transmission leading to data interception and manipulation	Message authentication code (MAC) and digital signatures. Secure shell (SSH)	[25, 26]
Authentication and authorization mechanisms compromise	Single sign-on (SSO), least privilege access control	[27, 28]
Malware and ransomware attacks	Intrusion detections and intrusion prevention (IDS/IPS) systems, next-generation firewall	[5, 29, 30]
Malicious device, operating system manipulation	ZTS, trust devices access, access control list, and whitelisting	[31, 32]

recovery (IR&DR) play pivotal roles in enhancing the security of telehealth systems. These techniques complement baseline approaches like encryption, authentication, access control, and secure communication protocols, which are fundamental for security augmentation. Together, they form a robust security framework aimed at mitigating the risk of unauthorized access, data interception, and tampering, thereby safeguarding patient information and preserving the confidentiality of healthcare transactions.

Among these techniques, encryption, authentication, authorization, audit, and awareness are widely utilized in literature (Table 9.2). Encryption ensures data confidentiality by encoding information, while authentication verifies the identity of users accessing the system. Authorization controls user access based on predefined permissions, while audit trails track system activities for accountability. Additionally, raising awareness among stakeholders about security best practices and risks is crucial for maintaining a secure telehealth environment. These combined efforts contribute to a comprehensive security posture that addresses the evolving threats and challenges in the telehealth sector.

Encryption cryptographic methods play a crucial role in safeguarding the confidentiality, integrity, and authenticity of data within telehealth systems, preventing unauthorized access, modification, or disclosure of sensitive information like patient records or medical images. Authentication procedures verify the identities and credentials of users and devices involved in telehealth services, mitigating risks of impersonation, fraud, or unauthorized system access [17]. Meanwhile, authorization mechanisms determine access rights based on user roles or policies, preventing inappropriate use of telehealth resources. Regular auditing of network and application activities ensures accountability, transparency, and the detection of security breaches, while raising awareness through training and education enhances user knowledge and fosters a culture of security and trust within telehealth services [5]. These measures align with emerging trends highlighted by the Health Sector Coordinating Council, indicating the ongoing evolution and importance of cybersecurity practices in telehealth [18].

9.2.2 Challenges in Ensuring Healthcare Security

The complexity of healthcare systems, coupled with the rapid evolution of cyber threats, presents numerous challenges in ensuring the security of patient data. Healthcare organizations must contend with diverse factors, including the proliferation of connected medical devices, the prevalence of legacy systems, and the need to comply with stringent regulatory requirements such as the Health Insurance Portability and Accountability Act (HIPAA). Additionally, the emergence of novel attack vectors, such as ransomware and insider threats, further complicates efforts to fortify healthcare data security.

Physicians, typically working as independent contractors within healthcare settings, often use personal devices, compounding challenges in endpoint device management, a key factor changing overall cybersecurity of organizations. The integration of many outdated monitoring devices into networks, coupled with limited support for patching security vulnerabilities, presents a landscape fraught with challenges. However, amidst these obstacles lies an opportunity for innovation and improvement in cybersecurity practices.

Major challenges of health care include social engineering, where individuals are manipulated into compromising security measures. Ransomware poses significant risks by encrypting data and demanding payment for its release. Incidents involving the loss or theft of equipment or data jeopardize patient privacy and data integrity. Insider threats, whether accidental or malicious, can result in data breaches. Additionally, attacks against network-connected medical devices exploit vulnerabilities, threatening patient safety and confidentiality. The most devastating attack so far taking place in health care is the ransomware attack, and researchers predicted that by 2031, ransomware attack cost around 256 million dollar per year [19].

The healthcare sector faces nearly three times the cost to remediate breaches compared to other industries, with an average of $408 per stolen healthcare record versus $148 per stolen non-health record [14]. Telehealth-enabled systems are facing integration issues, as legacy systems are not compatible with modern technology dynamics. Integration with existing healthcare infrastructure, including electronic health record (EHR) systems, medical devices, and other healthcare applications. Achieving seamless integration and interoperability while maintaining security and data integrity can be challenging, particularly due to differences in technology standards and protocols.

9.2.3 Strategies to Enhance Healthcare Data Security

Cyberattackers have capitalized on the rapidly evolving healthcare landscape during the COVID-19 pandemic. The transition to remote work, virtual care, and electronic consultations has introduced new vulnerabilities, providing cybercriminals with additional targets. Addressing the myriad challenges associated with healthcare data security requires a multifaceted approach encompassing technical, organizational, and regulatory measures. Healthcare organizations must prioritize the adoption of robust encryption protocols, access controls, and authentication mechanisms to protect patient data from unauthorized access and tampering. Moreover, fostering a culture of cybersecurity awareness and providing comprehensive training to healthcare personnel are integral components of a proactive security strategy. Additionally, adherence to regulatory frameworks such as HIPAA and the implementation of stringent data governance policies are essential for ensuring compliance and maintaining patient trust. Traditionally, organizations rely on network perimeter security to prevent intrusions, but this approach is increasingly outdated due to evolving cyber threats. Zero trust, a novel approach gaining traction, emphasizes scrutinizing all components including systems, users, and data for potential vulnerabilities rather than solely focusing on the network perimeter. By adopting a zero-trust model, organizations can better protect against sophisticated cyberattacks and minimize their impact. Strategies like secure communication over public network and automating the security operations and analysis can help to figure out the attack earlier. Security measures can be enhanced:

- By systematically hardening the digital attack surface with integrated, AI-driven processes and blockchain-based controls.
- Optimize and orchestrate enterprise security operations to improve investigations, decisions, and threat responses.
- Respond with agility and minimize the impact of evolving cyberthreats by adopting security standards.

Gartner has predicted that by 2023, up to 60% of enterprises will move from VPNs to zero-trust network access (ZTNA). ZTNA standards are being implemented using password-less and key-less secure tunneling, creating invisibility of the critical assets on the network from discovery using scanning networks. IBM research concluded that it takes seven months to identify a breach. In the current regime of cybersecurity attaining cyber-resilience is the need of the hour. European regulators advocate robust cybersecurity measures for IoT-based medical devices, aligning with general data protection rules (GDPRs) and data privacy regulations. While no single measure ensures absolute immunity, integrating cybersecurity into risk management and fostering cyber resilience is vital. This involves a comprehensive approach encompassing organizational culture, personnel, procedures, and technological safeguards. Cyber hygiene practices like secure backups and software updates are essential. Upholding confidentiality requires data anonymization and access limitations. Investment in secure data transfer systems and embedding security throughout the product lifecycle are imperative for holistic cybersecurity compliance.

In the healthcare industry, various best practices are being adopted to mitigate cybersecurity threats. These practices encompass a range of measures, including implementing email protection systems, endpoint protection systems, access management protocols, data protection and loss prevention strategies, and asset management procedures [33]. Additionally, network management, vulnerability management, and incident response protocols are crucial for maintaining security. Special attention is also given to securing network-connected medical devices and ensuring cybersecurity oversight and governance. These 10 practices collectively form a robust framework for safeguarding healthcare systems and patient data from cyber threats.

9.2.4 Key Security Features for Telehealth Systems

Security is a paramount concern in telehealth systems due to the sensitive nature of patient data and the potential consequences of security breaches. To address these concerns, telehealth systems implement a variety of security features and protocols to safeguard patient information, ensure data privacy, and maintain the integrity of healthcare services. Some of the key security features implemented in telehealth systems include the following.

9.2.4.1 Encryption

Encryption is used to secure the transmission of data between healthcare providers and patients, as well as between different components of the telehealth system. End-to-end encryption ensures that data stays confidential and protected from unauthorized access or interception by encrypting it during transmission and decrypting it only upon arrival at its intended destination [34]. Secured tunneling aligns with ZTNA principles and White Box Cryptography (WBC) standards to enhance security measures.

9.2.4.2 Authentication

Authentication mechanisms, such as usernames, passwords, biometric authentication, or multi-factor authentication (MFA) [35], are implemented to verify the identity of users accessing the telehealth system. Strong authentication helps prevent unauthorized access to patient data and ensures that only authorized individuals can view or modify sensitive information.

9.2.4.3 Access Control

Access control mechanisms are used to restrict access to patient data based on users' roles, permissions, and levels of authorization. Role-based access control (RBAC) or attribute-based access control (ABAC) systems are commonly employed to enforce least privilege principles, ensuring that users only have access to the data and functionalities necessary to perform their job responsibilities. Protecting the firewall from backdoor breaches is essential for maintaining security integrity.

9.2.4.4 Audit Trails

Audit trails are maintained to record and track all activities and interactions within the telehealth system, including user logins, data accesses, modifications, and deletions. Audit logs provide a detailed record of security-related events, enabling administrators to monitor system activity, detect unauthorized behavior, and investigate security incidents or breaches.

9.2.4.5 Data Integrity Checks

Data integrity checks, such as digital signatures, checksums, or hash functions, are used to verify the integrity of patient data and detect any unauthorized alterations or tampering. By comparing the checksum or hash value of the received data with the expected value, telehealth systems can ensure that data has not been modified or corrupted during transmission or storage.

9.2.4.6 Secure Communication Protocols

Secure communication protocols, such as transport layer security (TLS) or secure socket layer (SSL), are used to establish encrypted connections between clients and servers in the telehealth system. These protocols protect data in transit from eavesdropping, tampering, or interception by encrypting it before transmission and decrypting it upon receipt. Ensuring adherence to the Common Criteria (CC) for Information Technology Security Evaluation per the ISO/IEC 15408 standard, spanning Evaluation Assurance Levels (EAL) 1–7.

9.2.4.7 Security Audits and Penetration Testing

Regular security audits and penetration testing are conducted to assess the effectiveness of security controls, identify vulnerabilities or weaknesses in the telehealth system, and ensure compliance with security standards and regulations. By proactively identifying and addressing security risks, telehealth systems can mitigate the likelihood of security breaches and safeguard patient information.

Overall, implementing robust security features and protocols is essential to ensure the confidentiality, integrity, and availability of telehealth systems, protect patient privacy, and maintain trust in remote healthcare services. By adopting a multilayered approach to security and continuously monitoring and updating security measures, telehealth systems can effectively mitigate security risks and enhance the overall security posture of the healthcare organization.

9.2.4.8 Cyber Resilience

To enhance system resilience against cyberattacks, several strategies are employed. These include ensuring that the system does not have identifiable IP or MAC addresses, rendering it invisible within the network, and achieving zero visibility. Implementing a zero-trust approach is crucial, along with leveraging white-box cryptography for enhanced security. Additionally, deep packet inspection is utilized to detect communication anomalies in critical protocols. Moreover, the network adheres to zero-trust principles and complies with essential standards such as IEC62443.

9.2.4.9 Zero-Trust-Based Micro-segmentation

Zero-trust micro-segmentation represents a proactive and strategic method for enhancing security measures within an organization's network infrastructure. By implementing this approach, the attack surface is dynamically minimized, creating multiple isolated segments within the network. This segmentation prevents lateral movement of malware and unauthorized access, significantly enhancing overall security posture. Through meticulous control and monitoring of network traffic, zero-trust micro-segmentation ensures that each segment operates independently, mitigating the risk of potential breaches or intrusions. Ultimately, this strategy fosters a more resilient and secure network environment, safeguarding critical assets and data against evolving cyber threats.

Security is paramount in telehealth systems due to patient data sensitivity and potential consequences of breaches. Implementing encryption, authentication, access control, audit trails, data integrity checks, and secure communication protocols are key security measures. Regular audits and penetration testing help assess controls and identify vulnerabilities. Implementing strategies such as secure communication over public networks and automating security operations and analysis can aid in the early detection of attacks. To enhance security measures further, several steps can be taken. First, systematically hardening the digital attack surface through integrated, AI-driven processes can strengthen defenses against potential threats. Second, optimizing and orchestrating enterprise security operations can improve the efficiency of investigations, decision-making, and responses to threats. Finally, responding with agility to evolving cyber threats is crucial for minimizing their impact and ensuring continued protection of sensitive data and systems. Schemes like system resilience, zero-trust microsegmentation, and compliance with standards enhance security posture. A multilayered approach and continuous updates are essential to protect patient privacy and maintain trust in remote healthcare services.

9.3 Blockchain-Powered ZTS for Enhanced Security of Telehealth Systems

Technology has modernized the cyber security regime; conventional methods are being replaced by modern technology-based solutions. The traditional signature-based security is upgrading to ML-based context-aware firewalls-based security. Emerging technologies such as blockchain, ML, software-defined networking, and cloud computing offer promising solutions for enhancing the security and reliability of telehealth systems. However, these solutions still have loopholes and experience attacks, researchers principally consensus to emerging paradigm ZTS which so far offers optimal protection against the attacks.

Cybersecurity regimes can adopt two protection techniques, one is to develop self-resilient system that cannot be attacked or cracked through any means, second is to hide the system from the access of malicious actors. The first one is practically unworkable; however, the second approach is doable and can be achieved using advanced anonymizing and invisibility techniques. There are three areas where the data and security breach can occur: at the information/data generation stage, in transit and at the cloud or server stage. ZTS ensures protection at all stages by minimizing the attack surface.

9.3.1 Zero-Trust Security

Traditional perimeter based security strategies have been ineffective, due to a rise of zero-day attacks and insider attacks. Conventional perimeter security relays on the authentication and authorization at the entrance of a network perimeter, while trusting the entities inside the perimeter by default, which is the biggest threat to the network working inside the perimeter. In contrast, ZTS adopts a proactive approach to security by assuming that no entity, whether inside or outside the network, can be trusted. By implementing strict access controls, continuous monitoring, and real-time threat detection mechanisms, ZTS mitigates the risk of insider threats, unauthorized access, and data breaches in telehealth systems.

Implementing ZTS is not a one-time activity as well as not a single piece of program to be installed as a security measure, but it a new paradigm shift. Adopting ZTS reduces the chances of being victimized of the zero-day and other intrusion due to implicit trust. The principles of zero trust, as outlined in the NIST Special Publication 800-207 [36], encompass seven key areas. At the user level, continuous authentication and access monitoring are emphasized, with multifactor authentication and careful access governance being paramount. Devices are closely monitored in real-time, with features such as built-in memory and GPS tracking enabling immediate response to security threats. Applications and workloads are managed to ensure that only authorized users have access, while data security focuses on protecting information at rest, in transit, and in use. Network and environment security prioritizes data protection during transit, and automation facilitates rapid response to security incidents. Finally, visibility and analytics are leveraged to monitor and analyze events, enabling personalized and proactive security measures.

Implementing the anonymization to the ZTS provides an extra layer of security and facilitates secure data sharing and collaboration within a zero-trust environment. By combining blockchain's tamper-proof record-keeping with the foundational principles of ZTS, organizations can strengthen their overall cybersecurity posture while preserving data privacy and integrity.

Anonymization of the assets inside a ZTS-enabled network can be achieved by implementing an anonymization protocol, that is, Zero-Knowledge Proofs (ZKPs), Ring Signatures, and Coin Mixing, to obfuscate device identifiers or sensitive information being transmitted on the network through blockchain. These protocols employ cryptographic techniques to ensure data privacy while maintaining transparency and auditability. Figure 9.3 describes the enhanced version of the ZTS model being proposed by NIST as the baseline. In the enhanced model, blockchain is integrated at the core of the network so that critical information about the assets, devices, users, and applications inside the protected network is not accessible to outsiders. Attackers commonly used network-scanning tools to gather information about the devices and then launch the attack on the vulnerable device.

In conclusion, the integration of blockchain technology into ZTS frameworks holds immense promise for enhancing data privacy and security through anonymization. By leveraging blockchain's decentralized ledger and cryptographic features, organizations can establish a resilient security infrastructure that prioritizes trustless authentication, granular access control, and immutable audit trails. Blockchain enables the anonymization of user identities and sensitive data while ensuring transparency, integrity, and accountability within the ZTS ecosystem.

9.3.2 Blockchain

The proliferation of telehealth has introduced considerable convenience and accessibility to healthcare services, yet the issue of data security remains paramount. Blockchain technology offers a promising solution to significantly bolster

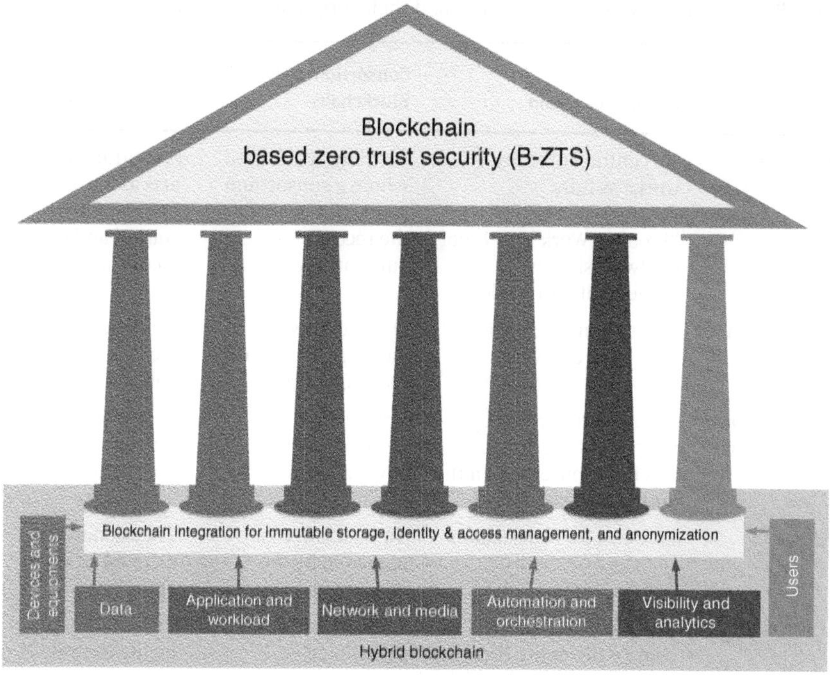

Figure 9.3 Blockchain-enabled zero-trust security framework for secure telehealth systems.

security within telehealth-enabled PMS. By leveraging blockchain technology, telehealth systems can ensure the immutability and integrity of patient records, enhance data transparency, and establish a tamper-resistant audit trail. Smart contracts can be utilized to automate and enforce data access controls, ensuring that only authorized parties can view or modify patient information.

Blockchain has the potential to share clinical trial launches and enrollments in real time to better match patients and prevent double enrollments, and smart contracts to connect different parties – such as providers, insurers, vendors, and auditors and automate transactions. There are three types of blockchain categories being used by industries, and each category has its own pros and cons, as summarised in Table 9.3.

Blockchain is capable to strengthen corporate reputation by providing transparency of materials used in products and services. It improves credibility and public trust in data shared and reduces potential public relations risk from supply chain malpractice. Due to this higher trust level, stakeholders' engagement increased.

Blockchain offers three fundamental business values: recordkeeping, value transfer, and smart contracts, each presenting compelling cases for adoption across various sectors. A smart identity system, utilizing distributed ledger

Table 9.3 Common types of blockchains and their core functionalities.

	Public Blockchain	Consortium / Hybrid Blockchain	Private Blockchain
Overview	No central authority works as fully decentralized under "Proof of Work" or "Proof of Ownership" authenticity principle	Partially centralized where a consortium of entities controls the record authenticity	A central authority acts as a trusted intermediary to control and ensure record authenticity
Permission	Permission less anyone c. join, read, and write	Only selected participant can join, read, and write	Write permissions are centralized to one entity
Transaction verification	Records are verified by majority of the "miners" reaching consensus on the authenticity	Transactions are verified by the consortium	Central authority verifies transactions
Costs	Low cost for transactions	Transaction cost agreed to by the consortium	Transaction cost dictated by one entity
Data storage	Records are distributed, a copy of the entire record is available to all users of the peer-to-peer network	Records are distributed throughout the consortium	Records are stored by the central Authority
Available blockchain	Ethereum, bitcoin, Vechain	Hyperledger, Quorum, c-rda, Enterprise Ethereum	Vechain, ripple, Ethereum

technology and smart contracts, can revolutionize identity management for customers, agents, and employees, ensuring efficiency and security within the ecosystem. Moreover, transfer pricing smart contracts address process fragmentation and information silos while upholding corporate governance and regulatory standards. Additionally, loyalty programs can leverage blockchain to tokenize reward points, facilitating their trade as assets and enhancing employee engagement or customer loyalty initiatives. Blockchain can be utilized for the security of the telehealth in following ways:

- **Secure Patients' Health Records:** IT systems holding and processing sensitive and confidential details about individuals form an attractive target for cybercriminals, protection of data both patient and staff data against theft, loss, or corruption is the utmost requirement of PMS.
- **Ensure Availability of Healthcare Services:** Disruption to healthcare services may have a devastating impact on patients' health, so using blockchain

reliability, resource-efficiency, resilience, operational continuity, and downtime can be minimized.

- **Maintain the Reputation:** The reputation of a healthcare facility is a non-negligible asset. A cyberattack – regardless of its nature – will damage credibility if disclosed to the public.

Healthcare providers are among the institutions in which we place the most trust; critical infrastructures in that they are vital to the well-being and safety of the public. Telehealth has a vast scope of accessibility, which makes it more attractive to malicious actors to be exploited. Modern laws also demand a foolproof security of personal data and health information to be protected with great care. The European Union (EU) has implemented the General Data Protection Regulation (GDPR) to protect the personal data of their residents, therefore, any application or system working in EU would have to comply with GDPR. GDPR demands that the owner of the data is the user any data utilization should take place with consent of the user. Blockchain offers temper-proof storage and code run as smart contract. If the personal and health information is stored on the blockchain, then the GDPR above clause is by default implemented.

In conclusion, the integration of blockchain technology into telehealth systems offers a transformative solution to address critical challenges in data security, patient privacy, and system integrity. By leveraging blockchain's decentralized ledger, cryptographic features, and smart contract functionality, telehealth platforms can enhance trust, transparency, and efficiency in healthcare delivery. From secure recordkeeping and identity management to streamlined transactions and incentivized loyalty programs, blockchain unlocks a wealth of possibilities to revolutionize telehealth services. The adoption of blockchain stands poised to drive innovation, improve patient outcomes, and redefine the future of remote medical care.

9.4 Cyber-resilient Telehealth-Enabled Patient Management System

Implementing a secure telehealth-enabled PMS requires careful planning, implementation, and ongoing operations to be aligned with the ZTS. The following framework would work as the baseline for defining the new strategies for telehealth-enabled PMS (Figure 9.4).

Achieving cyber-resilience in telehealth-enabled PMSs through blockchain-based ZTS entails leveraging blockchain for identity and access management (IAM), immutable provenance storage, and anonymization of devices, users, and applications. Blockchain technology provides a decentralized and tamper-proof

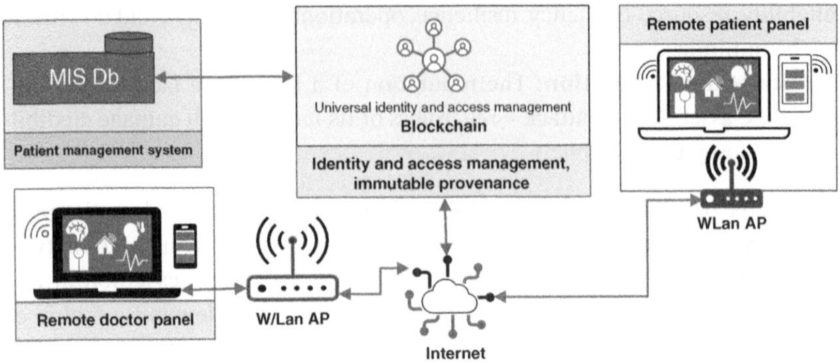

Figure 9.4 Zero-trust security for telehealth-enabled patient management system.

ledger for storing identity and access credentials, ensuring secure authentication and authorization processes. Additionally, blockchain's immutable nature enables the transparent tracking and auditing of data provenance, enhancing accountability and trustworthiness within the system. Moreover, blockchain facilitates the anonymization of sensitive information, such as patient data and device identifiers, thereby safeguarding privacy while maintaining data integrity. By integrating blockchain-based solutions for these critical security functions, telehealth systems can enhance their resilience against cyber threats and ensure the confidentiality, integrity, and availability of patient information.

Achieving ZTS in telehealth-based PMS is indeed a critical endeavor, necessitating fundamental transformations at the architectural and system levels. This entails conducting comprehensive assessments and planning exercises for architectural, hardware, and software components to ensure alignment with ZTS principles. Subsequently, rigorous implementation of security controls is imperative to establish a robust defense posture against cyber threats. Additionally, thorough testing and validation of ZTS standards are essential to verify the efficacy and resilience of the implemented security measures. Furthermore, providing training and raising awareness about cybersecurity among stakeholders is vital to foster a culture of vigilance and adherence to security protocols across the organization. By systematically addressing these aspects, telehealth-based PMS can enhance their security posture and mitigate risks associated with evolving cyber threats.

9.4.1 Assessment and Planning

Assessing and planning for the transformation of telehealth-enabled PMS involves several crucial steps. This includes evaluating the current infrastructure for compatibility issues and scalability considerations, followed by careful planning to migrate to a more secure and scalable architecture, such as distributed or cloud-based systems. Data migration strategies are essential to ensure a seamless transition while maintaining data integrity and security. Integration of security controls like encryption and access management is vital to enhance the system's security posture. Collaboration with stakeholders is key to aligning with organizational goals and regulatory requirements. Through comprehensive assessment and meticulous planning, telehealth-enabled PMS can transition to a more resilient and secure architecture, ensuring the delivery of high-quality healthcare services in a digital environment.

9.4.2 Infrastructure Setup

Implementing ZTS in healthcare systems needs substantial upgrades across hardware, software, and networking infrastructure to align with security standards and best practices. Hardware enhancements may include deploying advanced firewall appliances, intrusion detection/prevention systems, and secure routers or switches to enforce access controls effectively. Similarly, software improvements are vital, requiring endpoint security solutions for continuous monitoring, dynamic policy enforcement, and robust IAM. Networking infrastructure changes are also crucial, involving the establishment of micro-segmentation zones, VPN upgrades for secure remote access, and the implementation of logging and monitoring solutions to detect and respond to security incidents promptly. Overall, ensuring compliance with security standards across all components is essential for mitigating cybersecurity risks and enhancing the security posture of telehealth-enabled PMS.

9.4.3 Security Controls Implementation

Implementing ZTS compliance security controls is essential to safeguard patient data and ensure the integrity of telehealth-enabled PMS. This includes deploying encryption mechanisms to secure data transmission, authentication protocols to verify user identities, and access control measures to restrict unauthorized access to sensitive information. Additionally, intrusion detection systems are implemented to detect and respond to potential security breaches promptly. By incorporating these security controls, telehealth enabled PMS can enhance their resilience against cyber threats and maintain patient confidentiality and system integrity.

9.4.4 Training and Awareness

Effective training and awareness programs are essential for implementing ZTS. The focus of training should include educating healthcare professionals and IT staff about the principles and practices of ZTS, emphasizing the importance of continuous authentication, access control, and encryption in protecting data. Areas of focus should encompass identifying and mitigating cyber threats, recognizing phishing attempts, understanding the use of encryption tools, and adhering to ZTS policies and procedures. By providing comprehensive training and promoting awareness, healthcare organizations can ensure that staff members are equipped with the knowledge and skills needed to uphold the security of telehealth-enabled PMS and safeguard patient information effectively.

In Figure 9.5, a dedicated application on the remote patient's device collects vital signs, such as heart rate, blood pressure, and temperature. This application utilizes a single sign-on (SSO) facility powered by blockchain technology, ensuring secure authentication, anonymization, and access control for both the patient and healthcare providers. Once authenticated, the vital signs data is transmitted securely to the management information system (MIS) database, where it is stored for further analysis and review by healthcare professionals.

Additionally, a history of the vital signs data is maintained on the blockchain network. Each data entry is cryptographically linked and timestamped, creating an immutable record of the patient's health status over time. This ensures the integrity and traceability of the data, making it resistant to tampering or unauthorized

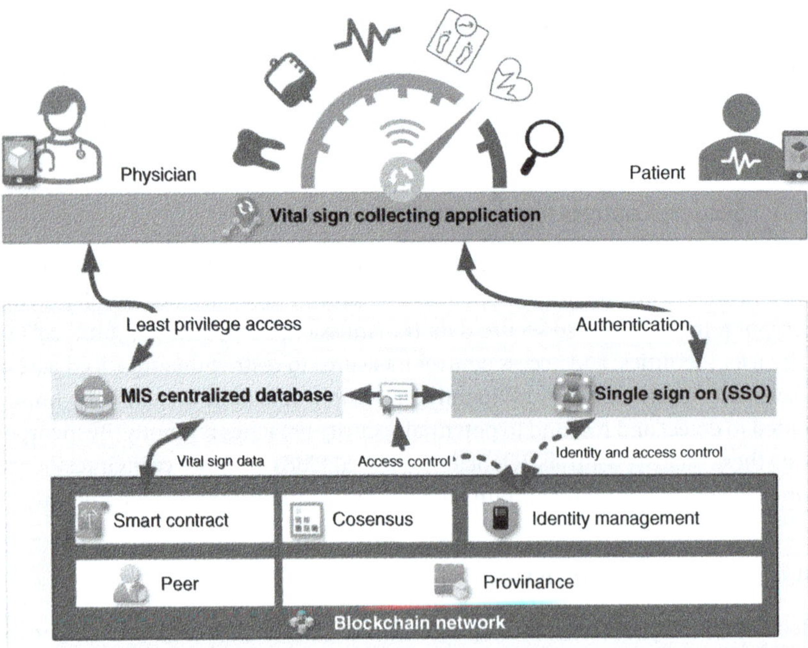

Figure 9.5 Blockchain-based zero-trust compliance for cyber resilience.

alterations. By leveraging blockchain for SSO authentication and data immutability, the diagram demonstrates a secure and transparent approach to collecting and managing vital signs data in telehealth applications. This not only enhances the privacy and security of patient information but also facilitates seamless access and historical tracking of health metrics for informed decision-making and personalized patient care.

9.4.5 Advantages and Limitations

The utilization of blockchain-powered ZTS in telehealth-enabled PMS offers several advantages and outcomes:

- Enhanced security and protection of patient data through decentralized and immutable blockchain technology.
- Improved integrity of patient data with transparent and verifiable transaction records.
- Increased trust and transparency for patients with accessible medical history and treatment plans.
- Streamlined interoperability across healthcare systems for seamless data exchange.
- Compliance with regulatory requirements such as HIPAA and GDPR.
- Cost savings through reduced risk of data breaches and operational efficiencies.
- Potential cost savings for healthcare organizations and patients.
- Real-time monitoring and early detection of health issues, leading to better patient outcomes.

The evolving nature of cyberattacks underscores the importance of continually adapting protection strategies to remain effective. Cybersecurity measures must evolve alongside emerging attack vectors to prevent unauthorized access and mitigate active threats effectively. While there is no perfect method to prevent attacks entirely, the zero-trust approach offers a comprehensive strategy by employing multiple layers of defense to enhance security posture and resilience.

- Security and privacy concerns related to the transmission and storage of sensitive patient data.
- Dependency on technology and internet connectivity, which may limit access for certain populations.
- Regulatory and legal challenges related to data protection, patient consent, and liability issues.
- Potential for misuse or abuse of telehealth systems, such as fraudulent activities or data breaches.

9.5 Summary

In conclusion, secure telehealth-enabled patient monitoring systems hold tremendous potential for improving healthcare accessibility, efficiency, and patient outcomes. By implementing robust security measures and leveraging advanced technologies such as blockchain-powered ZTS, healthcare organizations can mitigate security risks, protect patient privacy, and build trust in telehealth services. Given the criticality of healthcare data, it is imperative for healthcare systems to adopt a zero-trust approach. By embracing this proactive methodology and integrating zero-trust principles into the healthcare sector, organizations can initiate measures to mitigate the risk of unauthorized access. This shift toward a zero-trust model represents a crucial step forward in bolstering security within the healthcare industry and safeguarding sensitive patient information. The implementation of blockchain enables ZTS, facilitates secure identity access management, maintains an immutable audit trail, and enables anonymization of devices, users, and applications. Embracing a proactive zero-trust approach is essential to mitigate the risk of unauthorized access and bolster security within the healthcare sector. However, addressing security challenges requires a proactive and multi-faceted approach, involving collaboration between healthcare providers, technology vendors, regulators, and other stakeholders. With careful planning, implementation, and ongoing monitoring, secure telehealth systems can revolutionize the delivery of healthcare services and contribute to a healthier and more connected society.

References

1 Gulzar Ahmad, S., Iqbal, T., Javaid, A. et al. (2022). Sensing and artificial intelligent maternal-infant health care systems: a review. *Sensors* 22: 4362. https://doi.org/10.3390/S22124362.

2 Nait Hamoud, O., Kenaza, T., Challal, Y. et al. (2023). Implementing a secure remote patient monitoring system. *Information Security Journal* 32 (1): 21–38. https://doi.org/10.1080/19393555.2022.2047839.

3 Seigel, R., Lashway, S.T., Stein, M.M.K., and Rundell, C.J. (2022). Telehealth and digital health privacy regulations. *Diabetes Digital Health and Telehealth* 75–86. https://doi.org/10.1016/B978-0-323-90557-2.00020-0.

4 Houser, S.H., Flite, C.A., and Foster, S.L. (2023). Privacy and security risk factors related to telehealth services – a systematic review. *Perspectives in Health Information Management* 20 (1): 1f.

5 Upadrista, V., Nazir, S., and Tianfield, H. (2023). Secure data sharing with blockchain for remote health monitoring applications: a review. *Journal of Reliable Intelligent Environments* 9 (3): 349–368. https://doi.org/10.1007/S40860-023-00204-W.

6 Zheng, C., Li, J., and Yao, X. (2023). Design and implementation of trusted boot based on a new trusted computing dual-architecture. *Computer & Security* 127: https://doi.org/10.1016/J.COSE.2023.103095.

7 Obaid, O.I. and Salman, S.A.-B. (2022). Security and privacy in IoT-based healthcare systems: a review. *Mesopotamian Journal of Computer Science* 29–40. https://doi.org/10.58496/MJCSC/2022/007.

8 Jegadeesan, S., Dhamodaran, M., Azees, M., and Shanmugapriya, S.S. (2019). Computationally efficient mutual authentication protocol for remote infant incubator monitoring system. *Healthcare Technology Letters* 6 (4): 92–97. https://doi.org/10.1049/HTL.2018.5006.

9 Kumar, P., Kumar, R., Garg, S. et al. (2022). A secure data dissemination scheme for IoT-based e-health systems using AI and blockchain. *2022 IEEE Global Communications Conference, GLOBECOM 2022*, Rio de Janeiro, Brazil (4–8 December 2022) - *Proceedings*, 1397–1403. 10.1109/GLOBECOM48099.2022.10000801.

10 R. A. Alsemmeari, M. Y. Dahab, A. A. Alsulami, B. Alturki, and S. Algarni, "Resilient security framework using TNN and blockchain for IoMT," *Electronics (Switzerland)*, vol. 12, no. 10 May 2023, https://doi.org/10.3390/ELECTRONICS 12102252.

11 Xu, M., Guo, J., Yuan, H., and Yang, X. (2023). Zero-trust security authentication based on SPA and endogenous security architecture. *Electronics (Switzerland)* 12 (4): https://doi.org/10.3390/ELECTRONICS12040782.

12 A. Deshpande, K. Stewart, L. Lepetit, and S. Gunashekar (May 2017). "Distributed Ledger Technologies / Blockchain : Challenges, opportunities and the prospects for standards", the British Standards Institution (BSI).

13 Sassi, M. and Abid, M. (2022) Security and privacy protection in the e-health system: remote monitoring of COVID-19 patients as a use case. *Smart Innovation, Systems and Technologies* 237: 829–843. https://doi.org/10.1007/978-981-16-3637-0 58.

14 Clarke, M. and Martin, K. (2024). Managing cybersecurity risk in healthcare settings. *Healthcare Manage Forum* 37 (1): 17–20. https://doi.org/10.1177/08404704231195804.

15 Global ransomware damage costs predicted to exceed $265 billion by 2031 (2024). https://cybersecurityventures.com/global-ransomware-damage-costs-predicted-to-reach-250-billion-usd-by-2031 (accessed 06 March 2024).

16 Salem, O. and Mehaoua, A. (2022). A secure framework for remote healthcare monitoring using the internet of medical things. *IEEE International Conference on Communications*, Seoul, Korea (16–20 May 2022), 1233–1238. https://doi.org/10.1109/ICC45855.2022.9839025.

17 Samuel, O., Omojo, A.B., Mohsin, S.M. et al. (2023). An anonymous IoT-based e-health monitoring system using blockchain technology. *IEEE Systems Journal* 17 (2): 2422–2433. https://doi.org/10.1109/JSYST.2022.3170406.

18 Malasinghe, L.P., Ramzan, N., and Dahal, K. (2019). Remote patient monitoring: a comprehensive study. *Journal of Ambient Intelligence Humanized Computing* 10 (1): 57–76. https://doi.org/10.1007/S12652-017-0598-X.

19 Novikava, A. (2024). Gartner predicts 2023 to be the year of Zero Trust. *NordLayer Blog*. Nordlayer, https://nordlayer.com/blog/gartner-predicts-the-year-of-zero-trust/ (accessed 15 October 2024).

20 Saxena, A., Moghe, A.A., and Agarwal, R. (2022). IoT for health monitoring. In: *Advanced Sensing in Image Processing and IoT* (ed. R. Gupta, A.K. Rana, S. Dhawan, and K. Cengiz), 201–220. CRC Press https://doi.org/10.1201/9781003221333-11.

21 Gupta, K.D., Sharma, D.K., Dwivedi, R., and Al-Turjman, F. (2022). *Recent Advances in IoT and Blockchain Technology*. Bentham Science Publishers https://doi.org/10.2174/97898150516051220401.

22 Vardi, Y. (2024). The next gen of cybersecurity could be hiding in big tech. www.darkreading.com (accessed 16 March 2024). https://www.darkreading.com/cybersecurity-operations/next-gen-of-cybersecurity-could-be-hiding-in-big-tech.

23 Mishra, R. and Prasad, R. (2022). Towards efficient and secure framework for devices and informatics for internet of medical things. *International Symposium on Wireless Personal Multimedia Communications, WPMC*, Herning, Denmark (30 October–2 November 2022), 459–463. https://doi.org/10.1109/WPMC55625.2022.10014830.

24 Garg, N., Obaidat, M.S., Wazid, M. et al. (2021). SPCS-IoTEH: secure privacy-preserving communication scheme for IoT-enabled e-health applications. *IEEE International Conference on Communications* https://doi.org/10.1109/ICC42927.2021.9500388.

25 Srinivasan, C.R., Charan, G., and Babu, P.C.S. (2020). An IoT based SMART patient health monitoring system. *Indonesian Journal of Electrical Engineering and Computer Science* 18 (3): 1657–1664. https://doi.org/10.11591/IJEECS.V18.I3.PP1657-1664.

26 Hayajneh, T., Mohd, B.J., Imran, M. et al. (2016). Secure authentication for remote patient monitoring with wireless medical sensor networks. *Sensors (Switzerland)* 16 (4): https://doi.org/10.3390/S16040424.

27 Keerthana, D.A.K., Kiruthikanjali, N., Nandhini, G., and Yuvaraj, G. (2017). Secured smart healthcare monitoring system based on IOT. *SSRN Electronic Journal* https://doi.org/10.2139/SSRN.2941100.

28 Nidhya, R. and Karthik, S. (2019). Security and privacy issues in remote healthcare systems using wireless body area networks. *EAI/Springer Innovations in Communication and Computing* 37–53. https://doi.org/10.1007/978-3-030-00865-9_3.

29 Gopalan, S., Verma, R., and Jaswal, S. (2023). A secure and privacy preserving telehealth solution in fog based environment. *Proceedings – 2023 3rd International*

Conference on Smart Data Intelligence, ICSMDI, Trichy, India (30-31 March 2023), 38–47. https://doi.org/10.1109/ICSMDI57622.2023.00016.

30 Mohit, P. (2021). An efficient mutual authentication and privacy prevention scheme for e-healthcare monitoring. *Journal of Information Security and Applications* 63: https://doi.org/10.1016/J.JISA.2021.102992.

31 Pavana, B. and Prasad, S.K. (2022). Zero trust model: a compelling strategy to strengthen the security posture of IT organizations. *AIP Conference Proceedings* 2519: https://doi.org/10.1063/5.0110649.

32 Sultana, M., Hossain, A., Laila, F. et al. (2020). Towards developing a secure medical image sharing system based on zero trust principles and blockchain technology. *BMC Medical Informatics Decision Making* 20 (1): 256. https://doi.org/10.1186/s12911-020-01275-y.

33 Dhar, S. and Bose, I. (2020). Securing IoT devices using zero trust and Blockchain. *Journal of Organizational Computing and Electronic Commerce* 31 (1): 18–34. https://doi.org/10.1080/10919392.2020.1831870.

34 Khan, M.M., Alanazi, T.M., Albraikan, A.A., and Almalki, F.A. (2022). IoT-based health monitoring system development and analysis. *Security and Communication Networks* 2022: https://doi.org/10.1155/2022/9639195.

35 Rose, S., Borchert, O., Mitchell, S., and Connelly, S. (2024). *Zero Trust Architecture*, nvlpubs.nist.gov, https://nvlpubs.nist.gov/nistpubs/SpecialPublications/NIST.SP.800-207.pdf (accessed 20 October 2024).

36 Wang, Z., Chen, Q., and Liu, L. (2023). Permissioned blockchain-based secure and privacy-preserving data sharing protocol. *IEEE Internet of Things Journal* 10 (12): 10698–10707. https://doi.org/10.1109/JIOT.2023.3242959.

10

Advances in Multi-modal Remote Infant Monitoring Systems

Najia Saher[1], Omer Riaz[1], Muhammad Suleman[1], Dost Muhammad Khan[1], Nasira Kirn[2], Sana Ullah Jan[3], Rizwan Shahid[4], Hassan Rabah[5], and Naeem Ramzan[2]

[1] *Department of Information Technology, The Islamia University of Bahawalpur, Bahawalpur, Pakistan*
[2] *School of Computing, Engineering and Physical Sciences, University of the West of Scotland, Paisley, UK*
[3] *School of Computing Engineering and the Built Environment, Edinburgh Napier University, UK*
[4] *NHS Trust, Lincoln County Hospital, Lincoln, UK*
[5] *Institut Jean Lamour, Université de Lorraine, France*

With daily improvements in information and communication technologies (ICT) and related services, the field of remote patient monitoring (RPM) is expanding quickly. It seeks to improve treatment efficiency by remotely monitoring patients with the least amount of work, expense, and time. However, prompt medical attention is required in order to prevent major health hazards, such as neonatal mortality. This chapter offers a thorough analysis of recent research on RPM from the perspective of modified sensing technologies using both contact-based and contact-less sensors. On the other hand, monitoring of baby patients is the main application area. Through early interventions by healthcare providers and the integration of ICT with basic application software to create remote infant monitoring (RIM) systems that can efficiently monitor newborns' health at home, the goal is to lower the infant death rate by up to 50%. Additionally, significant problems with the current RIM systems are emphasized in order to open up new avenues for RIM research in the future.

Multimodal Intelligent Sensing in Modern Applications, First Edition.
Edited by Masood Ur Rehman, Ahmed Zoha, Muhammad Ali Jamshed, and Naeem Ramzan.

Published 2025 by John Wiley & Sons, Inc.

10.1 Remote Patient Monitoring

The advancement of technology has had a significant impact on the healthcare industry. By incorporating ubiquitous computing and the Internet of Things (IoT) into remote healthcare systems, doctors can effectively treat patients and closely monitor their health. The use of remote healthcare technology such as telemedicine and mobile health to monitor patients outside of hospitals reduces costs associated with patient care and improves patient outcomes.

It is crucial for Pakistan to focus on improving infant healthcare monitoring (Ministry of National Health Services [1]), as indicated by a Statista [2] report showing the country had the highest death rate between 2015 and 2020 (as shown in Figure 10.1). Implementing a remote infant health monitoring system with real-time alarm capabilities can provide emergency first aid, reduce infant hospitalizations, and prevent deaths at a lower economic cost.

Furthermore, premature newborns with underdeveloped organs and apnea issues, as well as neonates (defined as infants in the first 28 days after birth) with chronic illnesses, benefit from remote surveillance. Continuous monitoring is crucial for these infants, and research supports the idea that allowing babies to be in a family environment is preferable to expensive hospital stays. Various solutions

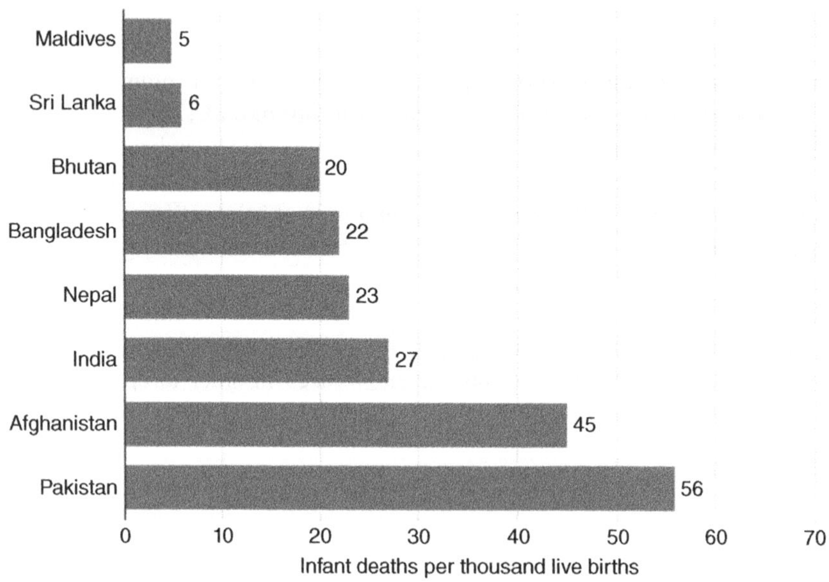

Figure 10.1 Infant mortality rates in South Asia between 2015 and 2020. *Source:* Statista [2]/Statista.

are being developed to support remote patient monitoring (RPM) using the latest technologies [3].

There are two types of sensors for infant monitoring: contact-based sensors and contactless sensors. Contactless sensors limit mobility as the patient must be within the communication range of the sensors, while contact-based sensors are more prone to causing skin infections [4]. It is important to address the limitations of RPM, including sensors, applications, and interfaces, to design a reliable system. Additionally, while there are circumstances that require invasive and contact-based approaches, the trend is toward noninvasive methods of obtaining physiological data [5, 6].

When designing remote infant patient monitoring systems, there are several challenges that need to be addressed. The first step is to select the type of sensors, algorithms, and communication protocols. Additionally, obtaining consent from both the medical staff and the patients is essential.

In this chapter, we examine the latest advancements in remote monitoring systems for infant healthcare. Our goal is to evaluate RPM systems designed for infants using both contactless and contact-based methods. These systems are further classified based on their application in respiratory diseases, sleep apnea, as well as heart and blood-related conditions. Additionally, we take a closer look at significant challenges such as sensor interference in the context of attacks on the technology used in these multi-modal wireless sensing networks.

10.2 Remote Infant Monitoring (RIM) System

The remote infant patient monitoring system typically collects physiological data from infant patients for medical monitoring purposes. Commonly used physiological data includes heart rate, blood oxygen levels, electroencephalogram (EEG), electrocardiogram (ECG), respiration rate, body temperature, blood pressure, blood sugar levels, and signals from the nervous system. Additionally, body weight and imaging data during sleep are gathered to enhance monitoring results [7]. Traditional systems collect data using contact-based sensors linked to the body, but they limit mobility and daily activities for infants [8]. This can impact the infant's comfort and may lead to inaccurate readings not caused by infant disturbance, but due to discomfort. Therefore, it is important to closely observe these effects to improve the quality of treatment. The features of physiological data monitoring devices for infants are outlined in Table 10.1.

Furthermore, the primary components of a remote infant patient monitoring system encompass data surveillance, data processing, and a communication network for data transmission. Data surveillance systems are developed using a variety of sensors equipped with wireless data transmission and reception capabilities.

Table 10.1 Features of physiological monitors for infant monitoring.

Brand name	Form	Pulse rate	Respiratory motion	Pulse oximetry
Baby Vida	Sock	Yes	No	Yes
Mimo	Onesie	No	Yes	No
MonBaby	Button	No	Yes	No
Owlet	Sock	Yes	No	Yes
Snuza Pico	Diaper Clip	No	Yes	No
Sprouting	Leg band	Yes	No	No

Notably, advancements in remote monitoring technologies have broadened the spectrum of sensors to include not only dedicated medical devices but also cameras and smartphones. Additionally, the subsequent sections will delineate contactless RPM systems and contact-based RPM systems.

10.2.1 Contactless Remote Patient Monitoring (RPM) Systems

Contactless RPM systems allow for the monitoring of patients without the need for physical contact between the sensor and the human body, using either a camera or radar. These systems can be divided into two categories: image-based RPMs and radar-based RPMs. Image-based systems detect patient illnesses or falls by analyzing images as data, while radar-based systems use radio frequencies to gather patient data. Radar-based systems are also sometimes used for patient localization. McDuff et al. [9] provide a review of photo-plethysmography imaging methods for contactless remote infant patient monitoring. This method uses a camera to obtain vital body data such as pulse rate variability, pulse rate, blood oxygen level, and respiratory rate (RR).

10.2.2 Contact-Based Remote Patient Monitoring (RPM) Systems

The contact-based sensors in wireless sensor networks (WSNs) can be categorized into wireless body area networks (WBANs) and personal area networks (PANs). These sensors comprise a data processing system with the ability to receive, transmit, and process data. The server terminal can be a computing device or a medical doctor's smartphone in a hospital [10]. Furthermore, communication networks link the data monitoring systems to the data processing systems and transmit the monitored data to a paramedic person connected to the system through the network.

Contact-based multi-modal RPM systems utilize software and hardware devices to monitor illnesses. Variations in heart rate and ECG are detected using off-the-shelf sensors (Bayoumy et al. 2021 [70]). Traditional RPM systems manage acute and chronic patient conditions using contact-based sensors for data collection. Different sensors, communication networks, data processing algorithms, and post-processing decision-making mechanisms are being employed. Therefore, contact-based RPM also targets different diseases and groups of patients. The variety in contact-based RPM systems arises due to the required outcomes of special features (low power, low cost, etc.) or the choices of system designers planned for its users. For instance, the designers may aim to target different age groups or develop a power-efficient system.

In a study conducted by Kang et al. in 2018, the development of patient health surveillance platforms through IoT-based smart devices and their security using Blockchain technology is discussed. The researchers focused on IoT-based smart devices being developed for patient monitoring. These devices collect real-time physiological data and process it to end-terminals such as doctors, parents, or healthcare providers for assessment. Park et al. [11] developed a technique using a flexible and transparent contact lens for detecting glucose levels in tears to monitor real-time diabetes wirelessly. They combined electrochemical sensors powered by batteries for digital signals monitoring and Bluetooth data transmission in wrist smart devices to monitor the concentration of potassium, lactate, and glucose in sweat [12]. Furthermore, they explored state-of-the-art RPM devices and potential ways to integrate Blockchain and cloud computing to efficiently transmit health-related data from multiple sources.

De la Iglesia et al. [13] focused on contextualizing data monitored by sensors. They presented a nonintrusive system effective for monitoring and detecting high-risk factors in the air that can lead to sudden infant death syndrome (SIDS). They used a context-aware framework named Big Data Context-Aware Monitoring (BDCaM) to enable personalized knowledge for users. Context data such as air quality and weather conditions enable sensitive systems for highly accurate analysis. An agent-based virtual organization coordinates with different sensors and integrates different communication protocols, deployed with sensors in the system.

Most health monitoring systems work offline, which can be crucial in emergency situations for patients, especially for infants. Parihar et al. [14] proposed a system to remotely monitor heart rate and body temperature through sensors in real time with Arduino ATmega328, temperature sensor LM35, Heartbeat sensor, and nRF24L01. The observed data is monitored and transmitted from remote locations through wireless systems and displayed to the doctor.

In 2018, Eddabbah et al. designed a web service-based gateway used as a bridge among heterogeneous body sensors. The purpose of the proposed system is to prove good scalability and low delay. This system can communicate with several communication modes depending upon RPM system objectives such as speed, quality of service (QoS), data rate, and range.

Despite limited applications of telemedicine and RPM in pediatric and neonatal settings, these technologies have outperformed in this field, as discussed by Sasangohar et al. [15]. Continued research and developments in telemedicine and RPM will further demand the need for work in pediatric health systems. Moreover, the security and accuracy of data are as important as the health of an infant. Schochet [16] emphasized the importance of security problems in WBANs and the necessity of secure and reliable distributed databases and distributed data access control. Data privacy involves integrity, accuracy, availability, authentication, data freshness, and secure localization and management, as discussed by Mainanwal et al. in 2015.

10.3 Disease-Specific Remote Infant Monitoring Systems

The vital parameters such as RR, ECG, heart rate, and body temperature of both term and preterm infants in incubator care are monitored using sensors or self-adhesive electrodes directly on the skin. This includes electrocardiography (ECG) and photoplethysmography (PPG), which are crucial for routine care in infant intensive care units to ensure their good health. However, these monitoring systems can be stressful for infants [17]. Therefore, it is important to measure vital parameters such as temperature, RR, pulse, and oxygen saturation without mechanical or conductive contact with the infants. As a contactless method of monitoring, an adaptive style of camera-based photoplethysmography imaging (PPGI) has been proposed to meet neonatal requirements [17]. The PPGI camera is placed directly above the incubator of the infant and is illuminated by an infrared light-emitting-diode (LED) array (850 nm). It records a five-minute video of each preterm infant and analyzes it post hoc. The infants are not limited to a specific position in front of the camera, allowing for accurate monitoring. This approach successfully records blood perfusion according to the heart rate. Additionally, wearable health monitoring devices enable continuous monitoring of health-related signs during daily life activities, such as walking, working, being at home, or engaging in sports, without causing discomfort or interference. In healthcare, wearable sensors, such as bands and smartwatches, are used to collect personal health data and send it to healthcare professionals in real time. The integration of mobile communication and wearable sensors has shifted healthcare services from clinic centered to patient centered and is known as telemedicine [18, 19]. The upcoming section will focus on the discussion of respiration and apnea monitoring systems and heart- and blood-related diseases in infants.

10.3.1 Respiration and Apnea Monitoring System

The World Health Organization [20] has reported an increase in respiratory diseases as leading causes of mortality. Common respiratory diseases in infants include asthma, apnea syndrome, and chronic obstructive pulmonary disease (COPD) [21]. The respiratory system consists of two phases: inhaling oxygen and exhaling carbon dioxide during energy-producing reactions [22]. Monitoring the RR is important for infant patients. RR is a key parameter measured in breaths per minute and helps prioritize necessary treatment in intensive care and operating theaters [23]. The normal range of RR is different for adults and preterm infants, with 12–20 breaths per minute for adults and 40–60 breaths per minute for infants [24].

Sleep apnea is a critical respiratory disease that occurs during a patient's sleep. Obstructive sleep apnea is a common breathing disorder during sleep in infants that occurs due to partial or complete obstruction in airflow. Sleep apnea disturbs sleep duration and quality, leading to chronic partial sleep deprivation. Continuous monitoring of RR and its variations is essential for premature neonates [25]. Apnea and bradycardia are common in infants and if not detected early, can lead to oxygen deprivation in the organs and result in a deficiency in neurodevelopmental outcome [26]. Tachypnea is also a crucial indicator of bacterial pulmonary infections [26]. Moreover, tachypnea is a sign of SIDS, a major cause of death in infants and adults [27].

Identifying critical diseases is essential for respiratory disorder diagnosis, and abnormal RR or atypical waveform are key indicators [28–30]. An increased RR, known as tachypnea, can be an early sign of lung and heart diseases, while a decreased RR, called bradypnea, can be caused by hypothermia, certain medications, or diseases affecting the central nervous system [31]. COPD is projected to be the fourth leading cause of death in 2030. An e-Health system with smart sensor-based devices for real-time monitoring of respiratory frequencies is considered a good approach for timely treatment and prevention of respiratory diseases [32], encouraging parents to adopt essential healthcare skills for improving the well-being of their infants [33].

10.3.1.1 Emerging Sensing Techniques for Respiratory Diseases

Many sensor-based technologies have been proposed for continuous respiratory monitoring. An accelerometer captures processed information to detect the RR from rib cage movements. Spirometry and capnography are common approaches to monitor respiratory and cardiac variables; however, these methods are uncomfortable for infants and newborns and can disturb the natural breathing pattern, which may affect health readings.

Detecting respiratory diseases and breathing abnormalities poses a challenge for remote monitoring systems because they must differentiate various breathing sounds produced by the human upper body, such as rhonchi, rales, bubbling, and wheezing. Respiratory inductive plethysmography (RIP) is used to measure the RR of the patient [34]. In RIP, the patient wears an elastic belt, and it measures the RR by detecting the change in its inductance. This approach is much invasive and causes discomfort for infants as an infant cannot manage the overtightening of the belt around the chest or abdomen and could cause suffocation. In contrast, the loosening of the belt can cause the loss of the signal. Hoffmann et al. [35]

reported new development in fabricating capacitive sensors for clothes to monitor respiration without being invasive. These may produce inaccurate results because of the infant patient's movements and crying. Other methods are introduced to measure RR from the unobstructed sensors and on-body wearable devices [36].

Naranjo-Hernández et al. [37] have discussed a technique for designing and implementing of a smart sensor-based device for respiration rate monitoring integrated with a smart vest. Such devices can be implemented for RR monitoring of COPD in infant patients during their rest time period. The requirements to design such devices are low cost and easy to use. Respiratory noise is an early sign of health abasement even though it is a vital sign for health measurement [38]. Further, a contactless respiratory system monitors respiration rate through camera sensors reliably. However, variations in breathing rate and shortness of breath are vital signs of health problems [39].

Researcher Nijsure et al. [40] developed a noninvasive, contactless, and wireless system for respiratory monitoring. This system can be utilized in polysomnography (PSG) studies, home respiratory systems, and physiotherapy applications. The researchers used ultrawideband (UWB) due to its large bandwidth, which facilitates location estimation, allowing precise ranging and high time resolution. The RPM system monitors an infant's respiration accurately, using the hidden Markov model (HMM) to jointly estimate the direction an infant is facing and received signal strength (RSS) propagation delay profile estimates the chest wall movements. A single segmentation algorithm was developed to estimate the infant's respiratory motion from backscattered signals at the ultra-wideband receivers. The researchers also estimated the number of apnea and hypopnea episodes.

Furthermore, Bartula et al. [39] deployed and implemented an efficient approach to extract raw breathing signals from video movements. In addition, a camera detects motion information in the explored scenes, improving subsequent breath-to-breath classification. The purpose of the proposed work is to develop an efficient image-processing algorithm that integrates low complexity and high sensitivity. This method uses an off-the-shelf camera without a light pattern projecting on the body. It provides directionality information to reconstruct inhale and exhale signals in real time without excessive post-processing.

Jiang et al. [41] developed a smart sensing strip for infants to noninvasively monitor their respiration rate in real-time. The monitoring system uses a flexible hot-film flow sensor integrated into a silicone case, along with a Bluetooth 4.0 LE module containing various components. This system allows respiratory data to be wirelessly acquired and processed on a personal mobile device, offering a simple wearable device to continuously monitor respiratory flow without the need for an uncomfortable nasal cannula. The flexible flow sensor consists of four elements of a Wheatstone bridge on a single chip, including a hot-film resistor, two balancing resistors, and a temperature-compensating resistor, all packaged in a silicon case. This smart strip can be attached to the upper lips of infants, providing a noninvasive and comfortable monitoring option.

Subsequently, Fekr et al. [42] addressed the issues with RPM systems for accurately detecting respiration rate and breath information. Their method utilizes accelerometer data to extract RR and Tidal Volume Variability (TVvar) estimation. They also designed an alert system to respond to tidal volume emergency situations, which is composed of wearable sensors, standard BLE (Bluetooth low energy), and a backend cloud storage system. These advancements aim to reduce diagnostic time, improve medical services, and provide important benefits in terms of convenience, cost, quality of service, and patient comfort.

Additionally, in 2016, Al-Naji et al. introduced a remote respiratory system designed to monitor an infant's RR and timing variables. The system is able to monitor the RR by utilizing a DSLR video camera to capture images of the infant's chest, even in regions of unclear visibility or when the infant is covered with a blanket. This method is effective in varying light conditions as it relies on motion magnification rather than skin color analysis. The research team utilized a magnification technique that involved elliptic filtering, wavelet decomposition, and frame subtraction based on motion detection to measure the RR of the infant's chest in the magnified videos. The system achieved a 99% accuracy rate in detecting RR and timing cycles across various infant positions. This system could potentially replace traditional contact-based sensors used for breath sensing.

Subsequently, in 2017, Elfaramawy et al. discussed a respiratory physiological monitoring system that employs both contactless and contact-based approaches for monitoring infant patients. The system involves a wearable device for detecting infant coughing and a wireless system for monitoring infant respiration. A micro-electro-mechanical-system-based microphone is used to record infant coughing sounds, while mutual measurement units are utilized to monitor thoracic and cavity motions for respiratory monitoring. The system is predominantly contactless but includes inertial measurement units (IMUs) placed on the infant's chest for certain measurements. The parameters are transmitted wirelessly, and the small and lightweight IMU units are designed with the comfort of the infants in mind.

Furthermore, in 2016 and 2018, Ferreira et al. and Jakubas et al. developed contact-based systems using web interface technologies for respiratory monitoring in infants. These systems utilize breathing rate sensors and textile sensors such as BabyTex, a knitted fabric made of electrically conductive strands. In 2019, Wang et al. proposed a contactless system using smart speaker hardware and the SimNewB infant simulator, which was effective for infants weighing a minimum of 3.5 kg.

Table 10.2 provides an overview of various studies that have proposed contact-based and contactless frameworks for remote infant respiratory monitoring systems.

Table 10.2 Respiratory-related disease observation in remote infant monitoring system.

Studies	Sensor type	Technology	Approach	Database	Limitations
[40]	Contactless	Ultrawideband (UWB) is used because of its large bandwidth that facilitates location estimation, allowing precise ranging and high time resolution	A single segmentation algorithm is developed to estimate the infant's respiratory motion from backscattered signals at the ultra-wideband receivers	Custom	Direction of an infant is significant
[39]	Contactless	Video movements	An efficient image processing algorithm that integrates low complexity and high sensitivity	Cloud storage	Single baby monitoring
[41]	Contact-based	Flexible silicone case and Bluetooth 4.0 LE module (containing a 12-bit ADC, an MCU part, a GPIO port, and Bluetooth 4.0 LE transceiver)	Noninvasively real-time monitoring of respiration rate acquired wirelessly through flow sensor	Custom	The proposed system can monitor single babies only
[42]	Contactless	Standard BLE (Bluetooth low energy)	A method is also proposed for Tidal Volume Variability (TVvar) estimation through accelerometer data validated by using Pearson correlation	Cloud storage	—
[43]	Contactless	Nikon D5300 DSL camera with a CMOS focal plane	A novel measuring method based on motion detection and motion magnification technique	Custom	The proposed system can monitor single babies only
[44]	Contact based	Breathing rate sensor, Baby Night Watch	The web interface was developed in PHP, SQL, JavaScript, HTML5, and CSS3	Cloud storage	The user has to be connected with the gateway to receive data

[45]	Contact-based/contactless	Micro-electro-mechanical-system-based microphone is utilized to record the coughing sound of infant patients and mutual measurement units are utilized to record the thoracic and cavity motions for monitoring respiration	A wearable device detects infant's coughing and a wireless system monitors the respiration of infants	Custom	—
[46]	Contact-based	Textile sensors, a knitted fabric made of electrically conductive strands, BabyTex	The APPA 505 multimeter	Custom	Sensors are dependent upon the quality of the stitch that is fabricated from electro-conductive yarn
[47]	Contactless	White noise	Smart speaker hardware, SimNewB infant simulator	Custom	The proposed system is effective only for infants with a minimum of 3.5 kg of weight. Underweight infants need to be focused too

10.3.2 Heart and Blood-Related Diseases Monitoring Systems

Heart monitoring systems are widely used to monitor cardiovascular problems. This is because the vital signs of the body linked with the heart can indicate many other diseases at the same time. Common diseases that can occur include chronic heart failure, blood clotting, strokes, cardiac arrhythmia, and high blood pressure. Various technologies such as ECG monitors and wearable devices are used to collect this data, although there may be chances of impurities in the collected data. This paper discusses various applications and devices that use smartphones for monitoring patients remotely and preventing cardiac diseases. However, sharing patient data can lead to security issues and compromise patient privacy. The Health Insurance Portability and Accountability Act (HIPAA) has listed measures to defend health information, but this information is still not enough to protect patient privacy against various health databases. According to recent reports, many RPM systems were designed for patients suffering from chronic illnesses such as cardiovascular diseases, cardiomyopathy, and cardiopulmonary diseases, which can cause heart failure in infant patients living in remote and urban areas and who cannot seek immediate medical assistance [48, 49].

10.3.2.1 Emerging Sensing Techniques for Heart and Blood Diseases

Remote heart monitoring systems are widespread due to heart-related diseases being the leading causes of death worldwide. These systems provide combined results for respiration and heart-related diseases. One crucial challenge is obtaining accurate readings from the patient's body. Traditional contact-based methods typically use ECG and photoplethysmographic (PPG) methods, which have proven to be useful. These methods use infrared light emitted by LED, which penetrates small veins. Some existing systems, such as Smart Vest (Pandian et al. 2008) and LOBIN [50], have a similar approach to monitoring remote patients using wearable clothes. The smart vest uses wireless data transmission from clothes to remote servers, while LOBIN uses wireless data transmission boards and distribution points. However, there are some disadvantages to these systems, such as discomfort for infants due to wearing unusual clothing.

Contactless physiological parameters using cameras are advanced technology these days. Recently, contactless systems have been developed in both offline and real-time mode by using different color spaces, such as RBG [51], Lab [52], and YCbCr [53]. Heart rate variability (HRV) is an important physiological parameter used to detect many fatal diseases.

In a telemedicine framework, multiple paramedics from an ambulance can communicate with a "Tele-EMS (Emergency Medical Service) physician" located in a "teleconsultation center." Research by Thelen et al. in 2016 suggests that simple devices and application interfaces are more effective than different specialized protocols.

Alves et al. in 2015 developed a system capable of detecting vital signs such as pulse rate and oxygen saturation. This system focuses on vital signs rather than monitoring joint angle and hydrotherapy. Bisio et al. in 2015 introduced the PAN upper-layer architecture for Android phones, which operates in a "hub + sensor + processor" manner. In 2015, Papon et al. presented a survey comparing smart cameras' applications for heart rate measurement. The comparison showed promising results in contact-based systems, while contactless approaches were affected by incorrect results due to ambient lighting. Donnelly et al. in 2015 minimized the disadvantages of deterioration monitoring with the wearable apparatus "patient guard 300 (PG300)," which measures heart rate, ECG, temperature, and respiration information. Mishra and Agarwal in 2015 presented an approach to measure physiological data related to pulmonary artery pressure (PAP) and ECG. They also encoded the detected data in a text message.

The researchers, Kakria et al. [54], proposed a real-time monitoring system that can detect medical parameters such as heart rate, body and skin temperature, and blood pressure simultaneously. The system is compatible with various sensors and aims to aid in the early detection of diseases such as high blood pressure, low blood pressure, hyperthermia, and arrhythmia through an alarming system based on preset threshold values. The monitoring system comprises two interfaces: one for medical professionals and one for the patient. The patient interface includes wearable sensors and transmission means to extract and transmit medical data via Bluetooth low energy to an Android-based listening port. This data is then transferred to a web server which extracts and transfers the data to an online MySQL database through Wi-Fi or GPRS/3G. Once processed, the data is displayed on the medical professional's interface along with the patient's location and identity information.

Zhang et al. [55] implemented a remote mobile health monitoring system based on smartphones and web browsers, targeting patients with chronic illnesses, especially infants. This system integrates vital sign sensors by using portable terminals to monitor physiological data such as heart rate, body temperature, and RR. A smartphone is used as an information transmission platform and an intuitive human–machine interface to continuously monitor the patient's health status. The patient's current location can be acquired by GPS for outdoor and Wi-Fi signals for indoor use. The remote server employs a browser/server architecture for data query, patient location, and observation curves. Notably, the portable terminal integrates various sensors, including heart rate, respiratory, temperature, and posture sensors. The Zephyr BioHarnessTM sensor is employed as a portable terminal, capable of detecting comprehensive physiological data like RR intervals, ECG, RR, heart rate, skin temperature, breathing wave altitude, vector magnitude peak acceleration, and posture. It weighs only 35 g and can be easily fixed on a patient's chest with a belt. However, it may be uncomfortable for infants due to the tightening belt. Furthermore, this system is capable of monitoring a patient's status in real-time, but not for professional analysis and instruction.

Ferreira et al. [44] developed a contact-based web interface in PHP, SQL, JavaScript, HTML5, and CSS3 to measure heart rate using baby night watch. This was followed by Blanik et al. [17] proposing a contactless Camera-based Photoplethysmograph-Phyimaging (PPGI) according to neonatal requirements which was found to be sensitive to the fast movement of infants.

Later, Rahman et al. [56] developed a contactless remote heart rate variability (HRV) monitoring system using facial video based on color disruption of facial skin caused by the cardiac pulse. The Lab color space is used to extract physiological data from facial video recording. Signal processing algorithms such as independent component analysis (ICA), principal component analysis (PCA), and fast Fourier transform (FFT) are employed to detect HRV. Inter-beat-interval (IBI)-based HRV features: time domain and frequency domain are extracted with a reference sensor system. The system is proposed for server-side languages PHP and MySQL and currently executes on a personal computer monitoring HRV features via a website.

Subsequently, Gogate and Bakal [57] presented a three-tier architecture for an infant healthcare monitoring system using a wireless sensor network (WSN) to constantly monitor the infant's body parameters. Various biosensors were attached to the Arduino Nano board to measure body oxygen level, heart rate, and temperature. The recorded signals are sent to the server using Node MCU wireless communication. Data is uploaded to a remote server and is available for paramedical staff and caregivers through an IoT application ThingSpeak. In case of an emergency, paramedics or caregivers are notified through smartphone alerts. The system is beneficial for cardiac patients and can be utilized for elderly care in homes and hospitals, with an accuracy of 99% and a 10-second response time.

Moreover, researchers Hussein et al. [58] proposed a cloud-based RPM system for monitoring HRV of infants and premature babies. This system is designed to offer better and timely healthcare monitoring to infants and newborns in rural areas with limited medical facilities due to the small doctor–patient ratio. In this study, the developed remote and diagnostic systems are discussed that are used to monitor an infant's heart conditions. This also helps patients in the prevention of heart diseases who are recovering from their illnesses. The researcher configured a wearable Electrocardiographic device to get HRV data. The data reflects the patient's health condition and helps to detect abnormalities in the heart like high blood pressure, low blood pressure, arrhythmia, and ischemia. The system's architecture is composed of wearable ECG devices, ECG system nodes, the main processing server, data files, and web graphical user interface (GUI). Two types of MIT Physionet databases are used for storing patient data: MIT-St. Petersburg and

MIT-BIH sinus rhythm. After analyzing the performance of the proposed system, the obtained results for specificity, sensitivity, and accuracy were 99.17%, 98.78%, and 99.02%, respectively. Achieved promising results concluded that the proposed system was quite reliable, valuable, and robust.

In a recent study Rudd et al. [59] remote home monitoring of infants with shunt-dependent single ventricle heart disease has been proposed to reduce mortality rate during the inter-stage period that is the time of discharge of a patient. In-home surveillance of infants detects physiological changes of inter-stage infants having single ventricle heart problems. This monitoring provides support to a pediatric cardiologist, nurses, care provider of infants during staged surgical palliation.

Table 10.3 summarizes the infant monitoring system for heart and blood-related diseases.

10.3.3 Infant Monitoring Systems for Various Diseases

There are several studies that propose different applications of infant disease monitoring aside from common respiratory and heart diseases. In 2014, researchers Greer et al. proposed a system for remotely monitoring physiological data from newborns. The system includes capture devices to monitor heart rate, respiration rate, and blood oxygen remotely, as well as a mobile application to view current and previous physiological data of patients. Thanks to multiple connectivity sources, the system demonstrated a high level of quality, accuracy, and reliability throughout the testing stages. Artemis is utilized by the system to capture and interpret data. Additionally, a web interface (web browser interface) was developed by Bauer and Mendes in 2015 to monitor the baby's incubator. Smartphone-assisted technology for eye care was presented by Lakshminarayanan in 2015. Similarly, in 2012, Benelli et al. presented a system to monitor various physiological measurements like body weight, blood pressure, glycemia, ECG, and spirometry, and these can be displayed on television. Table 10.4 summarizes the contactless and contact-based RPM systems for various infant diseases.

10.4 Challenges in Remote Infant Monitoring Systems

Although technology in multi-modal sensing is rapidly advancing, there are still significant shortcomings in research. For example, privacy and security issues in wireless sensing networks have not been adequately addressed. The most recent security issues involve the jamming of one or a few sensors during an attack. Sensor-based networks are vulnerable to various threats and attacks, which could potentially lead to the loss of an infant patient's life. Attackers could capture

Table 10.3 Summary of heart and blood-related diseases infant monitoring systems.

Studies	Sensor type	Technology	Approach	Database	Limitations
[54]	Contact-based	Bluetooth low energy to an Android-based listening port	Real-time monitoring system with various sensors	Cloud storage	Wearing a monitoring device continuously
[55]	Contact-based	Heart sensor, Zephyr BioHarnessTM sensor	Smartphone, which employs the Android-operating system	Remote server	Patients feel discomfort while continuously wearing a monitoring device
[44]	Contact-based	Heart rate sensor, Baby Night Watch	The web interface was developed in PHP, SQL, JavaScript, HTML5, and CSS3	Cloud storage	The user has to be connected with the gateway to receive data
[17].	Contactless	Camera-based photoplethysmograph-phyimaging (PPGI) according to neonatal requirements	Remote sensing	Custom	PPGI was found sensitive to the fast movement of infants
[56]	Contactless	Facial video recording by using a webcam	Remote heart rate variability (HRV) monitoring system using the facial video based on color disruption of facial skin caused by cardiac pulse	Custom	System is executing on a personal computer where HRV features are monitored by a website
[57]	Contact based	Biosensors	Node MCU ESP8266 wireless communication, IoT application ThingSpeak	Remote server	The proposed model should be tested on more patients
[58]	Contact based	MIT Physionet databases are used for storing patient data: MIT-St. Petersburg and MIT-BIH sinus rhythm	Wearable ECG devices, ECG system nodes, the main processing server, data files, and web GUI	Remote server	—
[60]	Contact based	Neo-SENSE, smart-mattress for newborn ECG monitoring	Electrometer	Custom	The device is limited to single infants only
[59]	Contact based	Shunt-dependent single ventricle heart disease	In-home surveillance of infants detects physiological changes of inter-stage infants	Custom	Single infants monitoring only

Table 10.4 Summary of remote infant monitoring systems for various infant diseases.

References	Sensor/Technology	Algorithm/Approach	Database	Limitations
[61]	BP sensor, temperature sensor, pulse sensor, wireless transmission	Develop application for smartphone	Custom database	The proposed system sends signals to the only Android-based smartphone and cannot send signals to medical devices
[62]	Microsoft Kinect v2 sensor, enhanced video magnification, motion-sensing technology	Eulerian video magnification technique	Custom database	—
[63]	Biomedical sensor nodes temperature and heart sensors, Arduino UNO as microcontroller, Bluetooth wireless technology, personal digital assistance (PDA), Raspberry Pi	Arduino UNO, Raspberry Pi 3	MySQL	Privacy and security are the issues in this system as any of medical staff access patient's health and locality status
[64]	Wearable calibrated accelerometer sensor, Bluetooth low energy (BLE)	Numeric integration algorithm	Cloud database	The system uses cloud database. Continues internet connectivity is required. It can be difficult for patients in rural areas
[39]	Monochrome camera, time-of-flight (ToF) camera	Proposed camera and image processing algorithms	Custom database	Camera sensor can produce inaccurate reading because of patient's movement
[65]	Thermal imaging sensor with uncooled infrared microbolometer	Black-Box Algorithm	Custom database	Thermal imaging is an expensive approach as compared to radar and other approaches
[66]	Web camera	Fast Fourier transform (FFT), principal component analysis (PCA), independent component analysis (ICA), Kanade–Lucas–Thomasi (KLD)	Cloud-based MySQL database	The developed system is not a real-time system. Multiple physiological parameters are not measured

(*Continued*)

Table 10.4 (Continued)

References	Sensor/Technology	Algorithm/Approach	Database	Limitations
[67]	Ultrawideband signals technology	Hidden Markov model (HMM)-based method, change point detection algorithm	—	If the procedure has converged, Marko-chain-Monte-Carlo (MCMC) techniques may produce erroneous results
[68]	Zephyr BT for heart rate, Omron wireless upper arm blood-pressure monitor for blood pressure, g-plus Bluetooth based for temperature	Fuzzy logic	SQLite internal database, online MySQL database	The delay in alarming the doctor about the patient's abnormal reading can be delayed due to weak internet signals and it can generate false alarm due to battery issues
[69]	Tailor-designed micro flow sensor monolithically integrated with a temperature-compensating resistor, two balancing resistors	A novel algorithm is proposed	Custom database	—
[57]	Heart rate sensor, temperature sensor, wireless sensor network	3-Tier architecture	Custom database	2-Way communication is not possible. So the doctor can advise patient in critical conditions

sensor data and send false information to doctors, resulting in life-threatening situations. Additionally, attackers could use nodes to block the entire network by jamming or disrupting the system by flooding it with excessive information, causing a Denial of Service (DoS) and compromising the network's ability to provide emergency services.

The following security threats need to be addressed in order to have a massive deployment and acceptability of the multi-modal infant monitoring sensing systems:

- **Data Confidentiality:** Patient data, such as health status, must be protected from unauthorized access using data encryption for secure communication.
- **Data Integrity:** Information should be protected from modification during transmission to prevent serious health concerns. This can be ensured through secure transmission and authentication protocols.
- **Network Availability:** Medical professionals should have continuous access to patient data, particularly for emergency cases involving newborns.
- **Data Authentication:** Implementing message authentication at each node of the network is essential to ensure trustworthiness.
- **Secure Management and Communication:** Secure encryption and decryption methods should be deployed, and secure routing protocols should be in place to manage group communication.
- **Flexibility:** Systems should allow controlled sharing of patient data and smooth transfer of control when a patient changes their physician.
- **Accountability:** Healthcare providers should be held accountable for securing patient data.

Another crucial issue is the usability of the system, as each system has a unique interface. Ensuring adaptability and ease of use for parents of infants is essential. Comfort is also a concern, as contact-based systems can cause discomfort for infants due to their sensitive skin. Contactless technologies such as radar and image sensing can alleviate this issue. Furthermore, error correction methods must be improved to gain the trust of the medical community. Overall, there is still a vast scope of research needed to assess the acceptance of technology in the medical field by patients and the medical community.

10.5 Summary

This chapter explores RIM and its feasibility and acceptability by parents and infants. RIM enables parents to monitor their infants without being physically present, offering flexibility and peace of mind. There are a number of RIM systems proposed dealing with infant respiration, apnea, heart diseases, and blood pressure, utilizing both contact-based and contactless technologies, this study primarily focuses on contactless radar and image-based technologies due to their favorable attributes. It underscores the significant impact of advancements in RPM systems for infants on society and researchers. It also emphasizes the importance of privacy-preserving methods to limit data availability and highlights the research challenges and opportunities associated with remote monitoring of infants.

References

1 Ministry of National Health Services, Regulations & Coordination (2021). *Pakistan: 2021 Monitoring Report: Universal Health Coverage. Moving Together to Build a Healthier Pakistan*, 1–84. Universal Health Coverage.

2 Statista (2021). Infant mortality rates in South Asia between 2015 and 2020, by country. https://www.statista.com/statistics/590050/infant-mortality-rates-in-south-asia (accessed 02 June 2024).

3 Mutanu, L., Gupta, K., Gohil, J. et al. (2020). Enhancing healthcare access through remote infant screening. *ACM International Conference Proceeding Series* 236–239: https://doi.org/10.1145/3411170.3411270.

4 Memon, S.F., Memon, M., and Bhatti, S. (2020). Wearable technology for infant health monitoring: a survey. *IET Circuits, Devices & Systems* 14 (2): 115–129. https://doi.org/10.1049/IET-CDS.2018.5447.

5 Beltrão, G., Stutz, R., Hornberger, F. et al. (2022). Contactless radar-based breathing monitoring of premature infants in the neonatal intensive care unit. *Scientific Reports* 12 (1): 1–15. https://doi.org/10.1038/s41598-022-08836-3.

6 Svoboda, L., Sperrhake, J., Nisser, M. et al. (2022). Contactless heart rate measurement in newborn infants using a multimodal 3D camera system. *Frontiers in Pediatrics* 10: https://doi.org/10.3389/FPED.2022.897961.

7 Teotia, M., Sharma, M., Dubey, K. et al. (2023). An approach on the development of a smart monitoring system of a baby cradle. *International Journal for Research in Applied Science & Engineering Technology (IJRASET)* 11: 2321–9653. www.ijraset.com.

8 Alam, H., Burhan, M., Gillani, A., and Haq, I. (2023). IoT based smart baby monitoring system with emotion recognition using machine learning. *Wireless Communications and Mobile Computing* 2023: 1–11. https://doi.org/10.1155/2023/1175450.

9 McDuff, D.J., Estepp, J.R., Piasecki, A.M. and Blackford, E.B. (2015). A survey of remote optical photoplethysmographic imaging methods. *2015 37th Annual International Conference of the IEEE Engineering in Medicine and Biology Society (EMBC)*, Milan, Italy, 6398–6404.

10 Shinde, S.K., Vijay, J.F., Meena, S.G. et al. (2023). IoT based smart intruder system for baby monitoring. *AIP Conference Proceedings* 2690 (1): https://doi.org/10.1063/5.0121400/2880462.

11 Park, J., Kim, J., Kim, S.Y. et al. (2018). Soft, smart contact lenses with integrations of wireless circuits, glucose sensors, and displays. *Science Advances* 4 (1): eaap9841.

12 Choi, J., Ghaffari, R., Baker, L.B., and Rogers, J.A. (2018). Skin-interfaced systems for sweat collection and analytics. *Science Advances* 4 (2): eaar3921.

13 De la Iglesia, D.H., De Paz, J.F., Villarrubia Gonzalez, G. et al. (2018). A context-aware indoor air quality system for sudden infant death syndrome prevention. *Sensors* 18 (3): 757.

14 Parihar, V.R., Tonge, A.Y., and Ganorkar, P.D. (2017). Heartbeat and temperature monitoring system for remote patients using Arduino. *International Journal of Advanced Engineering Research and Science* 4 (5): 237161.

15 Sasangohar, F., Davis, E., Kash, B.A., and Shah, S.R. (2018). Remote patient monitoring and telemedicine in neonatal and pediatric settings: scoping literature review. *Journal of medical Internet research* 20 (12): e295.

16 Schochet, L. (2017). *The Importance of Child Care Safety Protections*. Washington: Center for American Progress.

17 Blanik, N., Heimann, K., Pereira, C. et al. (2016). Remote vital parameter monitoring in neonatology–robust, unobtrusive heart rate detection in a realistic clinical scenario. *Biomedical Engineering/Biomedizinische Technik* 61 (6): 631–643.

18 Ping, W., Jin-Gang, W., Xiao-Bo, S. and Wei, H. (2006). The research of telemedicine system based on embedded computer. *2005 IEEE Engineering in Medicine and Biology 27th Annual Conference*, Shanghai, China, 114–117. IEEE.

19 Wu, X. and Lv, X. (2023). A sustainable energy strategy powered wireless sensor network system for monitoring child safety. *Sustainable Energy Technologies and Assessments* 57: 103183. https://doi.org/10.1016/J.SETA.2023.103183.

20 World Health Organization (2014). The 10 Leading Causes of Death in the World, 2000 and 2012. World Health Organization: Geneva, Switzerland. http://www.who.int/mediacentrefactsheets/fs310/en/ (accessed 05 June 2024).

21 Jain, C.D., Bhaskar, D.J., Kalra, M. et al. (2015). Recent improvements in the management of obstructive sleep apnea: the dental perception. *International Journal of Advanced Health Science and Technology* 1: 7–10.

22 Guyton, A. and Hall, J. (2011). *Guyton and Hall Textbook of Medical Physiology*. Philadelphia, PA: Saunders/Elsevier.

23 Cretikos, M.A., Bellomo, R., Hillman, K. et al. (2008). Respiratory rate: the neglected vital sign. *Medical Journal of Australia* 188 (11): 657–659.

24 Thomas, A. and Maxwell, L.J. (2016). Clinical assessment. In: *Cardiorespiratory Physiotherapy: Adults and Paediatrics*, 5e (ed. E. Main and L. Denehy). Amsterdam, The Netherlands: Elsevier, ch. 2.

25 Olivier, F., Nadeau, S., Caouette, G., and Piedboeuf, B. (2016). Association between apnea of prematurity and respiratory distress syndrome in late preterm infants: an observational study. *Frontiers in Pediatrics* 4: 105.

26 Tregoning, J.S. and Schwarze, J. (2010). Respiratory viral infections in infants: causes, clinical symptoms, virology, and immunology. *Clinical Microbiology Reviews* 23 (1): 74–98.

27 Fei, J. and Pavlidis, I. (2007). Virtual thermistor. *2007 29th Annual International Conference of the IEEE Engineering in Medicine and Biology Society*, Lyon, France, 250–253.

28 Kowalak, J.P. (2009). *Lippincott's Nursing Procedures*, 5e. Philadelphia: Lippincott Williams & Wilkins.

29 Lindh, W.Q., Pooler, M., Tamparo, C.D. et al. (2013). *Delmar's Comprehensive Medical Assisting: Administrative and Clinical Competencies.* Nelson Education.

30 Yoost, B.L., Crawford, L.R., and Castaldi, P. (2019). *Clinical Companion for Fundamentals of Nursing E-Book: Active Learning for Collaborative Practice.* Elsevier Health Sciences.

31 Shelledy, D.C. and Peters, J.I. (2014). *Respiratory Care: Patient Assessment and Care Plan Development*, 59. Jones & Bartlett Publishers.

32 Ambrosino, N., Makhabah, D.N., and Sutanto, Y.S. (2017). Tele-medicine in respiratory diseases. *Multidisciplinary Respiratory Medicine* 12 (1): 1–5.

33 Estrada, L., Torres, A., Sarlabous, L. et al. (2014). Respiratory rate detection by empirical mode decomposition method applied to diaphragm mechanomyographic signals. *2014 36th Annual International Conference of the IEEE Engineering in Medicine and Biology Society*, 3204–3207.

34 Iber, C., Ancoli-Israel, S., and Chesson, A. (2007). *The AASM Manual for the Scoring of Sleep and Associated Events.* Darien, IL: American Academy of Sleep Medicine.

35 Hoffmann, T., Eilebrecht, B., and Leonhardt, S. (2010). Respiratory monitoring system on the basis of capacitive textile force sensors. *IEEE Sensors Journal* 11 (5): 1112–1119.

36 Bianchi, A.M., Mendez, M.O., and Cerutti, S. (2010). Processing of signals recorded through smart devices: sleep-quality assessment. *IEEE Transactions on Information Technology in Biomedicine* 14 (3): 741–747.

37 Naranjo-Hernández, D., Talaminos-Barroso, A., Reina-Tosina, J. et al. (2018). Smart vest for respiratory rate monitoring of COPD patients based on non-contact capacitive sensing. *Sensors* 18 (7): 2144.

38 McGain, F., Cretikos, M.A., Jones, D. et al. (2008). Documentation of Clinical Review and Vital Signs after Major Surgery. *Medical Journal of Australia* 189 (7): 380–383.

39 Bartula, M., Tigges, T. and Muehlsteff, J. (2013). Camera-based system for contactless monitoring of respiration. *2013 35th Annual International Conference of the IEEE Engineering in Medicine and Biology Society (EMBC)*, Osaka, Japan, 2672–2675.

40 Nijsure, Y., Tay, W.P., Gunawan, E. et al. (2013). An impulse radio ultrawideband system for contactless noninvasive respiratory monitoring. *IEEE Transactions on Biomedical Engineering* 60 (6): 1509–1517.

41 Jiang, P., Zhao, S., and Zhu, R. (2015). Smart sensing strip using monolithically integrated flexible flow sensor for noninvasively monitoring respiratory flow. *Sensors* 15 (12): 31738–31750.

42 Fekr, A.R., Radecka, K., and Zilic, Z. (2015). Design and evaluation of an intelligent remote tidal volume variability monitoring system in e-health applications. *IEEE Journal of Biomedical and Health Informatics* 19 (5): 1532–1548.

43 Al-Naji, A. and Chahl, J. (2016). Remote respiratory monitoring system based on developing motion magnification technique. *Biomedical Signal Processing and Control* 29: 1–10.

44 Ferreira, A. G., Fernandes, D., Branco, S. et al. (2016). A smart wearable system for sudden infant death syndrome monitoring. *2016 IEEE International Conference on Industrial Technology (ICIT), Taipei,* Taiwan, 1920–1925.

45 Elfaramawy, T., Fall, C.L., Morissette, M. et al. (2017). Wireless respiratory monitoring and coughing detection using a wearable patch sensor network. *2017 15th IEEE International New Circuits and Systems Conference (NEWCAS),* Strasbourg, France, 197–200.

46 Jakubas, A. and Łada-Tondyra, E. (2018). A study on application of the ribbing stitch as sensor of respiratory rhythm in smart clothing designed for infants. *The Journal of the Textile Institute* 109 (9): 1208–1216.

47 Wang, A., Sunshine, J.E., and Gollakota, S. (2019). Contactless infant monitoring using white noise. *The 25th Annual International Conference on Mobile Computing and Networking*, Los Cabos, Mexico, 1–16.

48 Lee, C.D., Ho, K.I.J., and Lee, W.B. (2011). A novel key management solution for reinforcing compliance with HIPAA privacy/security regulations. *IEEE Transactions on Information Technology in Biomedicine* 15 (4): 550–556.

49 Mahmud, M.S., Wang, H., Esfar-E-Alam, A.M., and Fang, H. (2017). A wireless health monitoring system using mobile phone accessories. *IEEE Internet of Things Journal* 4 (6): 2009–2018.

50 López, G., Custodio, V., and Moreno, J.I. (2010). LOBIN: E-textile and wireless-sensor-network-based platform for healthcare monitoring in future hospital environments. *IEEE Transactions on Information Technology in Biomedicine* 14 (6): 1446–1458.

51 Rahman, H., Ahmed, M.U., and Begum, S. (2016). Non-contact physiological parameters extraction using camera. In: *Internet of Things. IoT Infrastructures: Second International Summit, IoT 360° 2015*, Rome, Italy (27–29 October 2015), Revised Selected Papers, Part I, 448–453. Springer International Publishing.

52 Rahman, H., Ahmed, M.U., and Begum, S. (2016). Non-contact heart rate monitoring using lab color space. *Studies in Health Technology and Informatics* 224: 46–53.

53 Zhang, Q., Xu, G.Q., Wang, M. et al. (2014). Webcam based non-contact real-time monitoring for the physiological parameters of drivers. In: *The 4th Annual IEEE*

International Conference on Cyber Technology in Automation, Control and Intelligent, 648–652. Hong Kong, China: IEEE.

54 Kakria, P., Tripathi, N.K., and Kitipawang, P. (2015). A real-time health monitoring system for remote cardiac patients using smartphone and wearable sensors. *International Journal of Telemedicine and Applications* 2015: 373474.

55 Zhang, Y., Liu, H., Su, X. et al. (2015). Remote mobile health monitoring system based on smart phone and browser/server structure. *Journal of Healthcare Engineering* 6 (4): 717–738.

56 Rahman, H., Ahmed, M.U., and Begum, S. (2018). Vision-based remote heart rate variability monitoring using camera. In: *Internet of Things (IoT) Technologies for HealthCare: 4th International Conference, HealthyIoT 2017*, Angers, France (24–25 October 2017), Proceedings 4, 10–18. Springer International Publishing.

57 Gogate, U. and Bakal, J. (2018). Healthcare monitoring system based on wireless sensor network for cardiac patients. *Biomedical & Pharmacology Journal* 11 (3): 1681.

58 Hussein, A.F., Burbano-Fernandez, M., Ramírez-González, G. et al. (2018). An automated remote cloud-based heart rate variability monitoring system. *IEEE Access* 6: 77055–77064.

59 Rudd, N.A., Ghanayem, N.S., Hill, G.D. et al. (2020). Interstage home monitoring for infants with single ventricle heart disease: education and management: A scientific statement from the American Heart Association. *Journal of the American Heart Association* 9 (16): e014548.

60 Aviles-Espinosa, R., Rendon-Morales, E., Luo, Z. et al. (2019). Neo-SENSE: a non-invasive smart sensing mattress for cardiac monitoring of babies. *2019 IEEE Sensors Applications Symposium (SAS)*, Sophia Antipolis, France, 1–5.

61 Ali, M.A.M., Tahir, N.M., and Ali, A.I. (2018). Monitoring healthcare system for infants: a review. In: *2018 IEEE Conference on Systems, Process and Control (ICSPC)*, Melaka, Malaysia, 44–47. https://doi.org/10.1109/SPC.2018.8704143.

62 Al-Naji, A. and Chahl, J. (2018). Detection of cardiopulmonary activity and related abnormal events using microsoft kinect sensor. *Sensors* 2018 (18): 920. https://doi.org/10.3390/s18030920.

63 Hassan, J. and Hassan, J. (2017). Implementation of wireless area network for patient monitoring system. *Iraqi Journal of Computers, Communication and Control & System Engineering (IJCCCE)* 17 (1): 1–9.

64 Fekr, A.R., Radecka, K., and Zilic, Z. (2015). Design and evaluation of an intelligent remote tidal volume variability monitoring system in E-health applications. *IEEE Journal of Biomedical and Health Informatics* 19 (5): 1532–1548. https://doi.org/10.1109/JBHI.2015.2445783.

65 Pereira, C. B., Yu, X., Goos, T., Orlikowsky, T., Reiss, I., Heimann, K., Venema, B., Blazek, V., Leonhardt, S., Teichmann, D. 2019, Noncontact monitoring of respiratory rate in newborn infants using thermal imaging, in *IEEE Transactions* *on Biomedical Engineering*, vol. 66, no. 4, pp. 1105–1114. doi: https://doi.org/10.1109/TBME.2018.2866878.

66 Rahman, H., Ahmed, M.U., and Begum, S. (2018). Vision-based remote heart rate variability monitoring using camera. In: *Internet of Things (IoT) Technologies for HealthCare. HealthyIoT 2017*, Lecture Notes of the Institute for Computer Sciences, Social Informatics and Telecommunications Engineering, vol. 225 (ed. M. Ahmed, S. Begum, and J.B. Fasquel). Cham: Springer https://doi.org/10.1007/978-3-319-76213-5_2.

67 Nijsure, Y., Tay, W.P., Gunawan, E. et al. (2013). An impulse radio ultrawideband system for contactless noninvasive respiratory monitoring. *IEEE Transactions on Biomedical Engineering* 60 (6): 1509–1517. https://doi.org/10.1109/TBME.2012.2237401.

68 Kakria, P., Tripathi, N.K., and Kitipawang, P. (2015). A real-time health monitoring system for remote cardiac patients using smartphone and wearable sensors. *International Journal of Telemedicine and Applications* 2015: https://doi.org/10.1155/2015/373474.

69 Jiang, P., Zhao, S., and Zhu, R. (2015). Smart sensing strip using monolithically integrated flexible flow sensor for noninvasively monitoring respiratory flow. *Sensors (Basel)* 15 (12): 31738–31750. https://doi.org/10.3390/s151229881.

70 Bayoumy, K., Gaber, M., Elshafeey, A. et al. (2021). Smart wearable devices in cardiovascular care: where we are and how to move forward. *Natural Reviews Cardiol* 18: 581–599.

11

Balancing Innovation with Ethics: Responsible Development of Multi-modal Intelligent Tutoring Systems

Romina Soledad Albornoz-De Luise[1], Pablo Arnau-González[1], Ana Serrano-Mamolar[2], Sergi Solera-Monforte[1], and Yuyan Wu[1]

[1] *Departament d'Informàtica, Universitat de València, València, Spain*
[2] *Departamento de Lenguajes y Sistemas, Universidad de Burgos, Burgos, Spain*

Intelligent tutoring systems (ITS) represent a paradigm shift in education, offering tailored learning experiences and continuous assessment. However, the integration of ITS raises profound ethical considerations. This chapter systematically explores the promise and perils of ITS, highlighting advantages such as personalized learning and accessibility, while also scrutinizing risks including algorithmic bias, privacy concerns, and socioeconomic disparities in access. Drawing on ethical frameworks and real-world examples, it underscores the importance of responsible development and deployment practices. By advocating for equity, transparency, and student-centered approaches, this chapter provides a comprehensive guide for navigating the multifaceted ethical landscape of ITS, ensuring the advancement of education while safeguarding the rights and dignity of learners.

11.1 Intelligent Tutoring Systems and Ethical Considerations

Intelligent tutoring systems (ITSs) have garnered significant attention in recent years as a promising tool to enhance education [1]. Powered by artificial intelligence (AI) and machine learning algorithms, these multi-modal systems offer

Multimodal Intelligent Sensing in Modern Applications, First Edition.
Edited by Masood Ur Rehman, Ahmed Zoha, Muhammad Ali Jamshed, and Naeem Ramzan.

Published 2025 by John Wiley & Sons, Inc.

personalized and adaptive learning experiences tailored to the individual needs and preferences of learners in a specific field. Through continuous assessment, immediate adaptive feedback, hints, and targeted instruction, ITS hold the potential to optimize learning outcomes and foster engagement in ways previously unattainable within traditional educational frameworks.

The main lure of ITS lies in its ability to transcend the limitations of one-size-fits-all instruction, providing learners with tailored pathways that accommodate diverse learning styles, paces, and abilities. By leveraging vast amounts of data and sophisticated algorithms, ITS can analyze and adapt to individual learning trajectories offering customized content, resources, and interventions to address specific areas of strength and weakness. This level of personalization not only enhances the learning experience but also promotes deeper understanding, retention, and mastery of the subject matter.

The advent of ITS signifies a paradigm shift in education, moving away from passive, lecture-based models toward more interactive, student-centered approaches. By placing learners at the center of the educational process, ITS empower individuals to take ownership of their learning journey, fostering autonomy, self-regulation, and critical thinking skills. However, besides the potential of ITS to influence and improve education, it is essential to recognize and address the complex ethical considerations inherent in their development, deployment, and use, as is the case in other contexts of AI application to education [2]. As with any technology, ITS bring with them a host of ethical challenges and dilemmas that demand careful scrutiny and deliberation.

One of the foremost concerns is the issue of algorithmic bias, wherein the algorithms powering ITS may inadvertently perpetuate or exacerbate existing societal biases present in the data on which they are trained. This can lead to unfair treatment, discrimination, or marginalization of certain groups of learners, perpetuating systemic inequities within educational systems. Additionally, the proliferation of ITS raises profound questions regarding data privacy, security, and consent. Given the vast amounts of sensitive student data collected and processed by these systems, there is a pressing need to establish robust safeguards and protocols to ensure the ethical and responsible handling of such information. Furthermore, the integration of ITS into educational settings has the potential to widen existing socioeconomic disparities, exacerbating the digital divide and perpetuating inequalities in access to quality education. Without careful consideration and proactive measures, ITS could inadvertently deepen educational inequities rather than mitigate them.

In light of these ethical challenges, it is imperative to adopt a proactive and conscientious approach to the development, deployment, and regulation of ITS in education. This entails not only addressing ethical concerns as they arise but also embedding ethical considerations into the design, implementation, and evaluation of ITS from the outset.

In this chapter, we aim to delve deeply into the ethical dimensions of ITS, exploring both the promises and perils they present. Through a comprehensive examination of the advantages offered by ITS, such as personalized learning and adaptive instruction, as well as the potential risks, including algorithmic bias and privacy concerns, we seek to elucidate the multifaceted ethical landscape in which these systems operate.

Drawing on ethical frameworks, theoretical insights, and empirical research, we will offer guidance and recommendations for navigating the ethical complexities of ITS in education. Our goal is to foster a deeper understanding of the ethical considerations at play and to empower educators, policymakers, developers, and other stakeholders to make informed decisions that prioritize the ethical use and deployment of ITS while safeguarding the rights, dignity, and well-being of all learners.

11.2 The Promise and Perils of ITS

Learning from machines in the form of multi-modal ITS has not only many advantages but also a number of concerns. This section details these benefits and challenges impacting the ITS.

11.2.1 Advantages of ITS in Education

ITSs have emerged as a transformative force in education, offering many advantages that have the potential to enhance the learning experience [3]. Intelligent tutors are capable of providing a tailored response to each student based on the individualized model (personalized learning) which is continuously updated by evaluating the student's proficiency at solving the problems (continuous assessment). Moreover, ITSs are software systems and therefore are not bound by geographical, or time restrictions (increased accessibility), which further accommodates the learning experience to the possibilities and necessities of each and every student.

11.2.1.1 Personalized Learning

Learning personalization in ITS can adopt multiple forms depending on the system, some systems might even present more than one specific type of adaptation. Commonly systems opt to tailor the instructions to each individual student [4]. This can be achieved by having multiple interaction models and allowing the system decide how to present the instructions according to the user model. Other adaptations consist of choosing the learning path or proposing problems [5]. Moreover, by adequately adapting the interaction between the user and the system, the user gains engagement [6] which positively affects the learning outcomes [7].

11.2.1.2 Increased Accessibility

ITSs are software systems, therefore their use is not bound by geographical barriers, and it is primarily limited only by the availability of computing devices. However, for internet-based ITS, internet access is essential. This accessibility allows the provision of high-quality education services to sparse or rural areas. Moreover, as software systems, ITS can be accessed at any given time, which enables asynchronous learning for students that need more time for developing their skills or simply have diverse schedules that are not compatible with attendance to traditional lessons.

11.2.1.3 Continuous Assessment and Feedback

Intelligent tutors are continuously evaluating the learners and providing feedback to them and the educators. This immediate feedback loop enhances the learning process, allowing for prompt corrections and adjustments. Beyond that, having a continuous assessment of the student capabilities allows for a granular tunning of the student model, and reduces significantly the response time human tutors need to adapt the tutoring to the specifics of the student. Moreover, an ITS can develop a unique tutoring strategy for each student, which is not typically possible with human tutors as normally, classes are formed by groups of 20–25 students with different particular needs. Moreover, the continuous assessment allows human tutors to adopt a data-driven approach to improve the learning outcomes of their students both individually, and collectively.

11.2.2 Potential Risks and Challenges

The implementation of intelligent tutors into education poses its own challenges and ethical concerns that must be addressed before proceeding to their implementation. ITSs is a highly experimental field [8], which means that most of the hypotheses are validated by conducting diverse experiments under different treatments to validate and provide a meaningful analysis. The main problem here is that in the particular case of ITS, the subjects are students, and students are often minors in the ages 7–18. Conducting behavioral and pedagogical research on humans requires to follow strict ethical guidelines, which are more strict when the population of the research are minors [9], which apart from being more protected by the law than adults, their consent to take place in academic studies is provided by their guardians rather than themselves. The main risks that arise from the use of ITS in education include bias in the algorithms and concerns regarding the capturing of sensitive data. However, certain aspects must be taken into account such as the socioeconomic disparities that might arise, or the dependency on technology that the democratization of ITS might generate.

11.2.2.1 Bias in Algorithms

Machine learning algorithms depend on the captured data to generate insights and make decisions. If the data-capturing strategy is not meticulously designed, algorithms tend to perpetuate societal biases present in the captured data. Although the presence of societal biases cannot be completely avoided, algorithms must be calibrated in order to operate beyond these cultural biases, as not doing so would violate principles such as equal opportunity. These biases have been extensively studied in the literature [10–12]. Particularly, in the field of ITS, Lin et al. [13] concluded that bias in the field of ITS produced harm, and eventually led to the derailment of the educational purposes. Moreover, they identified addressing bias in ITS as one of the open research questions in the field and proposed the use of explainable AI systems to deal with unfair biases.

11.2.2.2 Privacy Concerns

Most of the times the users provide demographic data to the system. This data is not especially sensitive. However, systems might collect many different types of data for creating the student model. As an example dialog between peers is being recorded in Koedinger et al. [14] and Bergner et al. [15] which exploit natural language capabilities to provide feedback based on the speech interaction between the peers. More examples include D'Mello et al. [16] which exploits gaze to enhance the student engagement. Gaze or chat interaction data are not inherently sensitive; however, in the context of ITS, students might have some sensitive conditions which are reflected in the user model that can be extracted from an analysis of the recorded data. The research conducted by Nerušil et al. [17] demonstrates the capability of detecting dyslexia by analyzing gaze data from different participants. Leontyev et al. [18] show that Attention Deficit Hyperactivity Disorder (ADHD) can be detected from mouse tracking information. Although the student models are not specifically designed to find these conditions, an ill-intentioned analysis of student models that incorporate these data modalities might reveal sensitive insights from the students.

11.2.2.3 Socioeconomic Disparities in Access

While ITS holds great promise, there is a risk of exacerbating existing socioeconomic inequalities if access to the necessary technology and infrastructure is not equitably distributed. According to the United Nations currently 3.6 billion people (roughly 50% of the world population) do not have internet access, although the objective is to reach 100% of world population with internet access by 2030 [19] often these objectives are not accomplished. Moreover, Statista[1] highlights that in

1 https://www.statista.com/statistics/748551/worldwide-households-with-computer

2019 in developing countries, only around a third of households had access to a personal computer. These figures show that while the technology that ITS develop has the potential to impact education globally, the real penetration of the developed systems is limited by the access of developing countries to the technology required to access the systems. Researchers must design intelligent tutors that can be operated using mobile devices, which have much higher penetration than personal computers[2] especially in developing countries where it is estimated that around 60% of the population has access to one.

11.2.2.4 Dependency on Technology

Development of ITS is a great complement to traditional teaching, where the student can get tailored tutoring outside the classroom. However, there is danger in developing the technology to a point where stakeholders might believe that human tutors are not needed anymore. This might be an insight taken from a data-driven approach in cases where education is not a right, but a commodity. Relying too much on technology might affect relations between the student and the tutor. Some researchers such as Keverne et al. [20] highlight the existing relationship between affect and learning, specially in early youth. More recently, Koca [21] showed that a positive teacher–student relationship improves the motivation, ultimately improving the learning outcome. It is essential that researchers find an equilibrium between contact and online learning when researching ITS methods.

11.3 Ethical Frameworks for ITS

11.3.1 Utilitarian Perspective

Utilitarianism defines that the outcome of an action is the criterion for determining whether the action is moral or not and that the right moral behavior brings the greatest happiness for the greatest number [22]. From this perspective, the use of technology is mainly to facilitate human life. Kulik and Fletcher's [23] study has revealed that online tutoring via ITS improves student's learning performance compared to traditional classroom methods.

The application of ITS improves the quality and efficiency of education. Aided by big data, AI, and cognitive science theories, the system provides personalized learning capabilities. In one-on-one instruction, the intelligent tutor can dynamically modify the learner's lesson plan in real time, tailoring it to the individual's characteristics and specific needs [24]. This personalized learning allows the learner to complete the task with the least amount of effort, acquire the most knowledge, and achieve higher scores on tests. ITS can be an effective solution to the disadvantages

---

2 https://www.statista.com/statistics/203734/global-smartphone-penetration-per-capita-since-2005

of the one-size-fits-all approach in traditional instruction [25]. Due to variations in student's strengths and weaknesses, employing a one-size-fits-approach would lead to substantial disparities in learning outcomes among individuals, undermining the motivation and engagement of students who are not comfortable with the approach.

The emergence of ITS has contributed to educational equality issue. Traditionally, the quality of education totally depended on the teacher's knowledge, experience, and teaching style. Different teachers can lead to inconsistencies in the quality of teaching. In addition, the educational resources in different regions are not balanced. Economically developed regions have more educational resources, while poor and backward regions have less. Therefore, people in different regions have unequal access to education and resources. An intelligent education system can reduce this difference to a certain extent [26]. It allows everyone to have the opportunity to learn knowledge at any time, anywhere, at a cost much lower than that of a real teacher.

Personalized recommendations enable students to access accurate instructional content, but they also simultaneously limit the learner's initiative to go out and get information. Students may become overly attached to ITS leading to a range of technology dependency issues. In human-computer interactions thismay hinder the development of students in nontechnical areas such as interpersonal communication, critical thinking, and practical experience. An over-reliance on the answers and guidance provided by an ITS reduces the student's ability to think independently and solve problems. This negatively impacts the purpose of education – to develop well-rounded individuals. Second, ITS must function with real entities, and in the event of system faults, bugs, or vulnerabilities, students may face risks such as compromised learning endpoints and data loss. These issues can lead to the system's inability to deliver suitable content or even render it inoperable.

Additionally, ITSs that emphasize utility may focus too much on goal orientation and outcomes, thereby neglecting the importance of student interest, creativity, and exploration. A prevalent example is ITS employing reinforcement learning methods. Because reinforcement learning algorithms are inherently designed based on the principle of goal maximization. This can lead to a lack of diversity in the learning experience and an inability to meet the individual requirements of students.

Finally, the widespread use of ITS will ultimately affect education practitioners. In certain programs, educational institutions will cease recruiting teachers, as ITS poses the capability and execute the duties of teachers. This shift will reduce the demand for education-related positions in society and may result in the unemployment of education professionals.

11.3.2 Deontological Perspective

Deontological ethics asserts that the justification of an individual's actions hinges on whether these actions inherently uphold obligations, duties, and powers. It emphasizes that these actions should be motivated by a commitment to adhere to ethical rules rather than being contingent on the outcomes they produce [27]. Examining ITS from a deontological perspective allows for an exploration of the responsibilities and duties it carries in terms of education and mentoring.

ITS take on the role of the teacher in teaching and learning. In order to ensure effective learning, the system needs to focus on several aspects. First, it should focus on imparting accurate knowledge and providing precise information to meet the individual learning needs of students. Second, the system has the responsibility to focus on the mental health of the students by monitoring learning performance and engagement, identifying potential problems in a timely manner, and providing appropriate support and guidance to ensure that the students have a positive experience in their learning.

Moreover, ITS bears the responsibility of advancing equitable and inclusive education. It must strive to eradicate disparities arising from factors such as race, gender, and socioeconomic status. By implementing personalized learning pathways and allocating resources appropriately, the system can effectively cater to the diverse needs of student groups, ensuring equal access to learning opportunities and resources for every student.

However, ITS needs to collect and analyze students' personal information in order to be able to provide personalized learning, and it is important that the designers comply with strict privacy regulations and data security when developing it to ensure that this information is not misused or compromised [28]. The study of Latham and Goltz [29] revealed that 77% of participants expressed concerns regarding the utilization of their data. Furthermore, almost 63% of the children involved in the study expressed opposition to the tracking of their data.

In addition, as AI technology continues to be updated, the behavior of AI becomes increasingly opaque and difficult to interpret when using massive amounts of data and large model parameters. In the event of algorithmic bias, teaching for intelligent tutors may become uncontrollable. For example, an intelligent tutor may exhibit behavior inconsistent with its motivation and input misleading knowledge to students.

11.3.3 Virtue Ethics

Virtue ethics assumes that essential individual characteristics, known as virtues, determine goodness. Ethical behavior demands a continual effort from individuals to embody these traits, allowing them to develop into virtuous persons. Unlike utilitarianism and deontology, which focus on the act itself, virtue ethics defines an action as right only if performed by a virtuous person. For example, Neubert [30] defined the basic virtues as prudence, temperance, justice, courage, faith, hope, and love.

ITSs do not maintain complete neutrality in their ethical standpoint. Initially, it is influenced by the moral values of the designer and will reflect the designer's value preferences. Then, because of the system's inherent adaptability and personalized learner characteristics, it will dynamically adjust itself to student's behaviors. The final moral values of the system will gradually tend to be closer to the students' moral views.

The main entities engaged in the research and development of ITS include academic researchers and businesses. While researchers work to drive technological innovation and educational advancement, it is really the companies that will make ITS available for long-term use. The ultimate goal of enterprises involved in ITS development is often the pursuit of profit. The profit-seeking nature of capital makes it possible for enterprises to pursue economic interests in system design and actual operation.

This motive of pursuing economic interests will inevitably affect the direction of ITS development. In the process, education, a cause that should be oriented toward public welfare, is likely to be impacted by commercialization and profit orientation. This, to a certain extent, makes it difficult for ITS to realize the essence of education emphasized by virtue theory, that is, the educational goal of fostering civic virtues and good personal character.

For example, the design of ITS may be more inclined to pursue high returns at the expense of the holistic and humanistic nature of education. This may lead to the system focusing more on superficial knowledge transfer in terms of content presentation and assessment methods, while neglecting the importance of cultivating students' ability to think deeply and make moral judgments. Under such circumstances, it is more likely that an intelligent education system will be reduced to a tool for the pure pursuit of economic interests rather than one that truly serves the public interest in education.

11.4 Bias and Fairness in ITS

Bias in ITS can manifest itself at different stages of development, such as data source, sampling, data cleaning, and choice of algorithm, as is the case in any area of machine learning application [31]. Suresh and Guttag [32] established six categories of bias that they mapped to the different stages of the machine learning process: historical bias, representation bias, measurement bias, aggregation bias, evaluation bias, and deployment bias. Whenever AI algorithms are created, they also create a dataset that represents historical and systemic biases in society, which ultimately become algorithmic biases. Subtle biases in the initial data or deviations during algorithm training, the choice of features used as inputs and changes in the distribution of training and test data can inevitably lead to manifestly unfair results. An example of bias produced during the data sampling phase with significant consequences was observed in the algorithm designed by Amazon for the recruitment of engineers, which evidenced a noticeable discrimination toward women and, as a result, was later abandoned.[3] Other aspects beyond the chosen data set, algorithm design aspects such as the initial assumptions of the model, or the selected optimization techniques may condition the behavior of the final model. Although the bias in the algorithmic model is not explicitly intended by the developers, numerous gender and racial biases can be found in different AI-based solutions [33]. The definition of algorithm bias has generated discussions with overlapping meanings of the term [34] while fairness can be defined as the lack of bias or favoritism of the intelligent system toward an individual or group based on their inherent characteristics. It is only recently that education researchers have begun to warn about the existence of cases of algorithmic bias in AI applied to education.

The main reason for the existing biases in the conversational models that integrate with an ITS is the datasets that are used to train these models. To fulfill one of their main functions, which is to understand the user's query and provide the best response, the models are trained with massive data sets and use various algorithms. The problem lies in the fact that, unless corrected, the age also learns and replicates human biases present in the dataset [35].

Take, for example, Microsoft's well-known bot, Tay, launched in 2016. Tay was designed to be a funny millennial girl and improve her conversational skills over time through informal chats with human users. However, in less than 24 hours, the bot became offensively sexist, misogynistic, and racist. The failure was due to a lack of sophisticated algorithms to overcome the inherent bias in the data entered into the program.

In 2017, researchers at the University of Massachusetts analyzed 52.9 million tweets and found that those containing African American slang and colloquial expressions were often not considered English. Twitter's sentiment analysis tools struggled and its "rudeness" filter tended to misinterpret and remove them [36].

On the other hand, another aspect to take into account in the interaction with the user is the gender that the system assigns to them and therefore uses in its interaction with them. Most designers look for interfaces that make users feel that the AI is conversing directly with them, and for them, an important aspect is to recognize their gender, especially when using gender-inflected languages, such as Spanish or French. When the algorithm does not take gender into account, it usually defaults to language that assumes the speaker is male, which excludes women and gender-diverse people. In their study, Prates et al. [37] demonstrate a notable inclination toward male defaults within widely used online translation

3 "2Amazon scraps secret AI recruiting tool that showed bias against women." https://www.reuters.com/article/us-amazon-com-jobs-automation-insight-idUSKCN1MK08G

services, particularly in domains traditionally characterized by imbalanced gender representation, such as STEM. This shows how AI models in language translation carry the societal biases and gender-specific stereotypes in the data.

Similarly, biases also affect when an algorithm must make decisions in the learning process: automated assessment, personalized learning, and predictive systems in the ITS. Although the main promise of machine learning models is greater accuracy and objectivity, numerous examples have revealed the opposite.

To prevent the generation of biased results in AI models integrated into an ITS, the best strategy is to take a proactive approach and avoid bias from the outset. While it is impossible to completely eliminate bias, steps can be taken to minimize it. This includes encouraging diversity in the development team, ensuring that developers come from different backgrounds to bring diverse perspectives. In addition, it is crucial that the data used by the model is varied and representative, avoiding exclusive reliance on a single source to avoid inherent bias. Continuous monitoring and feedback, as highlighted by Mitchell et al. [38], involve tracking AI system performance and fairness over time, alongside collecting user feedback to uncover biases, unfairness, and emerging issues not evident through traditional metrics. This ongoing process aids in identifying impacts arising from evolving data distributions and changing user contexts.

11.5 Privacy and Security Concerns

Being a branch of machine learning, ITSs generate their insights and models by analyzing amounts of data generated by the students. The core of an ITS lies in the correct modeling of the student. The systems produce this model by consuming different data sources that are specifically designed to model the student in the context of the system. Typically systems manage one or more of these sources

11.5.1 Self-Reported Sources

It includes the data that the student or the teacher explicitly introduces into the system through forms. This data often includes details such as the demographics of the student (age, sex, and region) and reports about perceptions during the use of the system. In this line, Mudrick et al. [39] present a system where students self-report their boredom while using the system. Similarly, Jaques et al. [40] propose an ITS that predicts the affect from captured data; however, in order to obtain affect information, the student fills out a form reporting their affect state. Similar approaches have been exploited by Myers [41] or D'Mello and Graesser [42].

11.5.2 Captured Data Sources

It includes data that is captured without direct input from the subject. The sources are diverse and unstructured. Captured data sources include interaction data, such as a number of clicks; typing patterns; or mouse movement, and observations such as gaze, facial expression, or other interaction with the system, such as conversations, or how well the user responds to the exercises.

The captured sources are used to create the internal representation of the student. Although the used data might not seem specially sensitive, it can often be exploited to correlate with some medical conditions, for example, the time taken to read the problem statement could be correlated with the probability of presenting dyslexia symptoms. This argument highlights the importance of treating all the retrieved data as sensitive data, since it can reveal medical conditions or other pieces of private information that the student might not consent to be divulged. The first step in this line would be to anonymize all the data that is gathered, that is, modify it so that it cannot be tracked back to the student's identity, not even by the researchers. In this direction, student accounts should not be identified by email, they rather should be identified by providing the student with a meaningless username. It must be taken into account that not all data can be anonymized without being severely altered, image data captured for facial expression analysis cannot be anonymized because of its nature. In these cases, researchers should implement adequate measures so that this data cannot be accessed without the participant's consent. According to best practices published by Princeton[4] and Imperial College London[5] sensitive data should be kept offline in detachable devices and encrypted. This is the preferable approach when dealing with a system that is under development. However, this is not possible for systems that provide service to many users, in this case, data should be kept encrypted as it is gathered and only perform the decryption when there is a need to retrain the user model, enforcing a security scheme for accessing the information.

Currently, the legislation framework varies among different countries. Countries belonging to the European Union abide by the General Data Protection Regulation (GDPR) and neighboring countries such as the United Kingdom follow similar principles. In the United States of America, the data protection regulation depends on each state, although child privacy is federally regulated in the "Family and Educational Rights and Privacy Act" (FERPA), similarly, Canada has federal and state laws regulating privacy, and the "Convention of the Rights of the Child" (CRC). However, most regulations protect similarly the private and sensitive data

4 https://ria.princeton.edu/human-research-protection/data/best-practices-for-data-a

5 www.imperial.ac.uk/research-and-innovation/support-for-staff/scholarly-communication/research-data-management/data-storage-and-security/storing-sensitive-and-personal-data

of minors. Besides carefully analyzing the ethical considerations of their work, researchers should at the very least comply with the data protection regulations that apply to the target population, but the target should be to excel at keeping the data private holding themselves to a higher standard than common applications.

Research and development in ITS should always follow a security-first approach ensuring that the privacy and rights of the students are preserved. Moreover, as systems mature, navigating from a research to a product environment, data requirements should be profoundly analyzed so that the amount of gathered information from the students is restrained, following the principle of data minimization [43] as the data protection agencies of many governments recommend.

11.6 Socioeconomic Disparities in Access

The advancement of personalized learning through computational tools is attributed to developments in ITS and Artificial Intelligence in Education (AIED). The development costs associated with creating tutoring systems encompass various factors, including comprehensive data-gathering processes, the availability of hardware, and the utilization of high-performance computers capable of processing large volumes of data at high speeds. Collaboration among multidisciplinary experts ensures that the system integrates insights from fields such as education, psychology, and computer science, resulting in a technologically advanced, pedagogically sound, and user-friendly platform.

Investments are essential for developing such systems, and they are often supported by research funding from government agencies. The research conducted by Guo et al. [44], analyzing the evolution and trends in ITS between 1963 and 2020, highlights the top five dominant ITS research countries/regions: the United States, Canada, Spain, Taiwan, and England. This trend highlights the global disparity in Research and Development (R&D) investment, where G20 countries account for 93% of the world's R&D spending, as reported by UNESCO in 2021. Such disparities underscore the link between a country's economic wealth, its capacity for research, and its investment in educational technologies.

Furthermore, the lack of agreements to share data, architecture, or even a common ITS hampers the ability to compare, modify, integrate, or utilize these resources among different institutions. This obstruction limits the accessibility of educational technologies to specific institutions, creating a barrier to the widespread adoption and adaptation of ITS.

The COVID-19 pandemic has starkly exposed the vulnerabilities in the educational systems of developing countries. On the one hand, inadequate electrical infrastructure and unreliable or insufficient internet connectivity, coupled with the absence or inadequacy of computer labs, and a lack of computers or laptops [45, 46], serve as significant barriers to access to such systems, especially in rural or remote areas. These technological constraints contribute to an inequality within the same country, further exacerbating the digital divide and hindering the potential for equitable education through advanced technologies.

In 2021, it was estimated that 53% of people in developing countries lacked access to the internet [47]. By January 2024, internet users comprised 66.2% of the global population.[6] By mid-2023, nearly 96% of the global digital population was estimated to use a mobile device for internet access,[7] highlighting a significant shift toward mobile connectivity across the world.

The surge in mobile device utilization underscores their critical role in educational innovation. Mobile applications (apps), in particular, have become invaluable tools in facilitating specific learning outcomes [48]. This evolution also highlights the potential to design ITSs optimized for mobile devices, offering accessible and flexible learning solutions that cater to the needs of a digitally connected populace.

Despite this, there are challenges that are computationally intractable, covering the socioeconomic difficulties students encounter, including family background, health status, resources at their disposal, and home living conditions [49]. These factors transcend the boundaries of the learning environment, thus presenting substantial hurdles for incorporation into ITS. The complexity of these issues exceeds the reach of existing computational models, highlighting the necessity for ITS to advance and become more adaptive and responsive. Such evolution is essential for the development of tutoring systems that can genuinely cater to the varied needs of learners, acknowledging the wide-ranging influences that impact their educational paths.

Recognizing the socioeconomic disparities in access to technological resources between developed and developing countries, as well as among different socioeconomic groups within countries, is essential, especially when considering the development and deployment of ITSs. These disparities significantly contribute to the widening technology divide, limiting the potential for these advanced educational tools to reach and benefit all segments of the population equally. Science and technology hold the promise of transforming communities and economies into more inclusive and equitable spaces. However, achieving this vision requires concerted global actions. It is imperative to develop strategies and policies that ensure equitable access to ITSs, thus leveraging their potential to enhance learning outcomes universally. Addressing these disparities is not just about providing

6 https://www.statista.com/statistics/617136/digital-population-worldwide
7 https://www.statista.com/topics/779/mobile-internet/#topicOverview

access to technology but about creating opportunities for all individuals to participate in and benefit from the digital revolution in education.

11.7 Dependency on Technology

Over the last few years, mobile devices such as laptops, tablets, and smartphones have become indispensable aspects of our daily lives. The integration of AI in education has revolutionized traditional instructional approaches, introducing ITS and virtual assistants that offer new opportunities for personalized and flexible learning experiences [50]. Specifically, mobile ITS have the potential for low-cost, personalized learning assistance beyond traditional classroom and computer lab settings [51]. The integration of these devices with a vast array of internet services has transformed educational landscapes, influencing student behaviors and learning processes [52].

While technology is celebrated for its numerous benefits – ranging from entertainment and business facilitation to cognitive skill development, social capital, and enhanced social interactions – the emergence of mobile learning has been particularly notable for its ability to significantly increase student motivation and engagement with academic material. However, this digital integration comes with its set of challenges, notably the issue of digital dependency. This condition is marked by an individual's inability to regulate the use of digital devices, leading to a high reliance on these technologies [53].

Digital dependency shares similarities with other forms of addiction, such as substance dependence [54, 55], through shared mechanisms like changes in dopaminergic and neuronal activity in the brain, alterations in brain structure, diminished impulse control, behavioral inhibition, and impaired cognitive functions. One notable manifestation of this dependency is nomophobia – the anxiety and nervousness or discomfort experienced in the absence of mobile phones, computers, or other communication devices [56].

Over the past decade, the reliance on and use of the internet has surged dramatically, along with the proliferation of web-based applications, which has directly impacted the educational context [52]. This shift has led to students perceiving their mobile devices as potential distractions, affecting their learning process. Studies have found that university students who show signs of nomophobia, or anxiety when separated from their mobile phones, struggle more with maintaining focus in class [57] and tend to perform poorly academically [58]. Educators are increasingly aware of technology's dual role in education – as both an enhancer of learning and a potential source of distraction, emphasizing the need for a balanced approach to technology use.

The growing concern over excessive mobile device usage highlights broader issues, including potential impacts on physical health and essential cognitive and social skills, such as critical thinking and problem-solving [59–61]. As technology continues to permeate the realm of education, particular concerns emerge about the potential weakening of human connection and the erosion of teacher–student rapport, with the consequent impact on the development of socio-emotional skills and a shared sense of classroom community [62].

This situation necessitates a thoughtful examination of our digital dependencies and the development of strategies that balance the advantages of technology with awareness of its possible adverse effects. By confronting these challenges, educators can cultivate an educational environment that not only utilizes technology to improve learning outcomes but also safeguards and enhances students' well-being and cognitive abilities.

11.8 Summary

There is no doubt of the potential of ITS to benefit students in general terms. ITSs increase the accessibility of education for students living in sparse areas, who must reconcile studying with other activities, or simply, with personal availability times that are not completely compatible with attending face-to-face lessons. Moreover, adaptation to the student allows for better learning outcomes, and improvement of the student's motivation to learn any specific subject, and having a system that automatically provides grading and evaluation significantly reduces tutors' workload.

However, it must be taken into account that the field overlays with sensitive areas such as dealing with underage students or managing and ensuring the privacy of sensitive data. As we move toward a more inclusive world, we, as researchers, must design our algorithms not to perpetuate cultural, racial, sexual, political, or other biases that are present in all individuals. Furthermore, as shown, all the information gathered in the system should be treated as if it were sensitive since it could be used to obtain sensitive information about the students, this should be done by encrypting the data and following a strong security scheme that grants access only to authorized stakeholders. It must be noted that while ITS have many advantages, they are not human. Relations among humans are vital in the development of the individual, and the context of learning is not an exception. ITS are great tools to support the learning of individuals, but they should not replace human teachers by any means. Furthermore, the implementation of ITS must ensure that the main objectives of the ITS are met. Working systems that cannot reach the target population are meaningless if the target population cannot access the system because they do not have the means, such as internet, or personal computer access.

References

1 Mousavinasab, E., Zarifsanaiey, N., Kalhori, S.R.N. et al. (2021). Intelligent tutoring systems: a systematic review of characteristics, applications, and evaluation methods. *Interactive Learning Environments* 29 (1): 142–163. https://doi.org/10.1080/10494820.2018.1558257.
2 Holmes, W., Porayska-Pomsta, K., Holstein, K., Sutherland, E., Baker, T., Shum, S. B., . . . others (2021). Ethics of AI in education: towards a community-wide framework. *International Journal of Artificial Intelligence in Education*, 32 1–23.
3 Burns, H.L. and Capps, C.G. (2013). Intelligent tutoring systems: an introduction. *Foundations of Intelligent Tutoring Systems* 3: 1–19.
4 Wray, R.E. and Woods, A. (2013). A cognitive systems approach to tailoring learner practice. *Proceedings of the Second Annual Conference on Advances in Cognitive Systems ACS* 21: 18.
5 Hong, C.-M., Chen, C.-M., Chang, M.-H., and Chen, S.-C. (2007). Intelligent web-based tutoring system with personalized learning path guidance. *Seventh IEEE International Conference on Advanced Learning Technologies (ICALT 2007)*,Niigata, Japan (18–20 July 2007), 512–516.
6 Chrysafiadi, K., Virvou, M., Tsihrintzis, G.A., and Hatzilygeroudis, I. (2023). Evaluating the user's experience, adaptivity and learning outcomes of a fuzzy-based intelligent tutoring system for computer programming for academic students in Greece. *Education and Information Technologies* 28 (6): 6453–6483.
7 Yu, Z., Yu, L., Xu, Q. et al. (2022). Effects of mobile learning technologies and social media tools on student engagement and learning outcomes of English learning. *Technology, Pedagogy and Education* 31 (3): 381–398.
8 McCalla, G.I. (1992). Intelligent tutoring systems: navigating the rocky road to success. In: *New Directions in Educational Technology*, 107–122. Springer Nature.
9 Flewitt, R. (2005). Conducting research with young children: some ethical considerations. *Early Child Development and Care* 175 (6): 553–565.
10 Cheng, L., Varshney, K.R., and Liu, H. (2021). Socially responsible AI algorithms: issues, purposes, and challenges. *Journal of Artificial Intelligence Research* 71: 1137–1181.
11 Johnson, G.M. (2021). Algorithmic bias: on the implicit biases of social technology. *Synthese* 198 (10): 9941–9961.
12 Vlasceanu, M. and Amodio, D.M. (2022). Propagation of societal gender inequality by internet search algorithms. *Proceedings of the National Academy of Sciences of the United States of America* 119 (29): e2204529119.
13 Lin, C.-C., Huang, A.Y., and Lu, O.H. (2023). Artificial intelligence in intelligent tutoring systems toward sustainable education: a systematic review. *Smart Learning Environments* 10 (1): 41.

14 Koedinger, K.R., Anderson, J.R., Hadley, W.H. et al. (1997). Intelligent tutoring goes to school in the big city. *International Journal of Artificial Intelligence in Education* 8 (1): 30–43.

15 Bergner, Y., Walker, E., and Ogan, A. (2017). Dynamic Bayesian network models for peer tutoring interactions. *Innovative Assessment of Collaboration* 249–268.

16 D'Mello, S., Olney, A., Williams, C., and Hays, P. (2012). Gaze tutor: a gaze-reactive intelligent tutoring system. *International Journal of Human-Computer Studies* 70 (5): 377–398.

17 Nerušil, B., Polec, J., Škunda, J., and Kačur, J. (2021). Eye tracking based dyslexia detection using a holistic approach. *Scientific Reports* 11 (1): 15687.

18 Leontyev, A., Yamauchi, T., and Razavi, M. (2019). Machine learning stop signal test (ML-SST): ML-based mouse tracking enhances adult ADHD diagnosis. *2019 8th International Conference on Affective Computing and Intelligent Interaction Workshops and Demos (ACIIW)*, Cambridge, UK (3–6 September 2019), 1–5.

19 Office of the Secretary-General's Envoy on Technology & International Telecommunication Union (2022). Achieving universal and meaningful digital connectivity. Setting a baseline and targets for 2030. https://www.itu.int/itu-d/meetings/statistics/wp-content/uploads/sites/8/2022/04/UniversalMeaningfulDigitalConnectivityTargets2030_BackgroundPaper.pdf

20 Keverne, E.B., Nevison, C.M., and Martel, F.L. (1997). Early learning and the social bond. *Annals of the New York Academy of Sciences* 807: 329–339.

21 Koca, F. (2016). Motivation to learn and teacher-student relationship. *Journal of International Education and Leadership* 6 (2): n2.

22 Sen, A. (1979). Utilitarianism and welfarism. *The Journal of Philosophy* 76 (9): 463–489.

23 Kulik, J.A. and Fletcher, J. (2016). Effectiveness of intelligent tutoring systems: a meta-analytic review. *Review of Educational Research* 86 (1): 42–78.

24 Grivokostopoulou, F., Perikos, I., and Hatzilygeroudis, I. (2017). An educational system for learning search algorithms and automatically assessing student performance. *International Journal of Artificial Intelligence in Education* 27 (1): 207–240. https://doi.org/10.1007/s40593-016-0116-x.

25 Yang, S., Tian, H., Sun, L., and Yu, X. (2019). From one-size-fits-all teaching to adaptive learning: the crisis and solution of education in the era of AI. *Journal of Physics: Conference Series* 1237: 042039.

26 Keleş, A., Ocak, R., Keleş, A., and Gülcü, A. (2009). Zosmat: Web-based intelligent tutoring system for teaching–learning process. *Expert Systems with Applications* 36 (2): 1229–1239.

27 Kant, I. and Schneewind, J.B. (2002). *Groundwork for the Metaphysics of Morals*. Yale University Press.

28 Lupton, D. and Williamson, B. (2017). The datafied child: the dataveillance of children and implications for their rights. *New Media & Society* 19 (5): 780–794.

29 Latham, A. and Goltz, S. (2019). A survey of the general public's views on the ethics of using AI in education. In: *Artificial Intelligence in Education* (ed. S. Isotani, E. Millán, A. Ogan, et al.), 194–206. Cham: Springer International Publishing.

30 Neubert, M.J. (2017). Teaching and training virtues: behavioral measurement and pedagogical approaches. In: *Handbook of Virtue Ethics in Business and Management*, 647–655. Springer.

31 Holmes, W. and Porayska-Pomsta, K. (2022). The ethics of artificial intelligence in education: practices, challenges, and debates. In: *The Ethics of Artificial Intelligence in Education: Practices, Challenges, and Debates*, 1–288. Routledge https://doi.org/10.4324/9780429329067.

32 Suresh, H. and Guttag, J.V. (2019). A framework for understanding unintended consequences of machine learning. *arXiv preprint arXiv:1901.10002, 2* (8).

33 Stahl, B.C. and Wright, D. (2018). Ethics and privacy in AI and big data: implementing responsible research and innovation. *IEEE Security & Privacy* 16 (3): 26–33. https://doi.org/10.1109/MSP.2018.2701164.

34 Baker, R.S. and Hawn, A. (2021). Algorithmic bias in education. *International Journal of Artificial Intelligence in Education* 2: 1–41.

35 Schlesinger, A., O'Hara, K.P., and Taylor, A.S. (2018). Let's talk about race: identity, chatbots, and AI. *Proceedings of the 2018 Chi Conference on Human Factors in Computing Systems*, Montreal, Canada (21–26 April 2018), 1–14.

36 Blodgett, S.L. and O'Connor, B. (2017). Racial disparity in natural language processing: a case study of social media African-American English. *arXiv preprint arXiv:1707.00061*.

37 Prates, M.O., Avelar, P.H., and Lamb, L.C. (2020). Assessing gender bias in machine translation: a case study with Google translate. *Neural Computing and Applications* 32: 6363–6381.

38 Mitchell, M., Wu, S., Zaldivar, A. et al. (2019). Model cards for model reporting. *Proceedings of the Conference on Fairness, Accountability, and Transparency*, Atlanta, USA (29–31 January 2019), 220–229.

39 Mudrick, N., Azevedo, R., Taub, M., and Bouchet, F. (2015). Does the frequency of pedagogical agent intervention relate to learners' self-reported boredom while using multi-agent intelligent tutoring systems? *The 37th Annual Meeting of the Cognitive Science Society,* Pasadena, USA (22–25 July 2015), 1661–1666.

40 Jaques, N., Conati, C., Harley, J.M., and Azevedo, R. (2014). Predicting affect from gaze data during interaction with an intelligent tutoring system. *Intelligent Tutoring Systems: 12th International Conference, Its 2014,* Honolulu, HI, USA (5–9 June 2014). *Proceedings 12,* 29–38.

41 Myers, M.H. (2021). Automatic detection of a student's affective states for intelligent teaching systems. *Brain Sciences* 11 (3): 331.

42 D'Mello, S. and Graesser, A. (2006). Affect detection from human-computer dialogue with an intelligent tutoring system. *International Workshop on Intelligent Virtual Agents*, Marina del Rey, USA (21–23 August 2006), 54–67.

43 Biega, A. J., Potash, P., Daumé, H. et al. (2020). Operationalizing the legal principle of data minimization for personalization. *Proceedings of the 43rd International ACM SIGIR Conference on Research and Development in Information Retrieval*, Virtual, China (25–30 July 2020), 399–408.

44 Guo, L., Wang, D., Gu, F. et al. (2021). Evolution and trends in intelligent tutoring systems research: a multidisciplinary and scientometric view. *Asia Pacific Education Review* 22: 1–21. https://doi.org/10.1007/s12564-021-09697-7.

45 Tadesse, S. and Muluye, W. (2020). The impact of covid-19 pandemic on education system in developing countries: a review. *Open Journal of Social Sciences* 8 (10): 159–170.

46 Zalat, M.M., Hamed, M.S., and Bolbol, S.A. (2021). The experiences, challenges, and acceptance of e-learning as a tool for teaching during the covid-19 pandemic among university medical staff. *PLoS One* 16 (3): 1–12. https://doi.org/10.1371/journal.pone.0248758.

47 Bamford, R., Hutchinson, G., and Macon-Cooney, B. (2021). *The Progressive Case for Universal Internet Access: How to Close the Digital Divide by 2030*. Institute Global.

48 García-Martínez, I., Fernández-Batanero, J.M., Cobos Sanchiz, D., and Luque de la Rosa, A. (2019). Using mobile devices for improving learning outcomes and teachers' professionalization. *Sustainability* 11 (24): https://doi.org/10.3390/su11246917.

49 Daniel, B.K. (2019). Big data and data science: a critical review of issues for educational research. *British Journal of Educational Technology* 50 (1): 101–113. https://doi.org/10.1111/bjet.12595.

50 Barrett, M., Branson, L., Carter, S. et al. (2019). Using artificial intelligence to enhance educational opportunities and student services in higher education. *Inquiry: The Journal of the Virginia Community Colleges* 22 (1): 11.

51 Brown, Q., Lee, F.J., Salvucci, D.D., and Aleven, V. (2008). The design of a mobile intelligent tutoring system. *Proceedings of the 9th International Conference on Intelligent Tutoring Systems*, Montreal, Canada (23–27 June 2008), Vol. 10, 978–983.

52 Artal-Sevil, J., Bernal-Agustín, J., and Domínguez Navarro, J. (2015). M-learning (mobile learning) in higher education. The impact of smartphone as interactive learning tool. In: *Edulearn15 Proceedings*, 8212–8221. Springer Heidelberg.

53 Gonçalves, L.L., Nardi, A.E., and King, A.L.S. (2023). Digital dependence in the past decade: a systematic review. *Journal of Addiction Research and Adolescent Behaviour* 6 (1): 1–18. https://doi.org/10.31579/2688-7517/059.

54 Andreassen, C., Billieux, J., Griffiths, M. et al. (2016). The relationship between addictive use of social media and video games and symptoms of psychiatric disorders: a large-scale cross-sectional study. *Psychology of Addictive Behaviors* 30: 252–262. https://doi.org/10.1037/adb0000160.

55 Griffiths, M. (1995). Technological addictions. *Clinical Psychology Forum* 76: 14–19. https://doi.org/10.53841/bpscpf.1995.1.76.14.

56 King, A., Valença, A., Silva, A. et al. (2013). Nomophobia: dependency on virtual environments or social phobia? *Computers in Human Behavior* 29 (1): 140–144. (Including Special Section Youth, Internet, and Wellbeing). https://doi.org/10.1016/j.chb.2012.07.025.

57 Lee, S., Kim, M.W., McDonough, I.M. et al. (2017). The effects of cell phone use and emotion-regulation style on college students' learning. *Applied Cognitive Psychology* 31 (3): 360–366.

58 Gutiérrez-Puertas, L., Márquez-Hernández, V.V., São-Romão-Preto, L. et al. (2019). Comparative study of nomophobia among Spanish and Portuguese nursing students. *Nurse Education in Practice* 34: 79–84.

59 Baciu, A.B. (2020). Medical and social consequences of digital addiction. *Proceedings of the Romanian Academy* 22: 141–147.

60 Domoff, S.E., Borgen, A.L., Foley, R.P., and Maffett, A. (2019). Excessive use of mobile devices and children's physical health. *Human Behavior and Emerging Technologies* 1 (2): 169–175. https://doi.org/10.1002/hbe2.145.

61 World Health Organization (2015). Public health implications of excessive use of the internet, computers, smartphones and similar electronic devices. Meeting Report, Main Meeting Hall, Foundation For Promotion of Cancer Research, National Cancer Research Centre, Tokyo, Japan (27–29 August 2014). World Health Organization.

62 Kamalov, F., Santandreu Calonge, D., & Gurrib, I. (2023). New era of artificial intelligence in education: towards a sustainable multifaceted revolution. *Sustainability*, 15 (16) 12451. https://www.mdpi.com/2071-1050/15/16/12451. https://doi.org/10.3390/su151612451

12

Road Ahead for Multi-modal Intelligent Sensing in the Deep Learning Era

Ahmed Zoha[1], Naeem Ramzan[2], Muhammad Ali Jamshed[1], and Masood Ur Rehman[1]

[1] *James Watt School of Engineering, University of Glasgow, Glasgow, UK*
[2] *School of Computing, Engineering and Physical Sciences, University of the West of Scotland, Paisley, UK*

In the field of computing and artificial intelligence (AI), multi-modality enables systems to process and integrate information from different sources, such as text, images, video, and audio. This capability enhances perception, allowing machines to understand their environment better and make improved decisions. Multi-modal systems utilize the unique strengths of each data type, creating robust solutions that overcome individual limitations, as demonstrated by autonomous vehicles using multiple sensor inputs for navigation. This approach also enhances contextual understanding, benefiting conditions monitoring and predictive maintenance. By integrating information from various modalities, machines can make better-informed decisions, which benefits a wide range of applications.

The rapid expansion of various networks, such as the Internet of Things, vehicular networks, and social networks, has resulted in a significant increase in the generation of big data. This extensive and diverse data often includes structured, semi-structured, and unstructured types, making it multi-modal. Data fusion, a key component of multi-modal data mining, aims to merge this diverse data into a unified space. This integrated approach can unveil intricate relationships and reveal deeper insights than analyzing individual data types in isolation from data silos.

While traditional methods have laid the groundwork for exploring relationships between different data types, they struggle to cope with the sheer scale and complexity of modern multi-modal big data. In the 1990s, traditional machine

Multimodal Intelligent Sensing in Modern Applications, First Edition.
Edited by Masood Ur Rehman, Ahmed Zoha, Muhammad Ali Jamshed, and Naeem Ramzan.

Published 2025 by John Wiley & Sons, Inc.

learning (ML) models were used to tackle multi-modal problems, extracting knowledge from diverse data and making decisions. However, these models heavily relied on feature engineering, a manual and domain-specific process often resulting in information loss. This limitation hindered their ability to fully capture the complex relationships between different data modalities, preventing them from achieving the ultimate goal of AI: mimicking or surpassing human performance. The challenge of automatically learning the complementary and redundant information in multi-modal data remained a significant obstacle in the traditional ML field.

Deep learning is a computational model that offers a promising solution for various applications, such as cross-modality retrieval, image annotation, and assistant diagnosis. It excels at learning multi-level abstract representations of data, using methods like convolutional neural networks (CNNs), recurrent neural networks (RNNs), and generative adversarial networks (GANs) to achieve breakthroughs in generative and discriminative tasks [1, 2]. For example, deep pyramidal residual networks have demonstrated superior pattern recognition accuracy by learning effective abstract representations [3]. Meanwhile, GANs, employing game theory, effectively capture and reconstruct input data. However, the field is still in its early stages and faces challenges. Traditional methods, relying on shallow models, struggle to capture the complex relationships and complementary correlations within multi-modal big data. Primary challenges include the complexity of data fusion (combining disparate data sources), aligning and synchronizing data from different modalities to represent the same real-world event, data quality and interoperability, and privacy and security concerns. Overcoming these obstacles is critical to unlocking the full potential of multi-modal systems across various fields.

In this chapter, we will explore the opportunities that existing systems present, such as improved accuracy and enhanced insights, as well as the challenges they face.

12.1 Future Challenges and Perspectives for Intelligent Multi-modal Sensing

12.1.1 Semantic Gaps and Cross-modality Representation

Multi-modal data contain information specific to each modality (inter-modality) and information arising from the relationships between different modalities (cross-modality). Deep learning models for multi-modal data fusion often struggle to capture the full semantic knowledge present in this data due to several limitations:

- **Limited Representation of Complex Relationships:** Linear concatenation of high-level representations extracted from each modality may not accurately reflect the complex, non-linear interactions between modalities.

- **Loss of Cross-modality Information:** Current approaches may not fully explain how cross-modality information is lost during the fusion process. When individual modality features are extracted and then combined, some information about how these features relate to each other can be lost, leading to a suboptimal understanding of the data.
- **Challenges with Multiple Modalities:** As the exploration of multi-modal data expands to incorporate three or more modalities, these limitations become even more pronounced. Capturing the relationships between a growing number of modalities becomes significantly more complex.

Different modalities inherently lead to varying data representations, dimensionalities, and semantic spaces, potentially resulting in information loss when directly correlating or fusing them. Addressing this challenge requires advanced strategies such as modality-specific feature extraction and representation learning, which can preserve unique semantic information within each modality before fusion, tailoring feature representations for the specific task [4]. Attention mechanisms and gating strategies can selectively highlight the most relevant features from each modality during fusion, thereby reducing information loss by focusing on the most critical aspects [5, 6]. Generative models, such as variational autoencoders and GANs, can learn a shared latent space that captures semantic relationships between modalities while preserving modality-specific information, enabling a lower-dimensional representation that retains the data's essence [7]. Additionally, ensemble and multi-stage fusion approaches can combine different fusion strategies at various levels (e.g., feature-level, decision-level), offering greater flexibility and potentially reducing information loss compared to early fusion techniques by optimizing how information from different modalities is integrated.

Capturing the rich semantic relationships and dependencies between different modalities is crucial for effective multi-modal data fusion, and integrating semantic fusion strategies with deep learning architectures presents a promising direction. Semantic fusion strategies like multiview fusion, transfer learning, and probabilistic dependency modeling aim to explicitly capture cross-modal correlations and interactions often overlooked by traditional techniques [4, 5]. For instance, multiview fusion approaches, such as co-training and co-regularization, leverage complementary information from different modalities to learn a shared representation that captures underlying semantics. Transfer learning techniques, like domain adaptation and cross-modal knowledge transfer, enable more effective fusion and knowledge sharing across modalities [6]. Probabilistic dependency models, such as Bayesian networks and graphical models, explicitly model the conditional dependencies and joint distributions between modalities, capturing semantic relationships more accurately [7]. Integrating these semantic fusion strategies with deep learning architectures, such as multi-modal attention mechanisms, cross-modal transformers, and probabilistic deep generative models, has shown promising results in capturing semantic relationships and improving fusion performance.

12.1.2 Concept Drift and Data Quality

Multi-modal data's dynamic nature, collected from changing environments, challenges traditional retraining methods. In practical applications, multi-modal data often encounters two key challenges: missing modalities and noisy modalities. The former refers to the absence of data from at least one modality within a sample, while the latter involves data that is corrupted, inaccurate, or misaligned. Notably, many state-of-the-art (SOTA) deep learning models for data fusion assume complete and clean data, limiting their effectiveness in real-world scenarios where data imperfections are expected. Strategies like creating synthetic data, imputation, and attention mechanisms can be employed to address missing modalities [8]. These approaches aim to compensate for the absence of specific modalities during the fusion process. For imbalanced data, resampling techniques, transfer learning, and modality-specific weighting or normalization can be used [9]. These methods help mitigate biases and ensure fair representation of all modalities in the fusion process.

The quality or distribution of multi-modal data can change over time due to various factors, a phenomenon known as data drift. Adaptive fusion techniques can help mitigate the issue by either updating the fusion model parameters or adapting the fusion strategy based on the changing data distributions. Online and incremental learning offer promising solutions for real-time adaptation but require further research for effective multi-modal data fusion. Instead of retraining the fusion model from scratch when data drift occurs, incremental learning techniques allow the model to continuously learn and adapt to the new data distributions without forgetting previously learned knowledge. This can be achieved through techniques like online gradient descent, reservoir sampling, or experience replay [9, 10].

Additionally, neural network architectures can incorporate domain adaptation techniques like adversarial training, maximum mean discrepancy minimization, or domain-invariant feature learning to address variations in data distributions across modalities [5, 9]. These methods aim to learn consistent representations across modalities, making the model more robust to changes in the input data. Attention and gating mechanisms within neural networks can further enhance adaptability by dynamically weighting the importance of each modality based on the specific input. This selective focus on relevant modalities helps mitigate the adverse effects of data drift [5].

12.1.3 Computational Demands and Model Scalability

Deep learning models for multi-modal data fusion often have many parameters, some of which may be redundant and contribute little to the task at hand. Training these models requires substantial amounts of data and computational power, which can be time-consuming and resource-intensive. While parallel computing techniques and high-performance hardware like GPUs and cloud platforms have improved training efficiency, the increasing volume and complexity of multi-modal data continue to challenge current computational capabilities. The sheer volume of multi-modal data, especially in applications like multimedia analysis, healthcare, and remote sensing, demands highly scalable fusion techniques capable of handling large-scale datasets efficiently. Traditional fusion methods may not scale well to these demands, necessitating the development of distributed and parallel computing approaches to tackle the training challenges of complex deep learning models.

Several distributed and parallel computing approaches have been explored for multi-modal data fusion to address the computational challenges and scalability issues:

- **MapReduce Frameworks:** Large-scale multi-modal datasets can be processed in parallel using frameworks like Apache Hadoop and Spark. For instance, researchers have developed MapReduce-based algorithms for fusing text and image data [9]. In these algorithms, the mapping phase extracts features from each modality, and the reduce phase performs the fusion.
- **Leveraging GPUs:** The parallel processing capabilities of modern GPUs can significantly accelerate multi-modal fusion tasks, especially those involving deep learning models. Techniques like splitting the data or the model across multiple GPUs (data parallelism and model parallelism) can be employed to distribute the computation workload [4, 11].
- **Distributed Deep Learning Frameworks:** Frameworks like TensorFlow, PyTorch, and Apache MXNet enable training large multi-modal deep neural networks in a distributed manner across multiple machines or compute nodes [4].
- **Federated Learning for Decentralized Data:** When multi-modal data is distributed across different locations (data silos) or devices, federated learning approaches can be beneficial. These techniques allow training a shared multi-modal model while decentralizing the data on individual devices, reducing communication costs and preserving privacy [5].
- **High-Performance Computing Resources:** High-performance computing clusters and cloud computing resources can be utilized to parallelize computationally intensive multi-modal fusion tasks, such as feature extraction, model training, and inference.

- **Real-time Stream Processing:** For real-time multi-modal data fusion applications, streaming frameworks like Apache Kafka, Apache Flink, and Apache Spark Streaming can be employed. These frameworks enable distributed and parallel processing and fusion of continuous multi-modal data streams.

These distributed and parallel computing approaches aim to utilize the computational resources available in modern computing infrastructures, enabling efficient processing and integration of large-scale, high-dimensional multi-modal datasets. Additionally, reducing the model size through parameter compression techniques, such as pruning (removing unimportant parameters) or quantization (reducing the number of bits used to represent parameters), offers a potential solution to enhance training efficiency and reduce computational demands. Future research could focus on developing new learning frameworks with more powerful computing architectures and exploring innovative parameter compression techniques tailored explicitly for multi-modal data fusion. This could involve, for example, designing compression methods that consider the specific characteristics and relationships between different modalities within the data.

12.1.4 Interpretability

Data-driven approaches, such as deep learning, have greatly impacted various fields; however, their "black box" nature can make it difficult to understand and interpret their predictions. There are two potential solutions to this challenge: First, combining statistical signal processing and deep learning can improve the scalability and interpretability of multi-modal data fusion. Statistical signal processing provides understandable mathematical representations that reflect the underlying physics, prior knowledge, and domain expertise, resulting in flexibility, versatility, scalability, and robustness. On the other hand, deep learning offers powerful capabilities for learning representations, enabling extracting complex patterns and relationships from data. By leveraging the strengths of both approaches, we can enhance the interpretation of the fusion process, making it a promising direction for future research.

Second, incorporating human cognitive strengths and expertise in heterogeneous multi-modal networks can improve decision quality and situational awareness. Critical applications like smart manufacturing, intelligent IoT, augmented reality, and remote diagnosis systems can benefit from seamless human-machine collaboration. Human judgment can complement automatic ML-based decision-making in high-stakes situations like natural disasters or pandemic response. However, modeling and analyzing human decision-making remains challenging, requiring consideration of cognitive biases, uncertainty handling mechanisms, and human-machine interactions.

The lack of interpretability in deep learning models, especially in multi-modal data fusion, poses a significant challenge. In critical applications like healthcare and finance, where transparency and explainability are paramount, the black-box nature of deep learning models can be a major drawback. Several approaches have been proposed to address this issue. For instance, attention mechanisms provide insights into the relative importance of different modalities and their contributions to the final decision [12]. Gradient-based interpretation techniques, like Grad-CAM and integrated gradients, generate visual explanations highlighting relevant input regions or features. Concept Activation Vectors (CAVs) and Multi-modal CAVs (MC-CAVs) capture high-level semantic concepts and cross-modal interactions, respectively, aiding in understanding the model's decision-making process [12]. Disentangled representations separate different factors of variation in the input data, improving interpretability. Explainable AI techniques such as LIME, SHAP, and rule extraction methods can be adapted to provide explanations for model behavior [5]. Incorporating domain knowledge into the model's architecture or training can align its behavior with human-understandable concepts and constraints.

Addressing the interpretability issue effectively is crucial for building trust, ensuring fairness, and fostering human-AI collaboration in critical applications. Combining interpretability techniques with robust fusion models allows us to harness the power of multi-modal data while maintaining transparency and explain-ability.

12.1.5 Ethical Considerations

The fusion of multi-modal data, particularly in sensitive fields like healthcare, raises significant ethical challenges. Privacy concerns arise from collecting and integrating personal data from diverse sources, potentially infringing on individ ual autonomy and informed consent due to the passive and continuous nature of data collection. Furthermore, biases present in training data can be perpetuated or amplified by multi-modal fusion models, leading to discriminatory outcomes. The lack of transparency and interpretability in deep learning models further complicates ethical considerations, making it difficult to understand and trust their decision-making processes. Over-reliance on automated systems without human oversight can exacerbate these issues, potentially leading to errors and erosion of trust [13].

Several strategies can be employed to address these ethical challenges. First and foremost, establishing clear data ownership and governance policies is essential. This includes implementing robust data anonymization, encryption, and access controls to safeguard sensitive information while adhering to relevant data protection regulations. Ensuring transparent consent processes, where individuals are fully informed about data collection and usage and can opt out or revoke consent, is crucial for upholding individual autonomy [14].

Furthermore, efforts to mitigate biases should be prioritized throughout the development and deployment of multi-modal fusion systems. This involves using diverse and representative training data, employing bias mitigation techniques like adversarial debiasing and fairness-aware machine learning, and conducting regular audits to identify and address any potential biases.

Promoting transparency and interpretability in multi-modal fusion models is also vital. Developing explainable AI techniques, such as attention mechanisms and concept activation vectors, can provide insights into the fusion process and the contributions of different modalities, making the decision-making process more understandable and trustworthy. Incorporating human-in-the-loop approaches, where human experts collaborate with AI systems, can further enhance transparency and ensure responsible decision-making.

Additionally, as multi-modal fusion technologies rapidly advance, it is crucial to keep ethical guidelines and governance frameworks up to date. Balancing innovation with ethical considerations is essential for the responsible and beneficial use of multi-modal data fusion in various domains

12.2 Summary

While promising for advancing various fields, multi-modal data fusion faces significant challenges in the age of big data and deep learning. The dynamic and often low-quality nature of multi-modal data necessitates the development of adaptive and robust models. Computational demands and scalability issues require innovative solutions like parallel processing and model compression techniques. Interpretability remains a crucial concern, prompting the exploration of hybrid approaches that combine statistical signal processing with deep learning, as well as incorporating human expertise for decision-making. Additionally, ethical considerations surrounding privacy, bias, and transparency must be carefully addressed through clear governance policies, bias mitigation techniques, and explainable AI methods. By tackling these challenges head-on, we can unlock the full potential of multi-modal data fusion and ensure its responsible and beneficial use in a wide range of applications.

References

1 Hügle, M., Kalweit, G., Hügle, T. et al. (2021). A dynamic deep neural network for multi-modal clinical data analysis. In: *Explainable AI in Healthcare and Medicine. Studies in Computational Intelligence*, vol. 914 (ed. A. Shaban-Nejad,

M. Michalowski, and D.L. Buckeridge). Cham: Springer. https://doi.org/10.1007/978-3-030-53352-6_8.

2 Shi, Y., Paige, B., and Torr, P. (2019). Variational mixture-of-experts autoencoders for multi-modal deep generative models. *Advances in Neural Information Processing Systems*, Vancouver BC, Canada (8–14 December 2019).

3 Han, D., Kim, J., and Kim, J. (2017). Deep pyramidal residual networks. *Proceedings of the IEEE Conference on Computer Vision and Pattern Recognition*, Honolulu, HI, USA (21–26 July 2017).

4 Gao, J., Li, P., Chen, Z., and Zhang, J. (2020). A survey on deep learning for multi-modal data fusion. *Neural Computation* 32 (5): 829–864.

5 Pawłowski, M., Wróblewska, A., and Sysko-Romańczuk, S. (2023). Effective techniques for multi-modal data fusion: a comparative analysis. *Sensors* 23 (5): 2381.

6 Zhang, Y., Morel, O., Blanchon, M. et al. (2019). Exploration of deep learning-based multi-modal fusion for semantic road scene segmentation. *VISAPP 2019 14th International Conference on Computer Vision Theory and Applications*, Prague, Czech Republic (25–27 February 2019).

7 Wang, S., Huang, L., Wang, Z. and Ren, H. (2023). Deep learning based fusion models for sensitive information identification. *2023 IEEE 6th Information Technology, Networking, Electronic and Automation Control Conference (ITNEC)*, Chongqing, China (24–26 February 2023). Vol. 6. IEEE.

8 Zhang, Q., Wei, Y., Han, Z. et al. (2024). Multi-modal fusion on low-quality data: a comprehensive survey. *arXiv preprint arXiv: 2404.18947*.

9 Gaw, N., Yousefi, S., and Gahrooei, M.R. (2022). Multi-modal data fusion for systems improvement: a review. In: *Handbook of Scholarly Publications from the Air Force Institute of Technology (AFIT)*, vol. 1, 2000–2020 (ed. A.B. Badiru, F.W. Ciarallo, and E.G. Mbonimpa), 101–136. CRC Press.

10 Wu, D., Zhong, X., Peng, X. et al. (2022). Multi-modal information fusion for high-robustness and low-drift state estimation of UGVs in diverse scenes. *IEEE Transactions on Instrumentation and Measurement* 71: 1–15.

11 Lahat, D., Adali, T., and Jutten, C. (2015). Multi-modal data fusion: an overview of methods, challenges, and prospects. *Proceedings of the IEEE* 103 (9): 1449–1477.

12 Hu, W., Meng, X., Bai, Y. et al. (2021). Interpretable multi-modal fusion networks reveal mechanisms of brain cognition. *IEEE Transactions on Medical Imaging* 40 (5): 1474–1483.

13 Mökander, J., Morley, J., Taddeo, M., and Floridi, L. (2021). Ethics-based auditing of automated decision-making systems: nature, scope, and limitations. *Science and Engineering Ethics* 27 (4): 44.

14 Shaik, T., Tao, X., Li, L. et al. (2023). A survey of multi-modal information fusion for smart healthcare: mapping the journey from data to wisdom. *Information Fusion* 102040.

Index

Multimodal Intelligent Sensing in Modern Applications, First Edition.
Edited by Masood Ur Rehman, Ahmed Zoha, Muhammad Ali Jamshed, and Naeem Ramzan.

Published 2025 by John Wiley & Sons, Inc.

Printed and bound by CPI Group (UK) Ltd, Croydon, CR0 4YY

17/09/2026

14964331-0005